D0073286

Fluid Mechanics and its applications

Fluid Mechanics and its applications

James W. Murdock, P.E.
Drexel University

HOUGHTON MIFFLIN COMPANY BOSTON

Atlanta Dallas Geneva, Illinois Hopewell, New Jersey

Palo Alto London

Printed in the U.S.A.

Library of Congress Catalog Card Number: 75–31024

ISBN: 0–395–20626–x

To my wife Audrey

Contents

2 Fluid Statics

3 Fluid Kinematics

4 Fluid Dynamics

Preface

Most textbooks are the outgrowth of lecture notes, and this is no exception. For over two decades I developed a series of lecture notes in fluid mechanics for evening college students at Drexel University. In 1971 a college-level version of this text was produced at Drexel University for use by evening college students.

This text differs from the original in several important aspects. To make it also suitable for use in technology programs, mathematical rigor of derivation was reduced in a very few cases, for clarity of concept. Although calculus is used, students need not be familiar with partial derivatives or vector calculus. Another important difference is the introduction of SI (metric) units on an equal basis with English units, to meet the dual unit demands of the foreseeable future.

A step-by-step procedure is followed throughout the text to eliminate the guessing game between the author and the student. Except in Chapter 10, each new concept is followed by a numerical example.

The first five chapters, "Basic Definitions and Values," "Fluid Statics," "Fluid Kinematics," "Fluid Dynamics," and "Compressible Flow," provide the basic theoretical foundation. Chapter 6, "Dimensionless Parameters," provides tools for engineering analysis of fluid mechanics problems. Chapter 7, "Forces on Immersed Objects," Chapter 8, "Flow in Pipes," and Chapter 9, "Flow in Open Channels," are practical presentations. Chapter 10, "Fluidics," is intended only as an introduction to a new and rapidly developing area of fluid mechanics.

I would like to thank the many individual students and faculty members at Drexel University, particularly Prof. Lester J. Stradling, who reviewed for thermodynamic aspects, as well as many former associates at the Naval Ship Engineering Center, Philadelphia Division, who helped make this book possible through their comments and suggestions. I was indeed fortunate to have the help of Mr. Clarence Gregory, Head of the Scientific Branch, who reviewed the text and checked the problems with meticulous detail, and Michele Ann Ariza, my former secretary, who typed the text and corrected my English and spelling with unerring accuracy.

I would also like to thank the following individuals who reviewed this text in one or more of its developmental stages: Edward O. Stoffel, California Polytechnic State University; G. D. Carnahan, Bradley University; Mark A. Fugelso, University of Wisconsin (Madison); H. L. Banton, Purdue University; Anthony Peters, New York City Community College; Robert S. Williams, State University Agricultural and Technical College at Farmingdale; Alden R. Hodgen, The Pennsylvania State University; Arnold R. Ness, Bradley University; Alvin T. Greenspan, Temple University; Joseph Kolczynski, Lafayette College; and Leo T. Smith, Drexel University.

James W. Murdock

Principal Symbols and Abbreviations

Symbols	Definitions and Units
a	acceleration, ft/sec^2, m/s^2
a_x	acceleration along the x axis
a_y	acceleration along the y axis
a_z	acceleration along the z axis
A	area, ft^2, m^2
A_p	projected area
A_s	shear area
°API	American Petroleum Institute gravity
°Be	Baumé gravity
BHP	brake horsepower
Btu	British thermal unit
c	acoustic velocity, ft/sec, m/s; chord length, ft, m
c_x	specific heat for process x, Btu/lbm-°R, J/kg · K
c_p	specific heat at constant pressure
c_v	specific heat at constant volume
C	Celsius
C	coefficient, discharge for pipeline meters; Chezy for open channel flow, dimensionless
C_D	drag coefficient
C_f	skin friction coefficient
C_L	lift coefficient
C_p	pipe coefficient
C_r	residual coefficient
C_s	shape coefficient
C_t	tube coefficient
CF	center of force
CFM	cubic feet per minute
CG	center of gravity
C	Cauchy number, dimensionless

$\mathbf{C}_p$	pressure coefficient, dimensionless
$\mathbf{C}_T$	thrust coefficient, dimensionless
d	primary element diameter, ft, m
D	diameter, ft, m
D_e	equivalent diameter
D_o	outside diameter
E	bulk modulus of elasticity, lbf/ft^2, N/m^2
E_n	polytropic bulk modulus of elasticity
E_s	isentropic bulk modulus of elasticity
E_T	isothermal bulk modulus of elasticity
E	efficiency, percent
E_p	propeller efficiency
E_s	system efficiency
E	specific energy, ft, m
$\mathbf{E}$	Euler number, dimensionless
f	friction factor, dimensionless
f	frequency, sec^{-1}, s^{-1}
f_n	natural frequency
f_w	wake frequency
F	Fahrenheit
F	force, lbf, N
F_b	bearing load force
F_B	buoyant force
F_d	downward force
F_D	drag force
F_E	elastic force
F_f	friction force, skin friction force, vibratory force
F_g	gravity force
F_i	inertia force
F_L	lift force
F_p	pressure force
F_r	residual force
F_s	shear force
F_u	upward force
F_x	x-component of force
F_y	y-component of force

F_z	vertical component of force
$F_{\Delta\,(\text{delta})}$	roughness force
$F_{\mu\,(\text{mu})}$	viscous force
$F_{\sigma\,(\text{sigma})}$	surface tension force
$F_{\omega\,(\text{omega})}$	centrifugal force
F	Froude number, dimensionless
g	acceleration due to gravity, ft/sec^2, m/s^2
$g_{\phi\,(\text{phi})}$	gravity at latitude ϕ
g_0	earth standard gravity
g_c	proportionality constant, lbm-ft/lbf-sec^2, kg $\cdot$ m/N $\cdot$ s^2
G	gravitational constant, lbf-ft^2/lbm^2, N $\cdot$ m^2/kg^2
GPM	gallons per minute
h	height, ft, m
h_c	height, vertical, from a liquid surface to the center of gravity of a submerged object
h	specific enthalpy, Btu/lbm, J/kg
H	geopotential altitude, weir head, ft, m
H	energy transfer per unit mass, ft-lbf/lbm, J/kg
H_f	energy lost due to friction between stations 1 and 2
$\overline{H}_f$	heat lost due to friction, ft, m
I	area moment of inertia, ft^4, m^4
I_G	area moment of inertia about the center of gravity of an object
I_O	area moment of inertia about a liquid surface
J	joules
k	ratio of specific heats, dimensionless
K	kelvin
K	coefficient, dimensionless
K	resistance coefficient, dimensionless
K_t	resistance coefficient, total
$\overline{\text{K}}$	flow coefficient, dimensionless
KE	kinetic energy, ft-lbf, J
lbf	pound-force
lbm	pound-mass
L	length, ft, m
L_w	length of weir crest

m	metre
m	mass, slug, kg
m_o	mass at rest
$\dot{m}$	mass flow rate, slug/sec, kg/s
M	mass, lbm, kg
$\dot{M}$	mass flow rate, lbm/sec, kg/s
$\dot{M}_i$	mass flow rate, ideal
M	moment, ft-lbf, N · m
$M_{\max}$	moment, maximum
M_o	moment about a liquid surface
M	molecular weight, lbm-mol, kg · mol
ME	mechanical energy, ft-lbf, J
M	Mach number, dimensionless
n	exponent describing the p-v relationship of an ideal gas, dimensionless
n	roughness factor, $ft^{1/6}$, $m^{1/6}$
N	newtons
N	piping schedule number, dimensionless
N	speed, rpm, s^{-1}
N_s	specific speed
p	pressure, lbf/ft^2, Pa
p_b	atmospheric pressure
p_c	critical pressure
p_i	pressure on inner wall; indicated (gage)
p_o	pressure on outer wall
p_s	pressure at liquid surface
p_S	stagnation pressure
p_v	vapor pressure
p_x	x-component of pressure
p_z	vertical pressure
psia	absolute pressure, lbf/in^2
psig	gage pressure, lbf/in^2
p'	difference between pipe internal and external pressures, lb/in^2
P	perimeter, ft, m
P	power, ft-lbf/sec, J/s

P_i	input power
P_o	output power
P_p	plotting parameter for bearings, dimensionless
PE	potential energy, ft-lbf, J
q	heat transfer, Btu/lbm, J/kg; volumetric flow rate per unit width, ft^2/sec, m^2/s
Q	volumetric flow rate, ft^3/sec, m^3/s
r	radius, ft, m
r_i	inner radius
r_o	outer radius
r	pressure ratio
r_c	critical pressure ratio
R	Rankine
R	gas constant, ft-lbf/lbm-°R, J/kg · K; distance along an inclined tube, ft, m
$\overline{R}$	universal gas constant, ft-lbf/lbm-mol-°R, J/kg · mol · K
R_h	hydraulic radius, ft, m
RPM	revolutions per minute
$\mathbf{R}$	Reynolds number, dimensionless
$\mathbf{R}_d$	Reynolds number based on primary element diameter
$\mathbf{R}_D$	Reynolds number based on sphere or pipe diameter
$\mathbf{R}_t$	Reynolds number in transition
$\mathbf{R}_X$	Reynolds number of a flat plate at X distance from leading edge
s	entropy, Btu/lbm-°R, J/kg · K; slope of an open channel, ft/ft, m/m; span of a lifting vane ft, m
S	scale reading, ft, m; specific gravity, dimensionless; surface area, ft^2, m^2
S	stress, lbf/in^2, Pa
S_c	circumferential stress
S_L	longitudinal stress
SE	specific energy, ft-lbf/lbm, m · N/kg
SKE	kinetic specific energy
SPE	potential specific energy
SFW	specific flow work
slug	unit of mass (U.S.)
SSU	kinematic viscosity, Saybolt seconds universal

S	Strouhal number, dimensionless
t	time, sec, s
t	thickness of pipe wall, in.
t_{m}	minimum thickness
t_{s}	schedule thickness
t	temperature, °F, °C
t_{C}	Celsius temperature
t_{F}	Fahrenheit temperature
T	absolute temperature, °R, K
T_c	critical temperature
T_{K}	kelvin temperature
T_{R}	Rankine temperature
T	time, sec, s
T	thrust, lbf, N
u	internal energy, Btu/lbm, J/kg
U	streamline velocity, ft/sec, m/s
U_i	ideal streamline velocity
U_m	maximum streamline velocity
v	specific volume, ft^3/lbm, m^3/kg
V	average velocity, ft/sec, m/s
V_o	velocity of a body
V_v	minimum velocity for vaporization
Vol	volume, ft^3, m^3
V	velocity ratio, dimensionless
W	work, ft-lbf, J
W_{nf}	nonflow shaft work
W_{sf}	steady-flow shaft work
$\overline{W}$	work in units of heat, ft, m
W	Weber number, dimensionless
x	distance along an abcissa, ft, m
x_c	location of a vertical force, ft, m
X	distance along a flat plate, ft, m
y	distance along an ordinate, ft, m
y	distance from a liquid surface to the bottom of an open channel (stage), ft, m
y	distance from a liquid surface, ft, m

y_c	distance to the center of gravity of an object, ft, m
y_F	distance to the center of pressure of an object, ft, m
Y	expansion factor, dimensionless
z	vertical distance, ft, m
Z	crest height of a weir, ft, m; compression factor, dimensionless
α (alpha)	angle of attack, °, rad; kinetic energy correction factor, dimensionless
β (beta)	ratio of primary element diameter to internal pipe diameter, dimensionless
γ (gamma)	specific weight, lbf/ft^3, N/m^3
γ_f	specific weight of fluid
γ_m	specific weight of manometer fluid
δ (delta)	thickness of a boundary layer, ft, m
ϵ (epsilon)	surface roughness, ft, m
η (eta)	efficiency, per cent
θ (theta)	angle, °, rad
μ (mu)	dynamic viscosity, lbf-sec/ft^2, Pa · s
ν (nu)	kinematic viscosity, ft^2/sec, m^2/s
π (pi)	3.14159 . . .
Π (Pi)	dimensionless ratio
ρ (rho)	density, slugs/ft^3, kg/m^3
σ (sigma)	surface tension, lbf/ft, N/m
τ (tau)	unit shear stress, lbf/ft^2, Pa
τ_o	unit shear stress at wall
ϕ (phi)	latitude, degrees
ω (omega)	rotational speed, radians/sec, rad/s

1 Basic Definitions and Values

1-1 INTRODUCTION

Ancient man dug wells and constructed canals. He built and operated crude water wheels and pumping devices. He overcame fluid resistance with well-faired ships propelled by oars and sails. He used feathered arrows for better flight stability. As his cities grew, he was forced to build ever larger aqueducts, which reached their zenith in the city of Rome. However, except for the establishment of the elementary principles of buoyancy and flotation by Archimedes (287–212 B.C.), ancient man contributed little to our current knowledge of fluid mechanics.

After the fall of the Roman Empire (A.D. 476), no known progress was made in the development of fluid mechanics until the time of Leonardo Da Vinci (1452–1519). From that time on, the accumulation of knowledge rapidly gained momentum. The first theories that were proposed were, in general, confirmed by crude experiments, but as time went on, the divergences that developed between theory and experimental data led to their separation.

Hydrodynamics is that branch of mathematics dealing with the motion of an idealized fluid. Because hydrodynamics neglected the very fluid properties that determine fluid motion, it has contributed little of practical value to the engineer.

Hydraulics is the practical science of determining fluid behavior, primarily of water, from experimental data. This approach resulted in many empirical equations obtained by fitting curves to experimental data. In many instances, the relationship between the physical facts and the equations was not apparent, so that experimental data could not be extrapolated to other fluids or conditions beyond those actually covered by the experiment.

Fluid mechanics is a rational approach to problems of fluid behavior combining the theoretical concepts of hydrodynamics with the practical approach of hydraulics. The study of fluid mechanics is divided into three major parts: *statics,* the forces of fluids at rest; *kinematics,* the geometry of fluid motion; and *dynamics,* the forces exerted on or by fluids in motion.

1-2 FLUIDS AND OTHER SUBSTANCES

Substances may be classified by their response when at rest to the imposition of a shear force. Consider the two very large plates, one moving, the other stationary, separated by a small distance y, as shown in Fig. 1-1. The space between these plates is filled with a substance whose surfaces adhere to the plates in such a manner that the upper surface of the substance moves at the same velocity as the upper plate and the bottom substance surface is stationary. As the result of the imposition of the shear force F_s, the upper surface of the substance attains a velocity U. As y approaches dy, U approaches dU and the rate of deformation of the substance becomes dU/dy. The unit shear stress τ (tau) $= F_s/A_s$, where A_s is the shear area. Deformation characteristics are shown in Fig. 1-2.

An ideal or *elastic solid* will resist the shear force and its *rate* of deformation will be zero regardless of loading and hence is coincident with the ordinate of Fig. 1-2.

Plastic will resist shear until its yield stress is attained, and then the application of additional loading will cause it to deform continuously or flow. If the deformation rate is directly proportional to the applied shear force less that required to start flow, then it is called an ideal plastic.

If the substance is unable to resist even the slightest amount of shear force without flowing, then it is a *fluid.* An *ideal fluid* has no

FIGURE 1-1 Flow of a substance between parallel plates

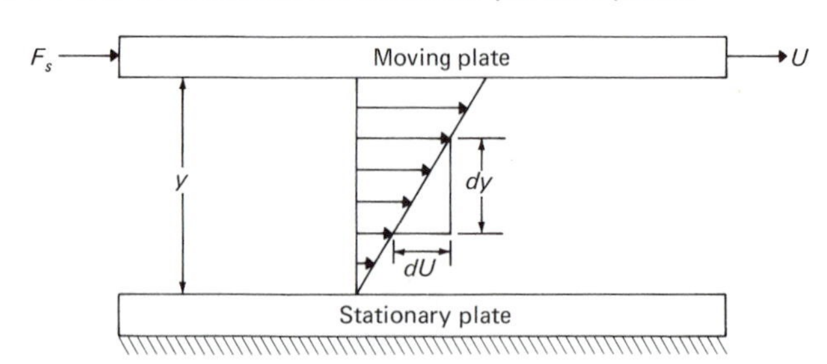

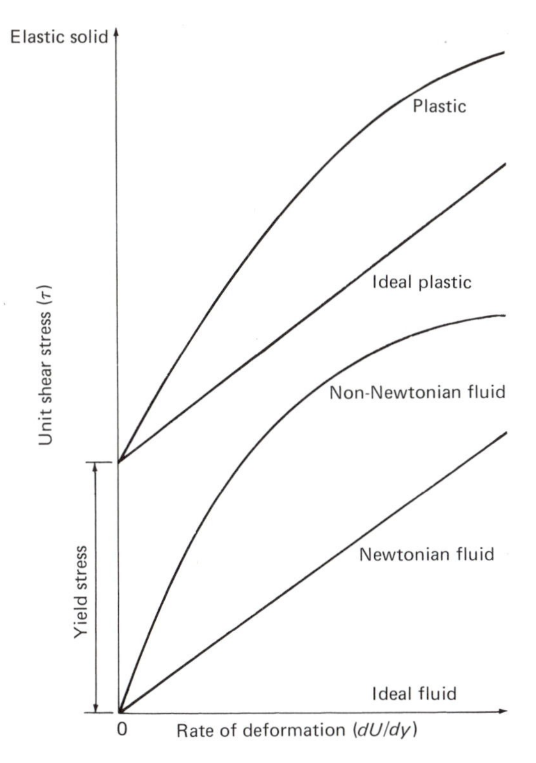

FIGURE 1-2 Deformation characteristics of substances

internal friction, and hence its deformation rate coincides with the abscissa of Fig. 1-2. All real fluids have internal friction, so that their rate of deformation is a function of the applied shear stress. If the rate of deformation is directly proportional to the applied shear stress, then it is a *Newtonian fluid* and if not, it is a *non-Newtonian fluid.*

Kinds of Fluids

For the purposes of the study of fluid mechanics two kinds of fluids are considered: compressible and incompressible. These characteristics are determined by molecular spacing and arrangement or *phase* of the substance. The phase relations of a pure substance are shown with respect to temperature and pressure in Fig. 1-3.

Liquids are considered to be incompressible except at very high pressures and unless otherwise specified will be treated as such throughout this text.

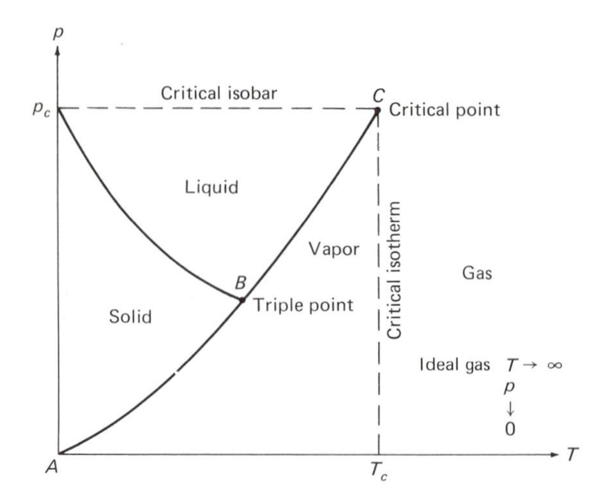

FIGURE 1-3 Phase diagram of a pure substance

Vapors are gases below their critical temperatures and are very compressible, but their temperature-pressure-volume relationships cannot be expressed by simple mathematical equations. Vapor properties are usually tabulated, as, for example, in Steam and Refrigeration Tables. Vapors are covered in the study of thermodynamics and are not considered in fluid mechanics.

Gases are compressible fluids. As the ratio of the temperature of the substance T to the critical temperature T_c approaches infinity and the ratio of the pressure p to the critical pressure p_c approaches zero, all substances tend to behave like ideal gases, i.e., their pressure-volume-temperature relations may be expressed by the equation of state for ideal gases (Sec. 1-14). No real gas follows this relation exactly, but for all practical purposes if $T/T_c > 2.5$ and $p/p_c < 5$ the substance may be treated as an ideal gas. Non-ideal gases are studied in thermodynamics; fluid mechanics considers only gases that may be treated as ideal.

1-3 UNITS

It is anticipated that the legislation now before the Congress to convert the United States to the use of the metric system will be passed. For the foreseeable future the United States will be faced with the problems involved in converting from its customary units (U.S.) of measure to the "Système Internationale d'Unites" (SI) units. During this long period, which will probably cover the professional life of those who study from this text, both systems will be employed. This makes it mandatory that engineers and technicians be proficient in the use of both systems.

One of the many problems arising from the use of a dual system of units is that individuals tend to think in the system that they learned first and their use of the other system becomes a mere mechanical conversion. Those engineers and technicians who are now in the field are of course used to the U.S. system, and those who enter in the future will be trained in the SI system. In addition to the "generation gap" that will exist between those now in the field and those who will enter in the future there will also be a "unit gap" to impede communications.

Both systems of units are used in this text. Although equal weight is given to each system, all the basic physical constants and standards are defined by international agreements in SI units. This sometimes results in the use of precise but inexact values for physical constants and standards in U.S. units.

Appendix B explains the SI system of units in regard to fluid mechanics and provides conversion factors for those who have already learned to think in the U.S. system.

The U.S. system is not really a system, since its units are based on customary use. Insofar as possible, the units used in this text are those traditionally used in mechanics, the foot (ft) for length, the second (sec) for time, and the pound-force (lbf) for force. Two units for mass are used, the slug (slug) and the pound-mass (lbm). The slug is the customary unit for mass used in fluid mechanics, but the pound-mass is used in general engineering practice and in thermodynamics in conjunction with the ideal gas laws. The common use of pressure in units of pounds per square inch and in terms of heights of liquid columns are handled by requiring their conversion to lbf/ft^2 before using them in problem solutions. In mechanics, energy (work) is expressed in units of ft-lbf, so that thermal energy terms normally expressed in British thermal units (Btu) must also be converted. In the step-by-step method of problem-solving followed in this text, the first step, Step A, is called data reduction, which includes the gathering and conversion as necessary to proper units of all the data required to solve the problem.

1-4 PRESSURE

Fluid forces that can act on a substance are *shear, tension,* and *compression.* By definition, fluids in a static state cannot resist any shear force without flowing. Fluids will support small tensile forces due to the property of *surface tension* (Sec. 1-18). Fluids can withstand compression forces, commonly called *pressure.*

Definition / Force per unit area
Symbol / p
Dimensions / FL^{-2} or $ML^{-1}T^{-2}$
Units / U.S.: lbf/ft^2 SI: N/m^2 or Pa

Atmospheric Pressure

The *actual* atmospheric pressure is the weight per unit area of the air above a datum and varies with weather conditions. Since this pressure is usually measured with a barometer, it is commonly called *barometric pressure.*

Standard Atmospheric Pressure

By international agreement the standard atmospheric pressure is defined as 101.325 kN/m^2. Converting to U.S. units using Eq. (B-3) of Appendix B, standard atmospheric pressure, for text calculation purposes, is

$$2116.216\,624 \text{ lbf/ft}^2 \qquad \text{or} \qquad 2116 \text{ lbf/ft}^2$$

Observed Pressures

Most pressure-sensing devices (Sec. 2-5) (the barometer is an exception) indicate the difference between the pressure to be measured and atmospheric pressure. As shown in Fig. 1-4, if the pressure being sensed is *greater* than atmospheric it is called *gage pressure,* if *lower* (negative gage) it is called a *vacuum.* The algebraic sum of the instrument reading and the actual atmospheric pressure is the true or absolute pressure. Thus

FIGURE 1-4 Pressure relations

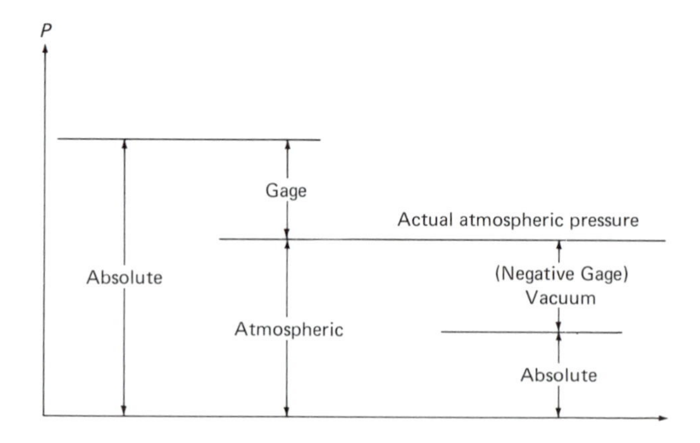

$$p = p_b + p_i \qquad \textbf{(1-1)}$$

where p is the absolute pressure, p_b the atmospheric (barometric) pressure, and p_i the instrument reading (positive for gage pressure, negative for vacuum). *All instrument readings must be converted to absolute pressure before they are used in calculations.* If the actual atmospheric pressure is not given in text problems, then standard conditions may be assumed.

Special note: This text follows the SI practice of leaving a space after each group of three digits, counting from the decimal point. (See Appendix B, especially Sec. B-2.) We do this (with metric units only) because, in many countries, a comma in a numeral is used to signify a decimal point. Examples: Instead of 5,720,626 we use 5 720 626; instead of 0.43875 we use 0.438 75; instead of 5,856 we write either 5856 or 5 856. (With English units, we use commas as usual.)

EXAMPLE *During a test of a steam turbine the observed vacuum in the condenser was 27.56 in. of mercury at 32°F and the actual atmospheric pressure was 14.89 psia. What was the absolute pressure in the condenser?

Step A *Data Reduction (Unit Conversions)*

(B-56) $p_i = -27.56\left(\frac{2.54}{100}\right) = -0.700\ 24 \text{ mm Hg} = -700.24 \text{ mm Hg}$

(B-108) $p_i = -700.24(2.784) = -1949 \text{ lbf/ft}^2$

(B-89) $p_b = 14.89(144) = 2144 \text{ lbf/ft}^2$

Step B *Compute Absolute Pressure*

(1-1) $p = p_b + p_i = 2144 + (-1949) = 195 \text{ lbf/ft}^2$

(B-87) $p = 195\left(\frac{1}{144}\right) = 1.354 \text{ psia}$

1-5 TEMPERATURE SCALES

Unlike the other properties studied in this text, temperature is based on a thermodynamic concept which is independent

* Here is where we first experience the difficulties of using a "pure" system of U.S. units. Conventional U.S. engineering practice is to use the unit lbf/in.^{2} (psi) for pressure. Gage pressures are indicated by psig and absolute pressures by psia. Vacuums are almost always reported in inches of mercury. Standard atmospheric pressure is 14.695 948 78 psia, or 14.70 for text calculation purposes. Standard atmospheric pressure is also 29.921 259 84 in. of mercury at 32°F and sea level, or 29.92 for calculations. The unit pound-force per square foot means little in engineering practice, so any problem solutions resulting in U.S. units for pressure should be converted to psia.

of the physical properties of any substance. The thermodynamic temperature can be shown to be related to the equation of state of an ideal gas (Sec. 1-14). The thermodynamic temperature is called an *absolute temperature* because its datum is absolute zero. The *thermodynamic temperature scale* has little practical value unless numbers can be assigned to the temperatures at which substances freeze or boil so that thermometers may be calibrated. The *International Practical Temperature Scale* is a document which defines and assigns numbers to fixed points (refer to Table 1-1) of selected substances and prescribes methods and instruments for interpolating between fixed points. Although the International Practical Temperature Scale is dependent on the physical properties of substances, it attempts to reproduce the thermodynamic temperature scale within the knowledge of the state of the art.

The Fahrenheit temperature scale is used in the United States for ordinary temperature measurements. It was invented in 1702 by Daniel Gabriel Fahrenheit (1686–1736), a German physicist. On this scale water freezes at 32°F (ice point) and boils at 212°F (steam point) and at standard atmospheric pressure.

The Celsius temperature scale (formerly centigrade) was first proposed in 1742 by Anders Celsius (1701–1744), a Swedish astron-

TABLE 1-1 *Primary and Secondary Fixed Points of the 1968 International Practical Temperature Scale*

Point	Substance	°C
	PRIMARY FIXED POINTS	
Triple	Hydrogen	−259.34
Boiling*	Hydrogen	−256.108
Boiling	Hydrogen	−252.87
Boiling	Neon	−246.048
Triple	Oxygen	−218.789
Boiling	Oxygen	−182.962
Triple	Water	0.01
Boiling	Water	100
Freezing	Zinc	419.58
Freezing	Silver	961.93
Freezing	Gold	1064.43
	SECONDARY FIXED POINTS	
Freezing	Tin	231.968
Freezing	Lead	327.502
Boiling	Sulfur	444.674
Freezing	Antimony	630.74
Freezing	Aluminum	660.37

*At $^{25}/_{76}$ atmospheres.

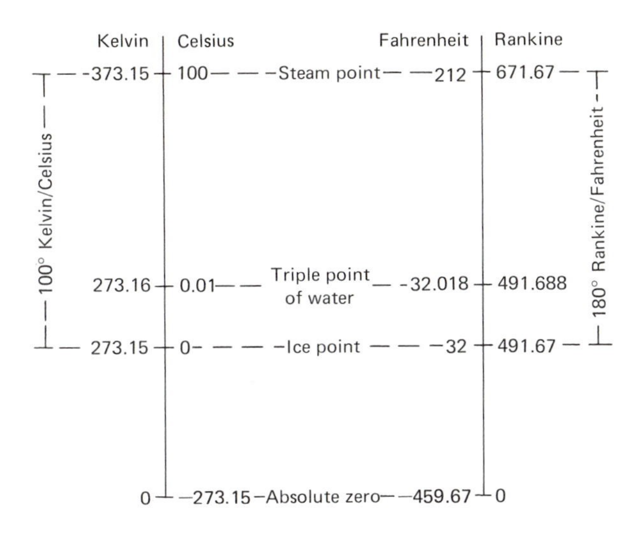

FIGURE 1-5 Temperature scales

omer. On the Celsius scale the ice point is 0°C and the steam point 100°C.

The Kelvin temperature scale is named in honor of the English scientist Lord Kelvin (William Thompson, 1824–1907) and is the absolute Celsius scale. The kelvin (K with no degree sign) is defined as the SI unit of temperature as 1/273.16 of the fraction of the thermodynamic temperature of the triple point of water. The International Practical Temperature Scale of 1968 assigns a value of 0.01°C to the triple point of water.

The Rankine temperature scale is named in honor of the Scottish engineer William J. Rankine (1820–1872) and is the absolute Fahrenheit scale.

Temperature scale relations are shown in Fig. 1-5. The Celsius scale has 100 degrees between the ice and steam points and the Fahrenheit 180. The relation between these scales may be shown as $\Delta t_F/\Delta t_C = 180/100 = 1.8$, where t_F is the Fahrenheit temperature and t_C the Celsius. At the ice point $t_F = 32$°F and $t_C = 0$°C, so that

$$\frac{\Delta t_F}{\Delta t_C} = \frac{t_F - 32}{t_C - 0} = 1.8$$

solving first for t_F and then t_C:

(B-25)* $$t_F = 1.8 t_C + 32$$ **(1-2)**

* Conversion factor equation numbers in Appendix B

(B-29)* $$t_C = \frac{t_F - 32}{1.8} \tag{1-3}$$

The triple point of water on the Celsius scale is 0.01°C, and on the kelvin 273.16K, so that $T_K = t_C + (273.16 - 0.01)$ or

(B-27)* $$T_K = t_C + 273.15 \tag{1-4}$$

where T_K is the temperature in kelvins.

Because the Kelvin and the Rankine scales are the absolute scales of the Celsius and Fahrenheit scales respectively, with the same differences between the ice and steam points, we can write: $\Delta t_F/\Delta t_C = \Delta T_R/\Delta T_K = 180/100 = 1.8$, where T_R is the temperature in degrees Rankine. Since both are to absolute zero:

$$\frac{\Delta T_R}{\Delta T_K} = \frac{T_R - 0}{T_K - 0} = 1.8$$

or

(B-32)* $$T_R = 1.8T_K \tag{1-5}$$

Substituting Eq. (1-4) in (1-5), we have $T_R = 1.8T_K = 1.8(t_C + 273.15)$ or

(B-26)* $$T_R = 1.8t_C + 491.67 \tag{1-6}$$

Substituting Eq. (1-3) in (1-6)

$$T_R = 1.8t_C + 491.67 = \frac{1.8(t_F - 32)}{1.8} + 491.67$$

or

(B-28)* $$T_R = t_F + 459.67 \tag{1-7}$$

EXAMPLE In a boiler furnace the gas temperature is 1,535°R and the water in the boiler tubes is at 486°F. Compute the temperature drop across the tube in kelvins (K).

Step A *Data Reduction* (Convert to K).

(B-32) $T_{gas} = 1535(1/1.8) = 852.8$ K

(B-28) $T_{water} = 486 + 459.67 = 945.67$°R

(B-32) $T_{water} = 945.67(1/1.8) = 525.4$ K

---

* Conversion factor equation numbers in Appendix B.

Step B *Compute Temperature Difference*

$$T_{gas} - T_{water} = 852.8 - 525.4 = 327.4 \text{ K}$$

1-6 MASS, FORCE, AND WEIGHT

A *mass* is a quantity of matter. Its value is the same any place in the universe. The opposition which a body offers to any change of motion is known as *inertia.* This property is common to all matter and leads to the concept of mass as inertia. *Force* and *mass* are related by Newton's second law of motion. *Weight* is the *force* exerted by a *mass* due to the acceleration of gravity.

Newton's second law of motion states that an unbalanced force acting on a body causes the body to accelerate in the direction of the force, and the acceleration is directly proportional to the unbalanced force and inversely proportional to the mass of the body. This law may be expressed mathematically as:

$$a \propto \frac{F}{M} \quad \text{or} \quad a = \frac{g_c F}{M} \quad \text{and} \quad F = \frac{Ma}{g_c} \tag{1-8}$$

where F is the unbalanced force, M is the mass of the body, a is the acceleration and g_c is the *proportionality constant.*

The numerical value of g_c depends on the units used in Eq. (1-8). The *newton* is defined as the force produced by the acceleration of the mass of 1 kg at a rate of 1 m/s^2. Solving Eq. (1-8) for these units,

$$g_c = \frac{Ma}{F} = 1\,\frac{(\text{kg})\,(\text{m/s}^2)}{\text{N}} = 1\,\frac{\text{kg}\cdot\text{m}}{\text{N}\cdot\text{s}^2} \tag{1-9}$$

The *slug* is a unit of mass named from the concept of mass as inertia or sluggishness. The slug is defined as that quantity of mass which when acted on by a force of 1 lb will undergo an acceleration of 1 ft/sec^2. The proportionality constant for the slug from Eq. (1-8) becomes

$$g_c = \frac{Ma}{F} = 1\,\frac{(\text{slug})(\text{ft/sec}^2)}{\text{lbf}} = 1\,\frac{\text{slug-ft}}{\text{lbf-sec}^2}$$

In U.S. units the *pound-mass* is defined by international agreement to be equal to 0.453 592 37 kg. The *pound-force* is defined as the *weight* of one pound-mass when subjected to the standard acceleration of gravity of 9.806 65 m/s^2. Converting this acceleration to U.S. units

$$\frac{9.806\,65 \text{ m/s}^2}{0.3048 \text{ m/ft}} = 32.174\,048\,56 \text{ ft/sec}^2$$

Again solving Eq. (1-8) for units of g_c,

$$g_c = \frac{Ma}{F} = \frac{1\ \text{lbm}(32.174\,048\,56\ \text{ft/sec}^2)}{\text{lbf}}$$

$$g_c = 32.174\,048\,56\ \frac{\text{lbm-ft}}{\text{lbf-sec}^2} \qquad \textbf{(1-11)}$$

For text calculations 32.17 should be used in Eq. (1-11). If Eqs. (1-10) and (1-11) are set equal to each other,

(B-98) $$1\ \text{slug} = 32.174\,048\,56\ \text{lbm} \qquad \textbf{(1-12)}$$

EXAMPLE A body requires the application of 500 N of force to produce an acceleration of 3 in./sec². Compute the mass of the body in (1) kilograms, (2) slugs, and (3) pounds-mass.

Step A *Data Reduction*

(1) *convert acceleration to SI units*

(B-56) $a = 3(2.54/100) = 7.620 \times 10^{-2}\ \text{m/s}^2 = 76.20\ \text{mm/s}^2$

(2) *obtain g_c for SI units*

(1-9) $g_c = 1\ \frac{\text{kg}\cdot\text{m}}{\text{N}\cdot\text{s}^2}$

Step B *Compute Mass in Kilograms*

(1-8) $M = \frac{Fg_c}{a} = \frac{500 \times 1}{76.20 \times 10^{-3}} = 6\,562\ \text{kg}$

Step C *Convert from SI to U.S. Units*

(1) *kilograms to slugs*

(B-99) $6\,562\frac{1}{14.59} = 449.8\ \text{slugs}$

(2) *kilograms to pounds-mass*

(B-94) $6\,562\frac{1}{0.453\,6} = 14{,}470\ \text{lbm}$

1-7 GRAVITY

Newton's Law of Universal Gravitation

Every particle in the universe attracts every other particle with a force that is directly proportional to the product of the masses of the particles and inversely proportional to the square of the distance between them. This may be expressed mathematically as

$$F_g \propto \frac{M_1 M_2}{L^2}$$

or

$$F_g = \frac{G M_1 M_2}{L^2} \tag{1-13}$$

where F_g is the attractive force, G is the gravitational constant ($G = 66.732\ pN \cdot m^2/kg^2$), M_1 is the mass of particle 1 and M_2 of particle 2, and L is the distance between the particles. Substituting F_g for F and g for a in Eq. (1-8) and equating with Eq. (1-13),

$$F_g = \frac{M_1 g}{g_c} = \frac{G M_1 M_2}{L^2}$$

Solving for g,

$$g = \frac{g_c G M_2}{L^2} \tag{1-14}$$

The variation of gravitational acceleration with distance from the surface of a body (planet, moon, asteroid, etc.) in space may be determined by letting g_ϕ be the surface acceleration at latitude ϕ, r the radius of the body, and z the height (altitude) above the surface of the body, substituting g_ϕ for g, r for L in Eq. (1-14),

$$g_\phi = \frac{g_c G M_2}{r^2} \tag{1-15}$$

And for altitude z above the surface, $L = r + z$:

$$g = \frac{g_c G M_2}{(r+z)^2} \tag{1-16}$$

Dividing Eq. (1-16) by Eq. (1-15),

$$\frac{g}{g_\phi} = \frac{r^2}{(r+z)^2} \tag{1-17}$$

Earth's Gravity

The standard acceleration due to gravity of the Earth is fixed as $g_0 = 9.806\ 65\ m/s^2$ by international agreement. For calculation purposes from Appendix B, $g_0 = 9.807\ m/s^2$ (B-51) and

TABLE 1-2 *Observed Acceleration of Gravity at Selected World Stations*

Station or Location	Latitude	Altitude		Deviation, %
	ϕ	m	ft	$100\frac{(g-g_0)}{g_0}$
Marigal, Africa	0°28′ N	1 036	3,399	−0.306
Dharwar, India	15°28′ N	728	2,288	−0.253
Chala, Peru	15°49′ S	14	46	−0.226
Santiago, Chile	33°27′ S	541	1,775	−0.126
Mount Wilson, Calif.	34°13′ N	1 719	5,640	−0.144
Ottawa, Ontario	45°24′ N	83	272	−0.004
Oslo, Norway	59°55′ N	28	92	+0.129
Arctic Ocean	81°48′ N	−3 402	−11,160	+0.247

Source: Smithsonian Physical Tables.

g_o = 32.17 ft/sec² (B-50). Standard gravity occurs at a latitude $\phi = 45°32'33''$. For other latitudes at sea level, g_ϕ may be calculated from

$$\frac{g_\phi}{g_o} = 999.95 \times 10^{-3}(1 - 2.6373 \times 10^{-3} \cos 2\phi + 5.9 \times 10^{-6} \cos^2 2\phi) \qquad \textbf{(1-18)}$$

Actual variation of gravitational acceleration from standard gravity on the Earth's surface is small, as shown in Table 1-2.

Solar System Gravities

The acceleration due to gravity of various bodies in the solar system varies greatly, as shown in Table 1-3.

EXAMPLE Calculate the acceleration of gravity at Marigal, Africa, and compare with the observed gravity deviation shown in Table 1-2.

Step A — *Data Reduction*

(1) *latitude and altitude of Marigal, Africa*

Table 1-2 — $\phi = 0°28'$ North and $z = 1\,036$ m $= 1.036$ km

(2) *radius of the Earth*

Table 1-3 — $r = 6\,371$ km

(3) *standard gravity*

(B-51) — $g_o = 9.807$ m/s²

TABLE 1-3 *Solar System Data*

Body	Distance From Sun, 10^6 miles	Mean Radius, *r*		Approximate Gravity, *g*	
		km	10^3 ft	m/s²	ft/sec²
Sun	0	696 000	2,283,000	273.7	897.0
Mercury	36	2 433	7,982	3.578	11.74
Venus	67	6 053	19,860	8.874	29.11
Earth	93	6 371	20,904	9.807	32.74
Moon	93	1 738	5,702	1.620	5.315
Mars	142	3 380	11,090	3.740	12.27
Jupiter	483	69 760	228,900	26.01	85.33
Saturn	886	58 220	191,100	11.71	36.65
Uranus	1,783	23 470	77,000	10.49	34.42
Neptune	2,793	22 720	74,540	13.25	43.42
Pluto	3,666	5 700	18,700	4 ?	14 ?

Source: *Handbook of Chemistry and Physics.*

Step B *Compute Acceleration*

(1) *effect of altitude*

(1-17)
$$\frac{g}{g_\phi} = \frac{r^2}{(r+z)^2}$$

(1-17)
$$\frac{g}{g_\phi} = \frac{(6\,371)^2}{(6\,371 + 1.036)^2} = 0.99967$$

or

$$g = 0.99967 g_\phi \qquad \text{(a)}$$

(2) *effect of latitude*

(1-18)
$$\frac{g_\phi}{g_o} = 999.95 \times 10^{-3}(1 - 2.6373 \times 10^{-3} \cos 2\phi + 5.9 \times 10^{-6} \cos^2 2\phi)$$

(1-18)
$$g_\phi = (9.807)(999.95 \times 10^{-3})\left(1 - 2.6373 \times 10^{-3} \cos \frac{2 \times 28}{60} + 5.9 \times 10^{-6} \cos^2 \frac{2 \times 28}{60}\right) = 9.780 \text{ m/s}^2$$

(3) *combine effects*

$$g = 0.99967 g_\phi = 0.99967(9.780) = 9.777 \text{ m/s}^2 \qquad \text{(a)}$$

Step C *Compare with Table 1-2*

Table 1-2
$$\frac{100(g - g_o)}{g_o} = -0.306 = \frac{100(g - 9.807)}{9.807},$$

solving for $g = 9.777$ m/s² which checks exactly with value obtained from Step B.

1-8 APPLICATIONS OF NEWTON'S SECOND LAW

Newton's second law may be used to establish the relationship between (a) force, mass, and acceleration, (b) work and energy, or (c) impulse and momentum.

Force, mass, and acceleration relationships were established in Sec. 1-6 by Eq. (1-8). This equation may be modified as follows:

(1-8) $$F = \frac{Ma}{g_c} = \left(\frac{M}{g_c}\right)a$$

Let $$m = \frac{M}{g_c} \qquad \textbf{(1-19)}$$

where m is the mass in units whose numerical value of the proportionality constant g_c is unity. In U.S. units the mass unit for m is the slug, Eq. (1-10), and in SI units the kilogram, Eq. (1-9). Substituting Eq. (1-19) in (1-8)

$$F = \left(\frac{M}{g_c}\right)a = ma \qquad \textbf{(1-20)}$$

or

$$Force = Mass \times Acceleration$$

EXAMPLE How much will a mass of 1000 lbm weigh on the surface of the planet Venus?

Step A — *Data Reduction*

(1) *proportionality constant*

(1-11) $g_c = 32.17$ lbm-ft/lbf-sec^2

(2) *unit conversion*

(1-19) $m = \frac{M}{g_c} = \frac{1000}{32.17} = 31.085$ slug

(3) *planetary gravity*

Table 1-3 Venus, $g = 29.11$ ft/sec^2

Step B — *Compute Weight*

Weight is the force due to gravitational acceleration:

(1-20) $F = ma = mg = (31.085)(29.11) = 904.9$ lbf

Work is defined as the amount of *energy* required to exert a constant force on a body which moves through a distance in the same direction as the applied force, or

$$work = force \times distance = mechanical\ energy$$

Mathematically,

$$\mathrm{W} = \mathrm{ME} = \int F\,dx \tag{1-21}$$

where W is the work, ME is the mechanical energy, and F is the force applied through the distance dx. Substituting Eq. (1-20) for force in Eq. (1-21),

$$\mathrm{ME} = \int F\,dx = \int ma\,dx = m\int a\,dx \tag{1-22}$$

Potential energy is defined as the energy required to lift a body to its present height from some datum. Substituting PE (potential energy) for ME, g for a and dz (elevation change) for dx in Eq. (1-22),

$$\mathrm{PE} = m\int a\,dx = m\int g\,dz \tag{1-23}$$

From Eq. (1-17),

$$g = \frac{g_\phi r^2}{(r+z)^2}$$

Substituting in Eq. (1-23),

$$\mathrm{PE} = m\int_0^z g\,dz = mg_\phi r^2 \int_0^z (r+z)^{-2}\,dz = mg_\phi z\left(1+\frac{z}{r}\right)^{-1} \tag{1-24}$$

Except for problems concerning the atmosphere (Chap. 2) the variation of gravity with elevation is so small that it may be assumed to be a constant and Eq. (1-23) integrated as follows:

$$\mathrm{PE} = m\int_0^z g\,dz = mg\int_0^z dz = mgz \tag{1-25}$$

EXAMPLE A Boeing 727 jet aircraft has a mass of 64 400 kg and is flying at an altitude of 10 km above sea level at a place where the gravitational acceleration is 9.806 m/s. Calculate the potential energy of the aircraft, assuming (a) constant gravity and (b) variation of gravity with elevation.

Step A *Data Reduction*

radius of the Earth

Table 1-3 $r = 6\,371$ km

Step B *Compute Potential Energy*

(1) *constant gravity*

(1-25) $\text{PE} = mgz = (64\,400)(9.806)(10 \times 10^3) = 6\,276 \text{ MJ}$

(2) *variable gravity*

(1-24) $\text{PE} = \dfrac{mg_\phi z}{1 + z/r} = \dfrac{(64\,400)\,(9.806)\,(10 \times 10^3)}{1 + 10/637\,1} = 6\,266 \text{ MJ}$

(3) *compare* Note: $g = g_\phi$ in Eq. (1-25)

(1-25)/(1-24) $\dfrac{\text{PE}_{g=c}}{\text{PE}_{g\neq c}} = \dfrac{mgz}{mg_\phi z(1 + z/r)^{-1}} = 1 + \dfrac{z}{r} = 1 + \dfrac{10}{637\,1} = 1.001\,570,$

which represents a deviation of 0.16%.

Kinetic energy is the energy of a body due to its motion. It is equivalent to the work required to impart this motion from rest in the absence of friction. Acceleration is the rate of change of velocity with time, or

$$a = \frac{dV}{dt} \qquad \textbf{(1-26)}$$

Substituting Eq. (1-26) in Eq. (1-20),

$$F = ma = \frac{m\,dV}{dt} \qquad \textbf{(1-27)}$$

And Eq. (1-27) in Eq. (1-22),

$$\text{ME} = m\int a\,dx = m\int\left(\frac{dV}{dt}\right)dx = m\int\left(\frac{dx}{dt}\right)dV \qquad \textbf{(1-28)}$$

Substituting KE (kinetic energy) for ME and $V = dx/dt$ in Eq. (1-28),

$$\text{KE} = m\int_0^V V\,dV = \frac{mV^2}{2} \qquad \textbf{(1-29)}$$

EXAMPLE A ball whose weight on the surface of Mars is 6 oz has a velocity of 20 ft/sec. What is its kinetic energy?

Step A *Data Reduction*

(1) *unit conversion*

(B-69) $F = 6(1/16) = 0.375$ lbf

(2) *planetary gravity* on Mars

Table 1-3 $g = 12.27$ ft/sec^2

Step B *Compute Mass of Ball* (weight is the force of gravity)

(1-20) $F = ma = mg$,

or

$$m = \frac{F}{g} = \frac{0.375}{12.27} = 30.562 \times 10^{-3} \text{ slugs}$$

Step C *Compute Kinetic Energy*

(1-29)
$$\text{KE} = \frac{mV^2}{2} = \frac{(30.562 \times 10^{-3})(20)^2}{2} = 6.112 \text{ ft-lbf}$$

Impulse and Momentum

Equation (1-27) may be written in the following form:

(1-27)
$$F\,dt = m\,dV$$

The *impulse* of a force is the integral of the left-hand side of this equation or

$$\text{Impulse of a force} = \int_{t_1}^{t_2} F\,dt \qquad \textbf{(1-30)}$$

For a *constant force* applied between t_1 and t_2,

$$\text{impulse of a force} = F\int_{t_1}^{t_2} dt = F(t_2 - t_1) \qquad \textbf{(1-31)}$$

Momentum is the product of mass times velocity and may be obtained by integrating the right-hand side of Eq. (1-27):

$$\text{Momentum change} = m\int_{V_1}^{V_2} dV = m(V_2 - V_1) = mV_2 - mV_1 \qquad \textbf{(1-32)}$$

Equating Eqs. (1-31) and (1-32),

$$F(t_2 - t_1) = m(V_2 - V_1) = mV_2 - mV_1 \qquad \textbf{(1-33)}$$

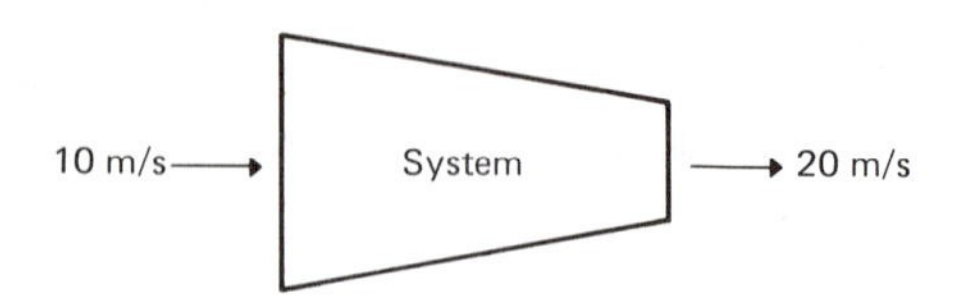

FIGURE 1-6 Jet propulsion system

or

impulse of a force = momentum change

For *constant mass* and *force* between t_1 and t_2 Eq. (1-33) may be written in the following form:

$$F = [m/(t_2 - t_1)][V_2 - V_1] = \dot{m}(V_2 - V_1) \quad \textbf{(1-34)}$$

where $\dot{m}$ is the mass flow rate.

EXAMPLE A stream of liquid flows steadily through the jet propulsion system shown in Fig. 1-6. It enters with a velocity of 10 m/s and leaves with a velocity of 20 m/s. During a 20-minute test period it was found that 22 Mg of liquid passed through the system. What is the thrust of this system?

Step A *Data Reduction*

(1) *unit conversion*

(B-68) $(t_2 - t_1) = 20 \times 60 = 1\,200$ s

Step B *Compute Thrust*

(1-34) $$F = \frac{m(V_2 - V_1)}{t_2 - t_1} = \frac{(22 \times 10^3)(20 - 10)}{1\,200}$$

$$F = 183.3 \text{ N}$$

1-9 DENSITY

Definition / Mass per unit volume
Symbol / ρ (rho)
Dimensions / ML^{-3} or FT^2L^{-4}
Units / U.S.: slugs/ft³ or lbf sec²/ft⁴ SI: kg/m³

Here again problems with "pure" U.S. units are encountered. Most heat transfer and all thermodynamic texts use the pound-mass for density. They also use the symbol ρ for density. The slug is used in fluid mechanics because its proportionality constant g_c has a numerical value of unity. In obtaining data from other sources in U.S. units students should use great care to make sure that they are in slugs/ft^3 and, if not, make the necessary conversions.

Values of density for selected liquids are given in Table A-1. Equations for the computation of the density of ideal gases are given in Sec. 1-13. Since density is the mass per unit volume, its numerical value is the same* any place in the universe because (Sec. 1-6) it represents a quantity of matter.

* Except where it is not. For example, we should not forget Albert Einstein's "energy = mass (velocity of light)^2". This may be restated in the following equation: $m = m_o/(1 - V^2/C_1^2)^{1/2}$, where m is the mass at velocity V, m_o the mass at rest, and C_1 the velocity of light. In any real engineering situation the ratio of fluid velocities to that of light is essentially zero. For this reason it will be assumed that mass (matter) will not change at any place in our universe.

EXAMPLE Check the value of density in SI units using the data given in Table A-1 for carbon tetrachloride at 68°F.

Step A *Data Reduction*

fluid properties, carbon tetrachloride at 68°F,

Table A-1 $\rho = 3.093$ slugs/ft^3

Step B *Convert to SI Units*

(B-101) $\rho = 3.093(515.4) = 1\,594$ kg/m^3

Step C *Compare Results*

Table A-1 $\rho = 1\,594$ kg/m^3 (checks exactly)

1-10 SPECIFIC WEIGHT

Definition / Weight (force) per unit volume
Symbol / γ (gamma)
Dimensions / FL^{-3} or $ML^{-2}T^{-2}$
Units / U.S.: lbf/ft^3 or slugs/(ft-sec)^2 SI: N/m^3

Relation to density / Eq. (1-20) related force to mass, and since both density and specific weight have the same volume units:

$$\frac{F_g}{\text{Vol}} = \frac{m}{\text{Vol}} \times g = \gamma = \rho g$$

or

$$\gamma = \rho g \tag{1-35}$$

EXAMPLE What is the specific weight of benzene at 30°C on the surface of (a) Earth and (b) of the planet Jupiter?

Step A *Data Reduction*

(1) *fluid properties,* benzene at 30°C

Table (A-1) $\rho = 868.0$ kg/m^3

(2) *planetary gravities*

Table 1-3 Earth, $g = 9.807$ m/s^2, Jupiter $g = 26.01$ m/s^2

Step B *Compute Specific Weight on Earth*

(1-35) $\gamma = \rho g = 868.0 \times 9.807 = 8.512$ kN/m^3

Step C *Compute Specific Weight on Jupiter*

(1-35) $\gamma = \rho g = 868.0 \times 26.01 = 22.58$ kN/m^3

1-11 SPECIFIC VOLUME

Definition / Volume per unit mass
Symbol / v
Dimensions / L^3M^{-1} or $F^{-1}L^4T^{-2}$
Units / U.S.: ft^3/lbm or ft^4/lbf − sec^2 SI: m^3/kg

Here we have problems with both definitions and U.S. units. Some fluid mechanics texts define specific volume as the reciprocal of specific weight. Thermodynamic textbooks and general engineering practice is to define specific volume as volume per unit mass. Tables of thermodynamic properties of substances report specific volume in units of cubic feet per pound-mass. There is little question that the U.S. customary unit is the pound-mass.

Relation to Density

Because in U.S. units the slug is used for mass in density and the pound-mass for specific volume, density can be stated as

$$\rho = \frac{m}{\text{Vol}} \quad \text{or} \quad \text{Vol} = \frac{m}{\rho} \quad \text{and} \quad v = \frac{\text{Vol}}{M}$$

From Eq. (1-19)

$$m = \frac{M}{g_c} \quad \text{or} \quad M = mg_c.$$

Substituting as follows,

$$v = \frac{\text{Vol}}{M} = \frac{m/\rho}{M} = \frac{m/\rho}{mg_c} = \frac{1}{\rho g_c} \tag{1-36}$$

Relation to Specific Weight

From Eq. (1-35)

$$\gamma = \rho g \quad \text{or} \quad \rho = \frac{\gamma}{g}.$$

Substituting in Eq. (1-36),

$$v = \frac{1}{\rho g_c} = \frac{1}{(\gamma/g)g_c} = \frac{g}{\gamma g_c} \tag{1-37}$$

EXAMPLE 1 What is the specific volume of fresh water at 68°F under 10,000 psia pressure (a) on the surface of the planet Neptune and (b) on the surface of the earth at a latitude of 45°?

Analysis Specific volume is the volume of a unit mass. Since mass is a quantity of matter, its value will be the same anywhere in the universe.

Step A *Data Reduction*

(1) *fluid properties,* water at 68°F and 10,000 psia

Table A-1 $\rho = 1.992$ lbf-sec²/ft⁴

(2) *proportionality constant* (U.S. units)

(1-11) $g_c = 32.17$ lbm-ft/sec²-lbf

Step B *Compute Specific Volume*

(1-36)

$$v = \frac{1}{\rho g_c} = \frac{1}{(1.992)(32.17)}$$

$$= 0.01560 \text{ ft}^3/\text{lbm}$$

EXAMPLE 2 A substance has a specific weight of 1 625 N/m^3 on the surface of the moon. What is its specific volume?

Step A *Data Reduction*

(1) *acceleration due to gravity* (moon surface)

Table 1-3 $g = 1.620 \text{ m/s}^2$

(2) *proportionality constant* (SI units)

(1-9) $g_c = 1 \text{ kg} \cdot \text{m/N} \cdot \text{s}^2$

Step B *Compute Specific Volume*

(1-37)
$$v = \frac{g}{\gamma g_c} = \frac{1.620}{1.625(1)}$$
$$= 996.9 \text{ cm}^3\text{/N}$$

1-12 SPECIFIC GRAVITY

Definition / Fluid density/reference fluid density
Symbol / S
Dimensions / Dimensionless ratio
Units / None
Reference fluids / Solids and liquids: water / Gases: air

Since the density of water changes with temperature and at high pressures with pressure (refer to Table A-1), for a precise definition of specific gravity, the temperatures and pressures of the fluid and the reference fluid should be stated. In practice, two temperatures are stated, for example, 60/60°F where the upper temperature pertains to the fluid and the lower to water. If no temperatures are stated, it must be assumed that reference is made to water at its maximum density. The maximum density of water at atmospheric pressure is at 3.98°C and has a value of 999.972 kg/m^3 (1.940 268 $slugs/ft^3$).

For gases it is common practice to use the ratio of the molecular weight of the gas to that of air (28.9644), thus eliminating the necessity of stating the pressure and temperature for ideal gases.

Hydrometer Scale Conversions

In certain fields of industry hydrometer scales are used which have arbitrary graduations. In the petroleum and chemical industries, the Baumé (°Be) and the American Petroleum Institute (°API) are used. Conversions are as follows:

Baumé Scale

Heavier than water $$S_{60/60°F} = \frac{145}{145 - °Be} \quad \text{(1-38)}$$

Lighter than water $$S_{60/60°F} = \frac{140}{130 + °Be} \quad \text{(1-39)}$$

American Petroleum Institute Scale

$$S_{60/60°F} = \frac{141.5}{131.5 + °API} \quad \text{(1-40)}$$

The Baumé scale for liquids lighter than water is very nearly the same as the American Petroleum Institute scales, both being 10° for a specific gravity of unity. The use of the American Petroleum Institute scale in preference to the Baumé scale for liquids lighter than water is recommended by the American National Standards Institute. Standardized hydrometers are available in various ranges from −1° API to 101° API for specific gravity ranges of 1.0843 to 0.6086 at 60/60°F.

EXAMPLE What is the American Petroleum Institute gravity of benzene?

Step A *Data Reduction*

fluid properties

Table A-1

Temperature	Water	Benzene	Units
50°F	1.940	1.726	slugs/ft^3
68°F	1.937	1.705	slugs/ft^3

Step B *Compute 60/60°F Gravities*

$$S_{50/50°F} = \frac{1.726}{1.940} = 0.88969$$

$$S_{68/68°F} = \frac{1.705}{1.937} = 0.88023$$

By linear interpolation

$$S_{60/60°F} = \frac{(0.088023 - 0.88969)(60 - 50)}{68 - 50} + 0.88969 = 0.8844$$

Step C *Compute °API*

(1-40) $$S_{60/60°F} = \frac{141.5}{131.5 + °API}$$

Solving for °API,

$$°API = \frac{141.5}{S_{60/60°F}} - 131.5 = \frac{141.5}{0.8844} - 131.5 = 28.50$$

1-13 THERMODYNAMIC PROCESSES

Because the study of fluid mechanics involves the behavior of compressible fluids, it becomes necessary to consider certain thermodynamic processes. For derivations and additional information, the student should refer to a thermodynamics textbook.

When one out of all the properties of a fluid changes, the fluid is said to have undergone a *process*. If the fluid can be made to return to its original state by exactly retracing its path then the process is said to be *reversible*. A reversible process is frictionless and cannot occur in nature, so that reversible processes serve as ideals.

All processes are *polytropic* processes, and the processes shown below are all special cases of the polytropic. For an ideal gas, the relation between pressure and specific volume is given by

$$pv^n = \text{a constant} \tag{1-41}$$

Values of n for various ideal gas processes are

**Isentropic*	$s = c$	*frictionless adiabatic*	$n = k = c_p/c_v$
Isothermal	$T = c$	*constant temperature*	$n = 1$
Isobaric	$p = c$	*constant pressure*	$n = 0$
Isometric	$V = c$	*constant volume*	$n = \infty$
Polytropic	–	*any*	$n = -\infty$ to $+\infty$

1-14 EQUATION OF STATE FOR AN IDEAL GAS

Definition

An ideal gas is one that obeys the equation of state (1-42) and whose internal energy is a function of temperature only.

* If a process takes place without heat transfer and is also reversible, it follows a path of constant entropy (s) and hence is called an isentropic process. Many fluid mechanics texts call this process a frictionless adiabatic or simply (and incorrectly) an adiabatic process.

The derivation of the equation of state from Boyle's and Charles' laws may be found in any thermodynamics textbook.

Equations

The equation of state for an ideal gas is

$$pv = RT \tag{1-42}$$

where

Unit	U.S.	SI
p = pressure	lbf/ft^2	N/m^2
v = specific volume	ft^3/lbm	m^3/kg
R = gas constant	ft-lbf/lbm-°R	J/kg · K
T = temperature	°R	K

Ideal Gas

Properties of selected gases are given in Table A-7. Consideration of Avogadro's principle leads to the concept of a universal gas constant:

$$\bar{R} = MR \tag{1-43}$$

where

Unit	U.S.	SI
M = Molecular weight	lbm-mol	kg mol
$\bar{R}$ = Universal gas constant	1545 ft-lbf/lbm-mol-°R	8 314 J/kg · mol · K

For computation of density, Eqs. (1-36) and (1-42) yield

$$\rho = \frac{p}{g_c RT} \tag{1-44}$$

Real gases do not obey the ideal gas relationships exactly. For engineering calculations, the equation of state gives satisfactory results if the pressure is 1/5 of the critical pressure and the absolute temperature is 2.5 times the critical temperature. As $T \to \infty$ and $p \to 0$ all gases approach the ideal gas state.

Other *p-v-T relations* for ideal gases may be obtained by combining the equation of state $pv = RT$ (1-42) with the polytropic process relation $pv^n = C$ (1-41) to produce the following:

For pressure

$$\frac{p_2}{p_1} = \left(\frac{v_1}{v_2}\right)^n = \left(\frac{T_2}{T_1}\right)^{n/(n-1)} \tag{1-45}$$

For specific volume $$\frac{v_2}{v_1} = \left(\frac{p_1}{p_2}\right)^{1/n} = \left(\frac{T_1}{T_2}\right)^{1/(n-1)} \quad \textbf{(1-46)}$$

For temperature $$\frac{T_2}{T_1} = \left(\frac{v_1}{v_2}\right)^{(n-1)} = \left(\frac{p_2}{p_1}\right)^{(n-1)/n} \quad \textbf{(1-47)}$$

EXAMPLE A tank with a fixed volume of 62.42 ft³ initially contains carbon monoxide at 15 psia and 70°F. Three pounds of carbon monoxide are added to the tank. If the final temperature is 75°F, what is the final pressure?

Step A *Data Reduction*

(1) *unit conversion*

(B-89) $p_1 = 144 \times 15 = 2160 \text{ lbf/ft}^2$

(B-28) $T_1 = t_1 + 459.67 = 70 + 459.67 = 529.67°\text{R}$

(B-28) $T_2 = t_2 + 459.67 = 75 + 459.67 = 534.67°\text{R}$

(2) *fluid properties,* carbon monoxide

Table A-7 $R = 55.17 \text{ ft-lbf/lbm-°R}$

Step B *Write the Equation of State in Terms of Mass*

(1-42) $pv = RT$

(1-36) $v = \dfrac{\text{Vol}}{M}$, substituting in Eq. (1-42),

$$p\left(\frac{\text{Vol}}{M}\right) = RT \quad \text{or} \quad M = \frac{p\text{Vol}}{RT} \quad \textbf{(a)}$$

Step C *Apply the Conservation of Mass Principle*

final mass = initial mass + mass added

$$M_2 = M_1 + M_a \quad \textbf{(b)}$$

Substituting from (a)

$$\frac{p_2\text{Vol}_2}{RT_2} = \frac{p_1\text{Vol}_1}{RT_1} + M_a \quad \textbf{(c)}$$

Solving (c) for p_2

$$p_2 = \frac{RT_2}{\text{Vol}_2} \times \frac{p_1\text{Vol}_1}{RT_1 + M_a} \quad \textbf{(d)}$$

Noting that $\text{Vol}_2 = \text{Vol}_1 = 62.42 \text{ ft}^3$

$$p_2 = \frac{55.17 \times 534.67}{62.42} \times \frac{2160 \times 62.42}{55.17 \times 529.67 + 3} = 3{,}598 \text{ lbf/ft}^2$$

(B-87) $p_2 = \dfrac{3598}{144} = 24.99 \text{ psia}$

1-15 BULK MODULUS OF ELASTICITY

Definition / Stress/volumetric strain
Symbol / E
Dimensions / FL^{-2} or $ML^{-1}T^{-2}$
Units / U.S.: lbf/ft^2 SI: N/m^2 or Pa

Derivation of Basic Equations

Consider the piston and cylinder of Fig. 1-7. A fluid originally under a pressure of p had a volume of Vol. An additional pressure of dp is imposed, resulting in a decrease of volume $-d\text{Vol}$. From the definition of bulk modulus,

$$E = \frac{\text{stress}}{\text{strain}} = \frac{dp}{-d\text{Vol}/\text{Vol}} = -\text{Vol}\left(\frac{dp}{d\text{Vol}}\right) \tag{1-48}$$

Substitution of $\text{Vol} = vM$ from Eq. (1-36) in (1-48) results in

$$E = -vM\left(\frac{dp}{dMv}\right) = -v\left(\frac{dp}{dv}\right) \tag{1-49}$$

Equation (1-49) cannot be integrated unless the pressure–specific volume relationship, and hence the process, is known, so that

$$E_n = -v\left(\frac{dp}{dv}\right)_n \tag{1-50}$$

FIGURE 1-7 Notation for bulk modulus

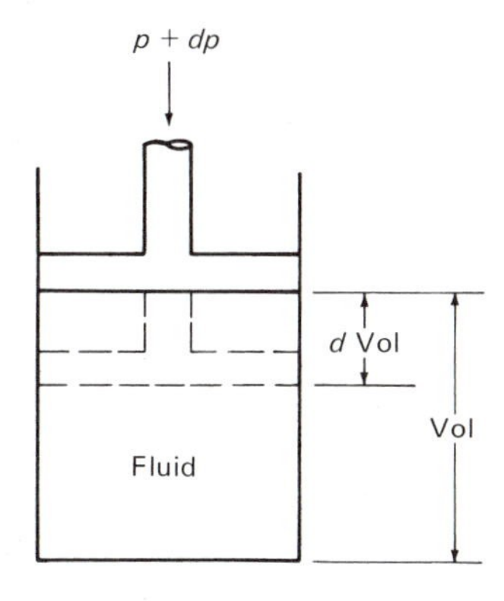

where E_n is the bulk modulus of elasticity for process n and $(dp/dv)_n$ indicates the pressure–specific volume gradient for that process. Although any number of processes are possible, conventional practice is to use only the *isothermal bulk modulus* (E_T) and the *isentropic bulk modulus* E_s.

Ideal Gases

If Eq. (1-41) $pv^n = C$ is written in logarithmic form ($\log_e p + n \log_e v = \log_e C$) and differentiated,

$$\frac{dp}{p} + \frac{n\,dv}{v} = 0, \quad \text{or} \quad np = -v\left(\frac{dp}{dv}\right) \tag{1-51}$$

Equating Eqs. (1-50) and (1-51),

$$E_n = -v\left(\frac{dp}{dv}\right)_n = np \tag{1-52}$$

For an ideal gas the bulk modulus of elasticity is the product of the process exponent and the pressure.

The Isothermal Bulk Modulus of Elasticity E_T

For an isothermal process, $n = 1$, so that from Eq. (1-52)

$$E_T = -v\left(\frac{dp}{dv}\right)_T = (1)p = p \tag{1-53}$$

The Isentropic Bulk Modulus of Elasticity E_s

For an isentropic process, $n = k = c_p/c_v$, so from Eq. (1-52)

$$E_s = -v\left(\frac{dp}{dv}\right)_s = kp \tag{1-54}$$

Liquids

At constant pressure the bulk modulus of most liquids decreases linearly with increasing temperature. Water is one exception and increases to a maximum value at 120°F (49°C) and decreases in value above that temperature. At constant temperature the value of the bulk modulus of elasticity increases with temperature for all liquids. No simple pressure–specific volume relationship similar to $pv^n = C$, for ideal gases, exists for liquids. Values of liquid bulk modulus must be determined experimentally. Values of E_T and E_s for selected liquids are given in Tables A-2 and A-3 respectively. For liquids Eq. (1-50) may be integrated over small intervals as follows:

$$E_n = -\int_1^2 v\left(\frac{dp}{dv}\right)_n \approx -v_1\left(\frac{\Delta p}{\Delta v}\right) = v_1\left(\frac{p_2 - p_1}{v_1 - v_2}\right) \tag{1-55}$$

Some handbooks and other data sources use Eq. (1-55) as a definition of the bulk modulus. In obtaining and using data from other sources the type of equation used to define the bulk modulus should be verified.

EXAMPLE How many standard atmospheres must be imposed on water at 20°C to reduce its volume 1%?

Step A — *Data Reduction*

(1) *standard atmospheric pressure*

(B-5) $p_b = 1.013 \times 10^5\ \text{N/m}^2$

(2) *fluid properties,* water at 20°C

Table A-2 $E_T = 21.78 \times 10^8\ \text{N/m}^2$

Step B — *Compute Pressure*

Given $\dfrac{\Delta v}{v_1} = -1\% = -\dfrac{1}{100}$

(1-55) $E_T = -v_1\left(\dfrac{\Delta p}{\Delta v}\right)$

Solving for Δp,

$$\Delta p = -E_T\left(\frac{\Delta v}{v_1}\right) = -(21.78 \times 10^8)\left(-\frac{1}{100}\right) \tag{a}$$

$$= 21.78 \times 10^6\ \text{N/m}^2$$

$$\frac{\Delta p}{p_b} = \frac{21.78 \times 10^6}{1.013 \times 10^5} = 215\ \text{atm} \tag{b}$$

1-16 ACOUSTIC VELOCITY

Definition / Speed of a small pressure (sound) wave in a fluid
Symbol / c
Dimensions / LT^{-1}
Units / U.S.: ft/sec SI: m/s

Derivation of Basic Equations

Consider an elastic fluid in a rigid pipe fitted with a cylinder as shown in Fig. 1-8. The pipe has a uniform cross-sectional area of A. As a result of the application of a force dF the piston is suddenly advanced with velocity of V for a time dt. The fluid pressure p is increased by the amount of dp which travels as a wave front with a velocity of c. During the application time dt, the piston moves a distance of $V\,dt$ and the wave front advances a distance of $c\,dt$. The result of this piston movement is to decrease the volume $c\,dt\,A$ by the amount of the volume $V\,dt\,A$.

From Sec. 1-15, Eq. (1-48), the bulk modulus of elasticity of a fluid is

$$E = \frac{\text{stress}}{\text{strain}} = \frac{dp}{d\,\text{Vol/Vol}} = -\frac{dp}{(V\,dt\,A)/(c\,dt\,A)} = -\frac{c\,dp}{V}$$

or

$$c = -VE/dp \tag{1-56}$$

The force dF imposed is $(p + dp)A - pA = dp\,A$. The mass of fluid accelerated in time dt is $\rho c\,dt\,A$, so that the mass rate is $\dot{m} = m/dt = (\rho c\,dt\,A)/dt = \rho cA$. The velocity change is from V to 0. From the impulse-momentum equation (1-34) $F = \dot{m}(V_2 - V_1)$:

$$dF = dp\,A = \rho cA(0 - V) \quad \text{or} \quad c = -dp/\rho V \tag{1-57}$$

Multiplying Eq. (1-56) by Eq. (1-57),

$$c^2 = \left(\frac{-VE}{dp}\right)\left(\frac{-dp}{\rho V}\right) = \frac{E}{\rho}$$

or

$$c = \sqrt{E/\rho} \tag{1-58}$$

FIGURE 1-8 Notation for acoustic velocity

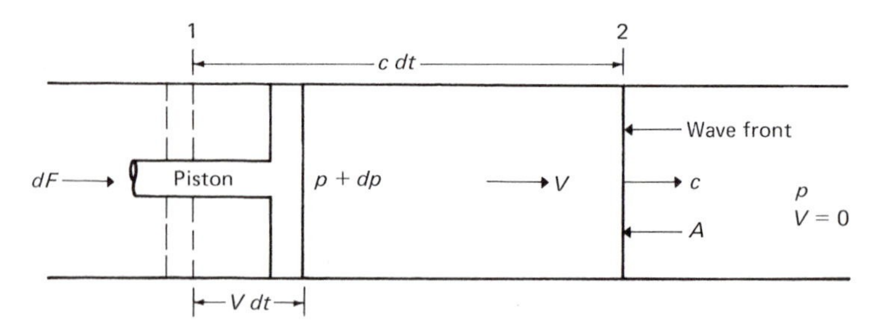

The numerical value of E depends upon the process. It is assumed that a small pressure (sound) wave will travel through the fluid without either heat transfer or friction so that the process is isentropic and Eq. (1-58) becomes

$$c = \sqrt{E_s/\rho} \tag{1-59}$$

Equation (1-59) may be used for any fluid whose value of E_s is known.

Ideal Gases

From Eq. (1-54) $E_s = kp$ and from Eq. (1-44) $\rho = p/g_c RT$. Substituting these values in Eq. (1-59),

$$c = \sqrt{E_s/\rho} = \sqrt{(kp)/(p/g_c RT)} = \sqrt{kg_c RT} \tag{1-60}$$

EXAMPLE 1 How long does it take a pressure wave to travel one mile through fresh water at 68°F and at atmospheric pressure?

Step A *Data Reduction*

(1) *unit conversion*

(B-65) 1 mile = 5280 ft

(2) *fluid properties,* fresh water at 68°F

Table A-1 $\rho = 1.937$ slugs/ft^3

Table A-3 $E_s = 45.81 \times 10^6$ lbf/ft^2

Step B *Compute Time of Travel*

(1-59) $c = \sqrt{E_s/\rho} = \sqrt{(45.81 \times 10^6)/(1.937)} = 4863$ ft/sec

$$\text{Time} = \frac{\text{distance}}{\text{velocity}} = \frac{5280}{4863} = 1.086 \text{ sec}$$

EXAMPLE 2 What is the velocity of sound in air at 20°C?

Step A *Data Reduction*

(1) *unit conversion*

(B-27) $T_K = t_C + 273.15 = 20 + 273.15 = 293.15$ K

(2) *fluid properties,* 20°C air

Table A-7 $R = 287.1$ J/kg · K, $k = 1.400$

(3) *proportionality constant*

(1-9) $g_c = 1$ kg · m/N · s^2

Step B *Compute Speed of Sound*

(1-60) $c = \sqrt{k g_c R T} = \sqrt{(1.400)(1)(287.1)(293.15)}$

$c = 343.3$ m/s

1-17 VISCOSITY

Definition

Viscosity is the resistance of a fluid to motion—its internal friction. A fluid in a static state is by definition unable to resist even the slightest amount of shear stress. Application of shear force results in the continual and permanent distortion known as flow.

Shear Stress

In Sec. 1-2, consideration of Fig. 1-1 led to the development of the unit shear stress τ (tau) $= F_s/A_s$, where F_s is the shear force and A_s is the shear area.

Rate of Shearing Strain

Also developed from consideration of Fig. 1-1 was the rate of deformation (shearing strain) as dU/dy, where U is the velocity and y is the distance perpendicular to the shear.

Newtonian Fluid

A Newtonian fluid is defined in Sec. 1-2 as a fluid whose rate of deformation is directly proportional to the applied shear stress, or

$$\tau = \frac{F_s}{A_s} \propto \frac{dU}{dy} = \frac{\mu \, dU}{dy} \tag{1-61}$$

where μ (mu) is the viscosity. In this form various texts call μ:

(a) coefficient of viscosity
(b) absolute viscosity
(c) dynamic viscosity (used in this text)

Dynamic Viscosity

Definition / Shearing stress/rate of shearing strain
Symbol / μ (mu)
Dimensions / FL^{-2} or $ML^{-1}T^{-1}$
Units / U.S.: lbf-sec/ft^2 or slugs/ft-sec SI: N · s/m^2 or Pa · s

---

Kinematic Viscosity

Definition / Dynamic viscosity/density
Symbol / ν (nu)
Dimensions / L^2T^{-1}
Units / U.S.: ft^2/sec SI: m^2/s
Relation to dynamic viscosity / $\nu = \mu/\rho$ **(1-62)**

Note that since the kinematic viscosity is a function of density it must be both temperature and pressure dependent. For ideal gases Eq. (1-44) $\rho = p/g_cRT$ may be substituted in Eq. (1-62). Thus

$$\nu = \mu g_c RT/p \qquad \text{(ideal gas only)} \tag{1-63}$$

Characteristics

In a flowing fluid tangential or shear stresses arise from two different molecular phenomena. The first is the cohesive or attractive forces of the molecules, which resist motion. The second is the molecular activity, which causes resistance to flow due to molecular momentum transfer. Molecular momentum transfer may be visualized by considering two trains made up of coal cars moving in the same direction on parallel tracks but at different speeds. As these trains pass each other coal is thrown from one train to the other and vice versa. From considerations of impulse and momentum, each will tend to resist the motion of the other to some degree and the slower train will tend to speed up and the faster to slow down.

Liquid Viscosities

In liquids cohesive forces predominate. Since cohesive forces decrease with increasing temperatures, so do liquid viscosities. The dynamic viscosity of selected liquids is given in Table A-4 and kinematic viscosity in Fig. 1-9.

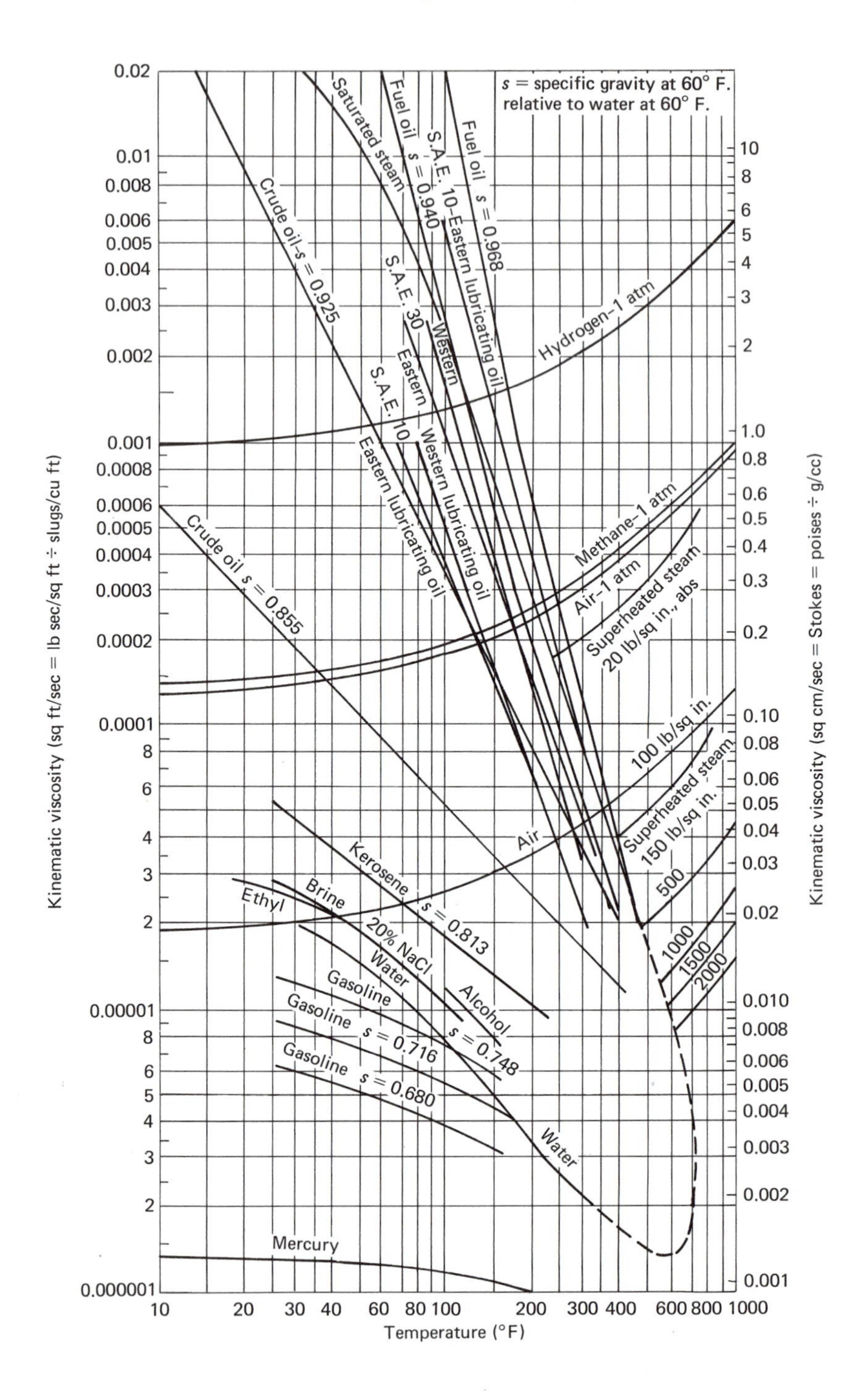

FIGURE 1-9 Kinematic viscosity of selected fluids (Adapted from *Pipe Friction Manual,* 3rd edition, 1951, published by Hydraulic Institute, Cleveland)

Ideal Gas Viscosities

In ideal gases, cohesive forces are absent. Molecular activity increases with temperature and so does viscosity. The dynamic viscosity of selected gases is given in Table A-8 and kinematic viscosity in Fig. 1-9.

Other Units of Viscosity

A unit of dynamic viscosity named after Jean Louis Poiseuille (1799–1869), a French scientist who performed meticulous tests on the resistance of flow through capillary tubes, a *poise* is defined as one dyne-second per square centimeter. In the SI system this is equal to $0.1\ N \cdot s/m^2$ or $0.1\ Pa \cdot s$. The viscosity of water at 20°C from Table A-4 is $10.02 \times 10^{-4}\ N \cdot s/m^2$ or 1.002×10^{-2} poises. Because of the magnitude of the poise the *centipoise* (1/100 poise) is used. The value of water at 20°C is thus approximately one centipoise. The conversion factor for U.S. units is 2.089×10^3 lbf-sec/ft^2.

A unit of kinematic viscosity named after George Gabriel Stokes (1819–1903), an English scientist who derived analytically various flow relationships ranging from wave mechanics to viscous resistances, particularly that for a falling sphere, a *stoke* is defined as one square centimeter per second. In SI units the stoke is equal to $1 \times 10^{-4}\ m^2/s$. In U.S. units the stoke is equal to 1.076×10^{-3} ft^2/sec. Like the poise, the *centistoke* is used because of size.

The standard viscometer for industrial work in the United States is the Saybolt universal viscometer. It consists essentially of a metal tube and an orifice built to rigid specifications and calibrated with fluids of known viscosity. The time required for a gravity flow of 60 cc is a measure of the kinematic viscosity of the fluid and is called SSU (*Saybolt Seconds Universal*). Approximate conversions of SSU to stokes may be made using the following equations:

(1) $\text{SSU} < 100$ seconds: $\text{stokes} = 0.00226\ \text{SSU} - 1.95/\text{SSU}$ **(1-64)**

(2) $\text{SSU} > 100$ seconds: $\text{stokes} = 0.00220\ \text{SSU} - 1.35/\text{SSU}$ **(1-65)**

EXAMPLE 1 An oil is tested in an industrial laboratory at a temperature of 60°F. It took 400 sec for 60 cc of this oil to flow from a standard Saybolt universal viscometer. An American Petroleum Institute hydrometer indicated that the oil had a gravity of 20°API. Compute the dynamic viscosity of this oil.

Step A *Data Reduction*

(1) *unit conversions*

(1-65) $\text{stokes} = 0.00220\ \text{SSU} - 1.35/\text{SSU} \quad \text{SSU} > 100$

(1-65) $$\text{stokes} = 0.00220(400) - \frac{1.35}{400} = 0.8766$$

(B-102) $$\nu = (1.076 \times 10^{-3})(0.8766) = 943.2 \times 10^{-6}\ \text{ft}^2/\text{sec}$$

(1-40) $$S_{60/60°\text{F}} = \frac{141.5}{131.5 + °\text{API}} = \frac{141.5}{131.5 + 20} = 0.9340$$

(2) *fluid properties*

Table A-1 $$\rho_w = 1.938\ \text{slugs/ft}^3 \text{ at } 60°\text{F (interpolated) (water)}$$

$$\rho_{oil} = S\rho_w = 0.934 \times 1.938 = 1.810\ \text{slugs/ft}^3 = \rho$$

Step B *Compute Dynamic Viscosity*

(1-62) $$\mu = \nu\rho = (943.2 \times 10^{-6})(1.810) = 1707 \times 10^{-6}\ \text{lbf-sec/ft}^2$$

EXAMPLE 2 What is the kinematic viscosity of carbon dioxide at 50°C and standard atmospheric pressure?

Step A *Data Reduction*

(1) *unit conversion*

(B-27) $$T_K = t_C + 273.15 = 50 + 273.15 = 323.15\ \text{K}$$

(2) *fluid properties,* carbon dioxide at 50°C

Table A-7 $$R = 188.9\ \text{J/kg} \cdot \text{K}$$

Table A-8 $$\mu = 16.20 \times 10^{-6}\ \text{N} \cdot \text{s/m}^2$$

(3) *proportionality constant*

(1-9) $$g_c = 1\ \text{kg} \cdot \text{m/N} \cdot \text{s}^2$$

(4) *standard atmosphere*

(B-5) $$p = 101.3\ \text{kN/m}^2$$

Step B *Compute Kinematic Viscosity*

(1-63) $$\nu = \frac{\mu g_c R T}{p} = \frac{(16.20 \times 10^{-6})(1)(188.9)(323.15)}{1.013 \times 10^5}$$

$$\nu = 9.762 \times 10^{-6}\ \text{m}^2/\text{s}$$

EXAMPLE 3 Castor oil at 20°C fills the space between two parallel plates 6 mm apart. If the upper plate moves with a velocity of 1.5 m/s and the lower plate is stationary, what is the unit shear stress in the oil if the velocity gradient is linear?

Step A	*Data Reduction*
	fluid properties, Castor oil at 20°C
Table A-4	$\mu = 9\,610 \times 10^{-4}\ \text{N} \cdot \text{s/m}^2$

Step B	*Derive an Equation*
Fig. 1-1	For a linear velocity gradient, by similar triangles,

$$\frac{U}{y} = \frac{dU}{dy} \qquad \textbf{(a)}$$

(1-61)

$$\tau = \mu\left(\frac{dU}{dy}\right)$$

Substituting from Eq. (a),

$$\tau = \frac{\mu U}{y} \qquad \textbf{(b)}$$

Step C *Solve Developed Equation*

$$\tau = \frac{\mu U}{y} = \frac{9\,610 \times 10^{-4} \times 1.5}{6 \times 10^{-3}} = 240.3\ \text{N/m}^2 \qquad \textbf{(b)}$$

EXAMPLE 4 A long vertical cylinder of 1.500-in. radius rotates inside a fixed tube. The tangential velocity of the outer surface of the rotating cylinder is 3 ft/sec. The uniform space of 0.0012 in. between the cylinders is filled with an oil which produces a resistance to motion of 450 lbf for a cylinder length of 6 in. What is the dynamic viscosity of this oil?

FIGURE 1-10 Notation for Example 4

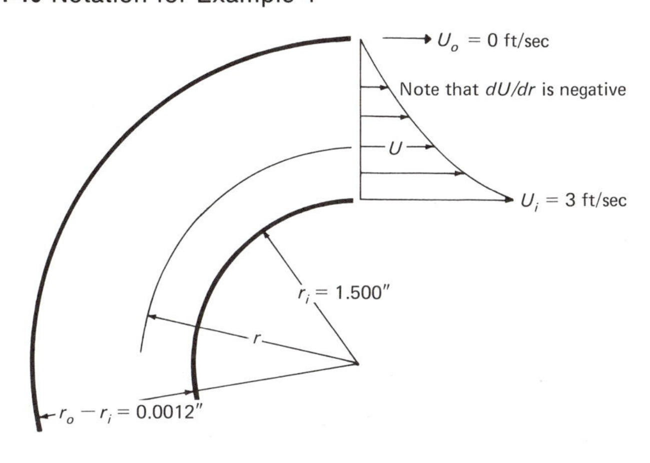

Step A	*Data Reduction*
	(1) *unit conversion*
(B-55)	Inner radius, $r_i = \dfrac{1.500}{12} = 0.125$ ft
(B-55)	Space, $r_o - r_i = \dfrac{0.0012}{12} = 1 \times 10^{-4}$ ft
(B-55)	Cylinder length, $L = \dfrac{6}{12} = 0.5$ ft
	(2) *geometric*
Fig. 1-10	$r_o = (r_o - r_i) + r_i = 1 \times 10^{-4} + 0.125 = 0.1251$ ft
	Shear area $A_s = 2\pi rL$ **(a)**

Step B *Derive an Equation for This Application*

(1-61) $\tau = \dfrac{F_s}{A_s} = \dfrac{\mu\, dU}{dy}$, substituting from Eq. (a)

$$\frac{F_s}{2\pi rL} = \frac{\mu\, dU}{dy}, \quad \text{or} \quad F_s = \frac{(2\pi rL)\mu\, dU}{dy} \qquad \textbf{(b)}$$

Fig. 1-10 As the radius increases the velocity decreases or the velocity gradient is negative, so that

$$\frac{du}{dy} = \frac{-dU}{dr}, \qquad \textbf{(c)}$$

substituting in Eq. (b)

$$F_s = 2\pi rL\mu\left(\frac{-dU}{dr}\right) \qquad \textbf{(d)}$$

which may be written as

$$\frac{-dr}{r} = \left(\frac{2\pi L\mu}{F_s}\right) dU, \qquad \textbf{(e)}$$

Integrating Eq. (e),

$$-\int_{r_o}^{r_i} dr/r = \int_{r_i}^{r_o} dr/r = \log_e(r_o/r_i) = (2\pi L\mu/F_s)\int_{U_o}^{U_i} dU$$

$$= (2\pi L\mu/F_s)(U_i - U_o),$$

simplifying and solving for μ, **(f)**

$$\mu = \frac{(F_s/2\pi L)(\log_e r_o/r_i)}{U_i - U_o} \qquad \textbf{(g)}$$

Step C *Solve Derived Equation*

$$\mu = \frac{(450/2\pi 0.5)(\log_e 0.1251/0.1250)}{3 - 0} \qquad \textbf{(g)}$$

$\mu = 0.03818$ lbf-sec/ft^2

1-18 SURFACE TENSION AND CAPILLARITY

Liquid surface characteristics are dependent upon molecular attraction. *Cohesion* is the attraction of like molecules, and *adhesion* the attraction of unlike molecules for each other. A liquid surface is able to support a very small tensile stress because of adhesion. *Surface tension* is the work done in extending the surface of a liquid one unit area. Liquid surfaces in contact with a solid surface will rise at the point of contact if adhesive forces predominate and will depress when cohesive forces are strongest, as shown in Fig. 1-11. *Capillarity* is the elevation or depression of a liquid surface in contact with a solid.

FIGURE 1-11 Notation for capillary study

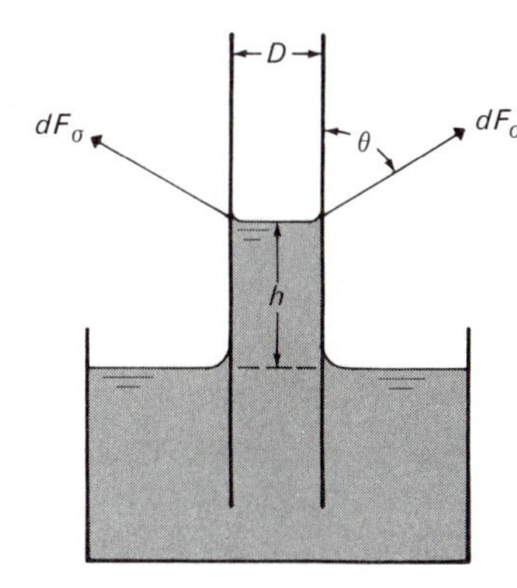

(a) Adhesive forces predominating

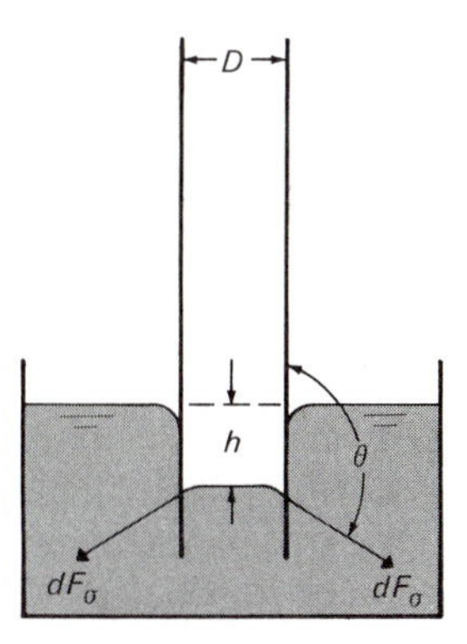

(b) Cohesive forces predominating

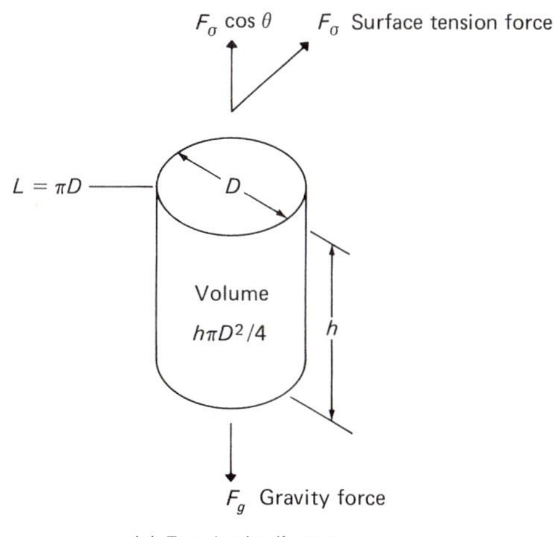

(c) Free body diagram

Surface Tension

Definition / Work per unit area	
Symbol / σ (sigma)	
Dimensions / $FL/L^2 = FL^{-1}$ or MT^{-2}	
Units / U.S.: lbf/ft SI: N/m	

Consider Fig. 1-12, which shows a soap film between a wire frame equipped with a frictionless slider which can move in the "x" direction. Moving the slider the distance dx to the right increases the surface S by the amount $L\,dx$. The force required to increase *both* (the film has two sides) surfaces $2L\,dx$ is $2F_\sigma$. The work of extension is $2F_\sigma\,dx$. From the definition of surface tension,

$$\sigma = \frac{\text{work}}{\text{unit area}} = \frac{dW}{dS} = \frac{2F_\sigma\,dx}{2L\,dx} = \frac{F_\sigma}{L}$$

or

$$F_\sigma = \sigma L \qquad \text{(1-66)}$$

The numerical value of surface tension depends upon the temperature and the fluids in contact at the interface, for example, water-air, water-steam, water-carbon dioxide, etc. Numerical values of surface tension for selected liquids is given in Table A-5.

Capillarity

Consider the tank and tube arrangements shown in Fig. 1-11. The angle θ (theta) is the contact angle between the liquid and solid surfaces. In Figure 1-11(a), the liquid has strong adhesive

FIGURE 1-12 Notation for surface tension

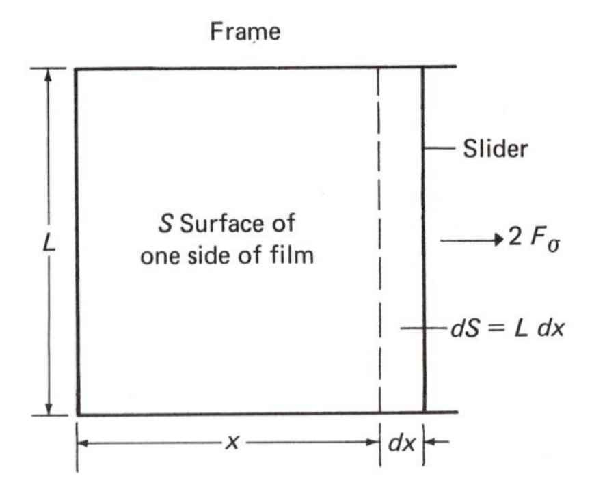

forces and rises at the liquid-solid interfaces. When the contact angle θ is less than 90° then the liquid is said to "wet" the tube walls. Figure 1-11(b), on the other hand, has a liquid with strong cohesive forces and depresses at the liquid-solid interfaces. The curved portion of the liquid surface in the tube is called the *meniscus*. The rise or fall h of a liquid column of diameter D can be derived from the free body diagram of Fig. 1-11(c). For equilibrium,

$$\Sigma F_z = F_\sigma \cos\theta - F_g = 0$$

or

$$F_\sigma \cos\theta = F_g \tag{1-67}$$

If the small mass of fluid above the meniscus is neglected, then the mass of fluid in the tube above the surface in the tank is $m = \rho h \pi D^2/4$. From Eq. (1-20) $F = ma$, and $F_g = mg$, so that the gravity force is $F_g = (\rho h \pi D^2/4)g = \rho g h \pi D^2/4$. The portion of the surface tension force acting in the "z" direction is $F_\sigma \cos\theta$. From Eq. (1-66) $F_\sigma = \sigma L$. The length L is the circumference of the tube, so that $F_\sigma = \sigma L = \sigma \pi D$. Substituting the above in Eq. (1-67),

$$F_\sigma \cos\theta = (\sigma\pi D)\cos\theta = F_g = \rho g h \pi D^2/4$$

or

$$h = 4\sigma \cos\theta/\rho_g D \tag{1-68}$$

From Eq. (1-68) it is evident that as the tube diameter D approaches zero the elevation (or depression) of the meniscus approaches infinity and there is no theoretical limit to the rise (or fall) of a liquid due to capillarity. Values of the contact angle θ for selected liquids when in contact with glass in air are given in Table 1-4.

TABLE 1-4 *Contact Angles of Selected Liquids When in Contact with Glass and Air*

Liquid	Angle θ
Alcohol, ethyl	0°
Benzene	0°
Carbon tetrachloride	0°
Glycerin	0°
Mercury	140°
Turpentine	0°
Water, fresh	0°

Source: *International Critical Tables of Numerical Data, Physics, Chemistry and Technology.*

EXAMPLE What is the minimum diameter of a glass tube in contact with water at 68°F in air required for an elevation of the meniscus of 1/10 in.?

Step A *Data Reduction*

(1) *unit conversion*

(B-35) $h = (1/12)(1/10) = 1/120$ ft

(2) *fluid properties,* water at 68°F

Table A-1 $\rho = 1.937$ slugs/ft^3

Table A-5 $\sigma = 49.85 \times 10^{-4}$ lbf/ft

Table 1-4 $\theta = 0°$

(3) *assume standard gravity*

(B-50) $g = 32.17$ ft/sec^2

Step B *Compute Required Diameter*

(1-68) $D = 4\sigma \cos \theta / \rho g h$

$$D = \frac{4(49.85 \times 10^{-4})(\text{Cos } 0°)}{(1.937)(32.17)(1/120)}$$

$D = 0.03840$ ft or $12 \times 0.03840 = 0.4608$

$D \approx 1/2$ in.

1-19 VAPOR PRESSURE

Definition / The pressure exerted when a solid or liquid is in equilibrium with its own vapor.
Symbol / p_v
Dimensions / FL^{-2} or $ML^{-1}T^{-2}$
Units / U.S.: lbf/ft^2 SI: N/m^2 or Pa

Vapor pressure is a function of the substance and its temperature. The temperature-pressure relation is shown as line *ABC* of Fig. 1-3. Numerical values of p_v for selected liquids are given in Table A-6.

Cavitation

If at some point in the flow of a liquid the existing fluid pressure is equal to or below p_v, the liquid will vaporize and a cavity or void will form. Fluctuations of liquid pressures above and below the vapor pressure result in the formation and collapse of vapor bubbles. The combination of sometimes violent collapse of these bubbles and related chemical reactions results not only in poor performance but also at times in severe damage to equipment. It is necessary in the design of fluid equipment to avoid this phenomenon.

Cavitation Velocity

The velocity at which cavitation will take place in a steady flow system may be determined by considering Fig. 1-13. The fluid mass shown has a length of dx, an area normal to its motion of dA, and the movement of this mass is horizontal. The mass of this fluid element is $\rho\, dA\, dx$. For frictionless movement the pressure forces opposing the motion must be equal to the fluid mass times its acceleration, or from Eq. (1-20) $F_x = ma_x$ and from Eq. (1-26) $a_x = dV/dt$. Substituting these values in Eq. (1-20),

$$F_x = ma_x = (\rho\, dA\, dx)\left(\frac{dV}{dt}\right)$$

Noting that by definition $V = dx/dt$, and again substituting in Eq. (1-20),

$$F_x = (\rho\, dA)\left(\frac{dx}{dt}\right) dV = \rho\, dAV\, dV \tag{1-69}$$

From Fig. 1-13 the sum of the pressure forces

$$\Sigma F_p = F_{p1} - F_{p2} = F_x = p\, dA - (p + dp)\, dA = -dp\, dA,$$

or

$$F_x = -dp\, dA.$$

Substituting in Eq. (1-69),

$$F_x = -dp\, dA = \rho\, dAV\, dV$$

FIGURE 1-13 Notation for cavitation study

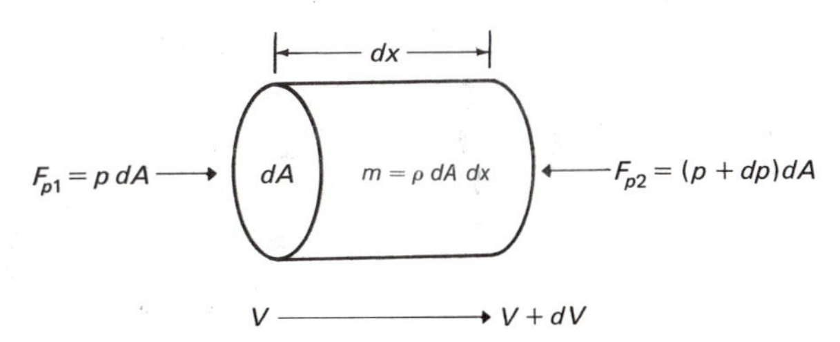

which reduces to

$$dp + \rho V\, dV = 0 \tag{1-70}$$

Integrating Eq. (1-70) for a liquid ($\rho =$ constant) between the limits of p_s to p_v and 0 to V_v,

$$\int_{p_s}^{p_v} dp + \rho \int_0^{V_v} V\, dV = 0 = (p_v - p) + \frac{\rho(V_v^2 - 0^2)}{2}$$

which reduces to

$$V_v = \sqrt{2(p_s - p_v)/\rho} \tag{1-71}$$

where p_s is the pressure of the fluid at rest, and V_v is the minimum velocity at which the liquid will begin to vaporize.

EXAMPLE Water at 50°C when at rest has a pressure of 101 kPa. At what velocity will vaporization start?

Step A	*Data Reduction*
	fluid properties, water at 50°C
Table A-1	$\rho = 998.0\ \text{kg/m}^3$
Table A-6	$p_v = 12.33 \times 10^3\ \text{N/m}^2$
Step B	*Compute Velocity*
(1-71)	$V_v = \sqrt{2(p_s - p_v)/\rho} = \sqrt{2(1.01 \times 10^5 - 12.33 \times 10^3)/998.0}$
	$V_v = 13.33\ \text{m/s}$

PROBLEMS

1. When the atmospheric pressure is 101 kPa a gage attached to a closed tank indicates 91.4 kPa. What will the gage indicate if the atmospheric pressure increases to 103 kPa and all other conditions remain the same?
2. On a day when the atmospheric pressure is at standard conditions the vacuum in a steam condenser is 28.92 in. of mercury at 32°F. What is the absolute pressure in the condenser?
3. A "Scientific Absolute Temperature Scale" is proposed where the ice point is fixed as 1000°S. What would be the temperature of the steam point in °S?
4. Convert 45°F to (a) degrees Rankine, (b) degrees Celsius, and (c) kelvins.

5. How much force will 25 kg of mass exert on the Earth's surface at a location where the gravitational acceleration is standard?
6. What acceleration is produced on a body whose mass is 15 slugs when it is subjected to a force of 100 lbf?
7. What is the acceleration due to the Earth's gravity at 100,000 ft above sea level at a latitude of 45°32′33″?
8. What is the sea level acceleration due to Earth's gravity at a latitude of 15°?
9. What force is exerted by a pound of mass on the surface of the planet Mars?
10. A quantity of matter exerts a force of 500 lbf on the surface of the planet Jupiter. What is its mass in (a) slugs, (b) pounds, and (c) kilograms?
11. An object exerts a force of 1000 lbf on the surface of the planet Neptune. If this object is raised 1000 ft vertically, what is the change in its potential energy?
12. What is the kinetic energy of a 140,000 lbm aircraft cruising at a speed of 300 knots?
13. What is the thrust produced when 20 lbm/sec of fluid flows through a jet propulsion system if its inlet velocity is 100 ft/sec and its exit velocity is 400 ft/sec?
14. When rocks were collected on the surface of the moon it was found that a mass of 165 lbm of them would fit into a box one cubic foot in volume. Estimate the minimum density of these rocks.
15. Check the density in U.S. units using the SI data given in Table A-1 for turpentine at 50°C.
16. Estimate the specific weight of mercury at 86°F on the surface of the planet Mercury.
17. Compute the specific weight of benzene at 10°C in Oslo, Norway.
18. What is the specific volume of carbon tetrachloride at 10°C?
19. What is the specific volume of linseed oil at 68°F?
20. For glycerin at 60°F compute the following specific gravities: (a) based on water at maximum density, (b) water at 60°F, (c) °Be, and (d) °API.
21. Compute the density and specific weight of air at atmospheric pressure and 120°F.
22. Carbon dioxide expands isentropically from 310 K and 700 kPa. Compute the temperature and density of the carbon dioxide at the point where the pressure reaches 350 kPa.
23. A cylinder contains 100 ft^3 of an ideal gas at 120°F and 40 psia. The gas is compressed isentropically to 20 ft^3 and 300 psia. What is the isentropic bulk modulus of elasticity of this gas at 300 psia?

24. At a depth of about 8 km in the ocean the pressure is 82.75 MPa. Estimate the density of 20°C sea water at this depth.
25. Compute the acoustic velocity of the following liquids at atmospheric pressure and at 122°F: (a) ethyl alcohol, (b) mercury, and (c) fresh water.
26. Compute the acoustic velocity of air at standard atmospheric pressure and at 15°C.
27. An oil whose specific gravity is 0.936 has a Saybolt viscosity of 150 SSU. What is its kinematic and dynamic viscosity in (a) U.S. units and (b) SI units?
28. A cylinder of 127 mm radius rotates concentrically inside a fixed tube whose inner radius is 133 mm. The annular space between the tube and the cylinder is filled with linseed oil at 50°C. If the cylinder length is 305 mm, what torque must be applied to maintain a rotational speed of 2 rps?
29. Manometer tables recommend a minimum bore of 10 mm for mercury manometer tubes to avoid capillary error. Estimate the depression of the meniscus for mercury at 20°C with this bore.
30. Octane at 122°F has a pressure when at rest of 5 psia. The octane is accelerated until its velocity is 30 ft/sec. Will cavitation take place?

REFERENCES

Benedict, Robert P., *Fundamentals of Temperature, Pressure, and Flow Measurement,* John Wiley and Sons, New York, 1969.

Benedict, Robert P., "International Practical Temperature Scale of 1968," *Instruments and Control Systems,* October 1969, pp. 85–89.

ASTM Hydrometers, American Society for Testing and Materials Standard Specification E-100-66.

Handbook of Chemistry and Physics, 52d ed. Edited by Robert C. Weast. The Chemical Rubber Company, 1971–1972.

International Critical Tables of Numerical Data, Physics, Chemistry and Technology, vols. 1–7. Published for the National Academy of Sciences National Research Council by McGraw-Hill Book Company, New York, 1926–1929.

Manometer Tables, Instrument Society of America Recommended Practice RP 2.1, 1952.

Smithsonian Meteorological Tables, 6th rev. ed. Smithsonian Institution, Washington, D.C., 1966.

Smithsonian Physical Tables, 9th rev. ed. Smithsonian Institution, Washington, D.C., 1954.

2 Fluid Statics

2-1 INTRODUCTION

Fluid statics is that branch of fluid mechanics that deals with fluids that are at rest with respect to the surfaces that bound them. The entire fluid mass may be in motion, but all fluid particles are at rest with each other.

There are two kinds of forces to be considered: (1) *surface forces,* forces due to direct contact with other fluid particles or solid walls (forces due to pressure and tangential, that is, shear stress), and (2) *body forces,* forces acting on the fluid particles at a distance (e.g., gravity, magnetic field, etc.). Since there is no motion of a fluid layer relative to another fluid layer, the shear stress everywhere in the fluid must be zero and the only surface force that can act on a fluid particle is normal pressure force. Because the entire fluid mass may be accelerated, body forces other than gravity may act in any direction on a fluid particle.

2-2 PASCAL'S PRINCIPLE

The great French philosopher and mathematician Blaise Pascal (1623–1662) is given credit for the first definite statement that the pressure in a static fluid is the same in all directions. Pascal's principle may be illustrated as follows.

Consider the prism of fluid shown in Fig. 2-1. The fluid is at rest and has a thickness of unity normal to the page. The prism has a mass of $\rho \Delta x \Delta z/2$. This prism is a small part of a container of liquid

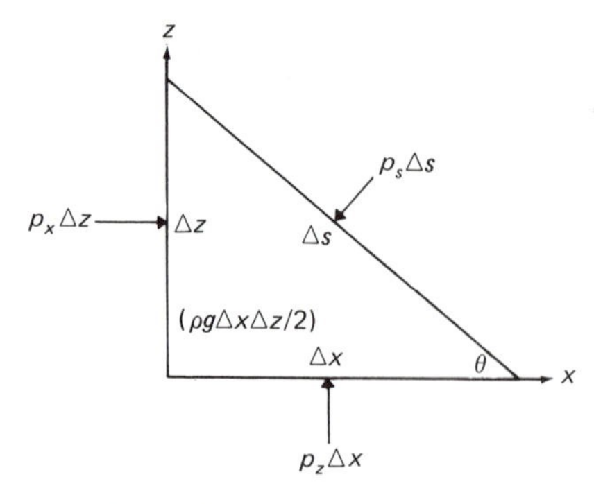

FIGURE 2-1 Notation for Pascal's principle

at rest. Since the entire mass is not being accelerated, the only body force acting is that of gravity in the negative "z" direction.

For a "free body" the sum of the forces in each direction must be zero for equilibrium, so that we may write

x direction $\qquad \Sigma F_x = 0 = p_x\Delta z - p_s\Delta s \sin\theta$

where $\Delta s \sin\theta = \Delta z$ and thus $p_x = p_s$

z direction $\qquad \Sigma F_z = 0 = p_z\Delta x - \dfrac{\rho g\Delta x\Delta z}{2} - p_s\Delta s\cos\theta$

where $\Delta s \cos\theta = \Delta x$ and thus

$$0 = \frac{p_z\Delta x - \rho g\Delta x\Delta z}{2 - p_s\Delta x}$$

which reduces to

$$0 = \frac{p_z - \rho g\Delta z}{2 - p_s}$$

As $\Delta z \to 0$, $p_z \to p_s$

And finally $p_x = p_z = p_s$

In a like manner, it may be demonstrated that for all other planes, the pressure is the same in all directions.

2-3 BASIC EQUATION OF FLUID STATICS

Body Forces

The infinitesimal fluid cube shown in Fig. 2-2 has a mass of $\rho\, dx\, dy\, dz$. This cube is a particle in a large container of

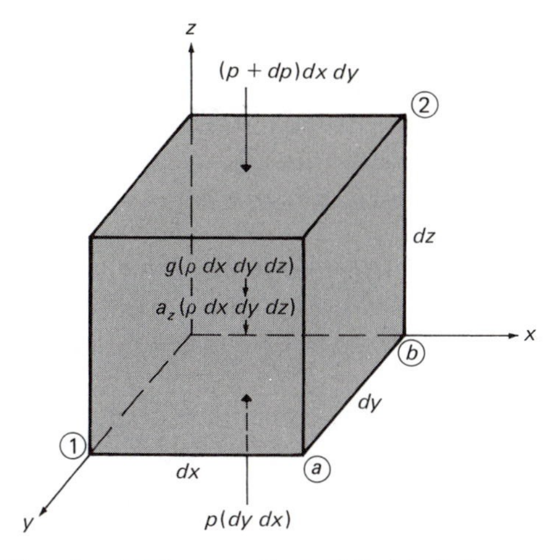

FIGURE 2-2 Notation for basic equation of fluid statics

fluid where all the particles are at rest with respect to each other. The entire fluid mass is subject to body force accelerations of a_x, a_y, and a_z opposite the directions of x, y, and z respectively. In addition, the acceleration due to gravity g acts opposite to the direction of z. Although, for clarity, only the z direction forces are shown in Fig. 2-2, forces also act in the x and y directions. From Eq. (1-20), $F = ma$, the body forces are: $F_{bx} = (\rho\ dx\ dy\ dz)a_x$, $F_{by} = (\rho\ dx\ dy\ dz)a_y$, $F_{bz} = (\rho\ dx\ dy\ dz)a_z$ and the gravity force $F_g = (\rho\ dx\ dy\ dz)g$.

Vertical Forces

By definition of pressure $F = pA$, the upward pressure force is $F_u = p\ dx\ dy$ and the downward pressure force is

$$F_d = (p + dp)dx\ dy$$

Considering the cube of Fig. 2-2 to be a free body and only vertical components acting:

$$\Sigma F_z = F_u - F_d - F_{bz} - F_g = 0$$

or

$$p\,dx\,dy - (p + dp)dx\,dy - \rho\,dx\,dy\,dz\,a_z - \rho\,dx\,dy\,dz\,g = 0$$

which reduces to:

$$dp = -\rho(a_z + g)\ dz \qquad (x,y \text{ constant}) \tag{2-1}$$

Combined Forces

In a like manner, it may be shown that with *only "y" direction forces acting,*

$$dp = -\rho a_y \, dy \qquad (x,z \text{ constant}) \tag{2-2}$$

and with only "x" direction forces acting,

$$dp = -\rho a_x \, dx \qquad (y,z \text{ constant}) \tag{2-3}$$

Forces may be combined by considering the pressure difference between points 2 and 1 of Fig. 2-2. In path $1 \rightarrow a$, x is varied and y and z are held constant so that Eq. (2-3) applies to the difference between a and 1. In a like manner, Eq. (2-2) may be used for path $a \rightarrow b$ and Eq. (2-1) for path $b \rightarrow 2$. The total difference is the sum of each component or

$$dp = -\rho a_x \, dx - \rho a_y \, dy - \rho(a_z + g)dz$$

or

$$dp = -\rho[a_x \, dx + a_y \, dy + (a_z + g)dz] \tag{2-4}$$

Equation (2-4) is the basic equation of fluid statics.

2-4 PRESSURE-HEIGHT RELATIONS FOR INCOMPRESSIBLE FLUIDS

For a fluid at rest and subject only to gravitational force, a_x, a_y, and a_z are zero, reducing Eq. (2-4) to

$$dp = -\rho g \, dz \tag{2-5}$$

Integrating Eq. (2-5) for an incompressible fluid in a field of constant gravity* and substituting $\gamma = \rho g$ from Eq. (1-35),

$$\int_1^2 dp = -\rho g \int_1^2 dz = (p_2 - p_1) = -\rho g(z_2 - z_1)$$
$$= -\gamma(z_2 - z_1) = h$$

* If Eq. (1-17) is solved for z, $z = r(\sqrt{g_\phi/g} - 1)$. For a 0.1% change in the earth's gravitational attraction, $z = 20{,}904{,}000\ (\sqrt{1.001} - 1) \approx 10{,}000 \approx 3$ km. In any practical engineering application involving liquid columns, constant gravity may be assumed.

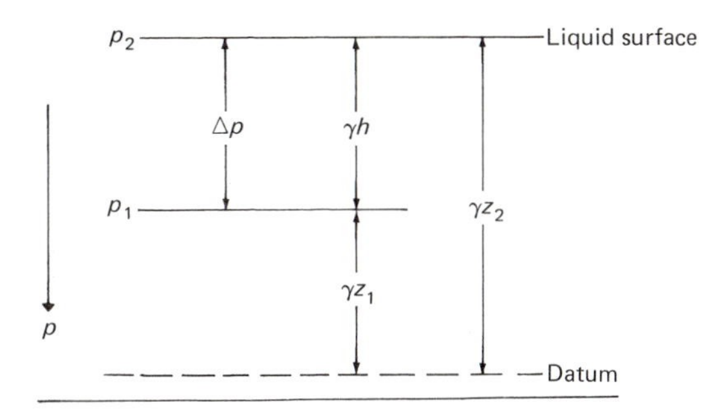

FIGURE 2-3 Pressure equivalence

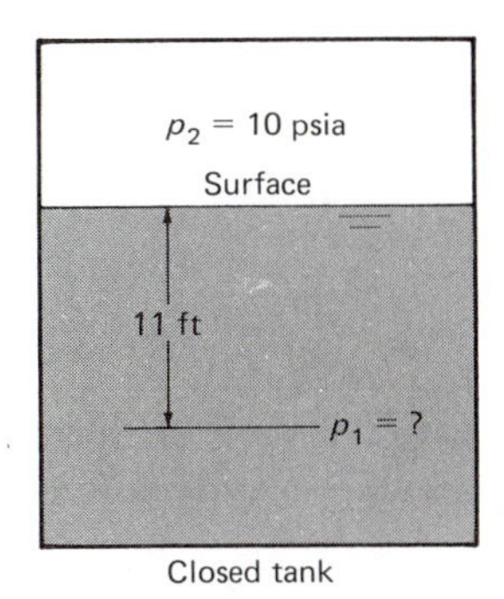

FIGURE 2-4 Notation for closed tank example

which reduces to

$$(p_1 - p_2) = \Delta p = \gamma h \tag{2-6}$$

where $h = (z_2 - z_1)$, or the height of a liquid column. The relationships of Eq. (2-6) are shown in Fig. 2-3.

EXAMPLE A large closed tank is partly filled with benzene at 68°F. If the pressure on the surface is 10 lbf/in² absolute, what is the absolute pressure in the benzene at a depth of 11 ft below its surface? (See Fig. 2-4.)

Step A *Data Reduction*

(1) *unit conversion*

(B-87) $p_2 = 144 \times 10 = 1440$ lbf/ft²

(2) *standard gravity* (assume)

(B-50) $g = 32.17$ ft/sec²

(3) *fluid properties,* benzene at 68°F

Table A-1	$\rho = 1.705 \text{ slugs/ft}^3$
(1-35)	$\gamma = \rho g = 1.705 \times 32.17 = 54.85 \text{ lbf/ft}^2$

Step B *Solve for p_1*

(2-6)
$$(p_1 - p_2) = \gamma h \text{ or } p_1 = \gamma h + p_2$$
$$p_1 = (54.85)(11) + 1440 = 2\ 043 \text{ lbf/ft}^2$$

(B-87)
$$= \frac{2\ 043}{144} = 14.19 \text{ psia}$$

2-5 PRESSURE-SENSING DEVICES

Bourdon tube gages are used for measuring pressure differences. The essential features are shown in Fig. 2-5. The Bourdon tube is made of metal and has an elliptical cross section. The tube is fixed at A and free to move at B. As the difference between the internal and the external pressures increases, the elliptical cross section tends to become circular, and the free end of the tube (point B) moves outward, moving the pointer C through suitable linkage. It should be noted that when the outside pressure is

FIGURE 2-5 Bourdon tube gage (Adapted from *Pressure Measurement,* Fig. 5-1, PTC 19.2-64, American Society of Mechanical Engineers, New York, 1964)

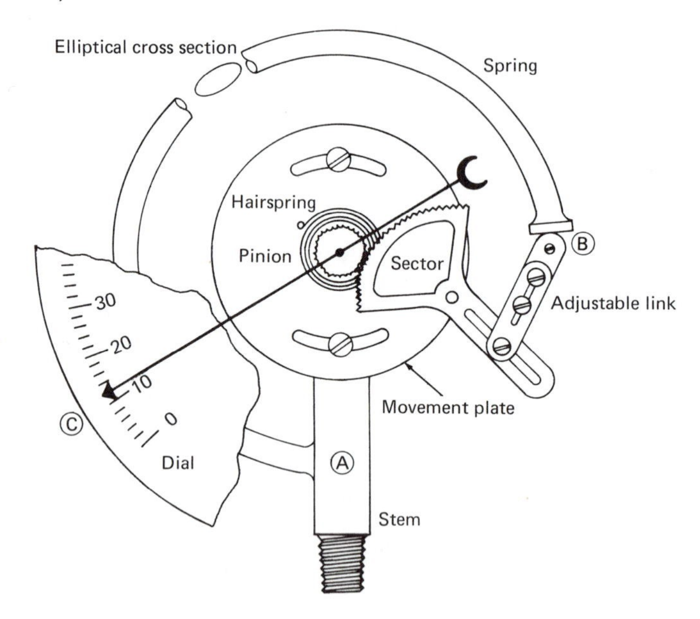

atmospheric, the Bourdon tube indicates *gage* pressure and when the internal pressure is less than the atmosphere, then a negative gage pressure or *vacuum* is sensed. Refer to Fig. 1-4 for these relationships.

Credit for the discovery of the *barometer* is given to Evangelista Torricelli (1608–1647), an Italian scientist who related barometric height to weight of the atmosphere. Fig. 2-6 shows the essential features of an elementary barometer. In its most primitive form, the barometer is made by filling a long glass tube with mercury and inverting it in a pan of mercury. If the height of the mercury column is less than the tube, then mercury vapor will form at the top of the tube. Application of Eq. (2-6) yields

$$p_b = \gamma h + p_v \tag{2-7}$$

The vapor pressure of mercury is very small; from Table A-6 we find that vapor pressure of mercury is

$$5.151 \times 10^{-4}\ \text{lbf/ft}^2\ (2.466 \times 10^{-2}\ \text{N/m}^2)$$

at 32°F. For all practical purposes, $p_b = \gamma h$.

Manometers are one of the oldest means of measuring pressure. They were used as early as 1662 by Robert Boyle to make precise measurements of steady fluid pressures. Because it is a direct application of the basic equation of fluid statics and also because of its inherent simplicity, the manometer serves as a pressure standard in the range of 1/10 in. of water to 100 psig.

The arrangement of the *U-tube manometer* is shown in Fig. 2-7. The manometer is acted upon by a pressure p_1 on the left and p_2 on

FIGURE 2-6 Barometer

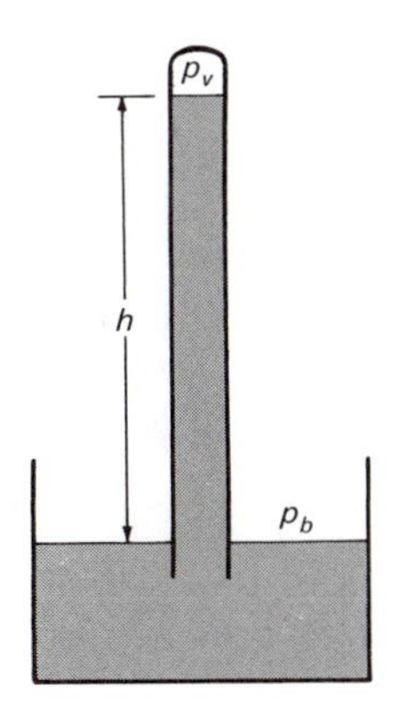

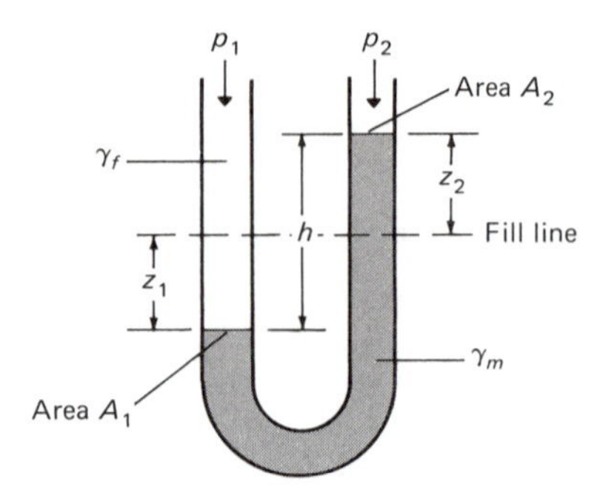

FIGURE 2-7 U-tube manometer

the right. If $p_1 > p_2$, then the fluid in the left leg of the manometer will be displaced to the right by a volume of z_1A_1, resulting in an increase of volume of z_2A_2 in the right leg. Application of Eq. (2-6) for equilibrium in the U-tube manometer results in

$$p_1 + \gamma_f\,(z_1 + z_2) = p_2 + \gamma_m\,(z_1 + z_2)$$

which reduces to

$$(p_1 - p_2) = (\gamma_m - \gamma_f)\,(z_1 + z_2) = (\gamma_m - \gamma_f)h \qquad \textbf{(2-8)}$$

where γ_m is the specific weight of the manometer fluid and γ_f that of the fluid whose differential is being sensed.

One of the disadvantages of the U-tube manometer is that unless $A_1 = A_2$ exactly, then both legs must be observed simultaneously. For this reason, the *well* or *cistern* type shown in Fig. 2-8 is some-

FIGURE 2-8 Well or cistern type manometer

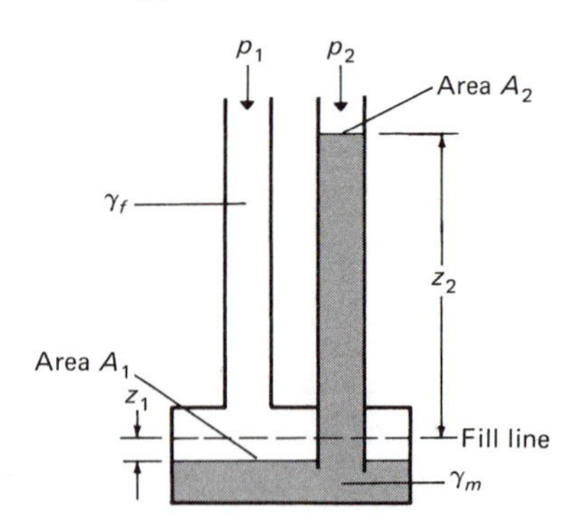

times used. In the well or cistern type of manometer, the areas A_1 and A_2 are controlled to give a maximum deflection of z_2 and a minimum for z_1. From consideration of volumetric displacement of the liquid from one leg to the other:

$$z_1 A_1 = z_2 A_2$$

or

$$z_1 = \frac{z_2 A_2}{A_1}$$

Substituting in Eq. (2-8),

$$\begin{aligned} p_1 - p_2 &= (\gamma_m - \gamma_f)\left(\frac{z_2 A_2}{A_1} + z_2\right) \\ &= (\gamma_m - \gamma_f)\left(1 + \frac{A_2}{A_1}\right) z_2 \end{aligned} \qquad \text{(2-9)}$$

Note that as $A_1 \to \infty$, $A_2/A_1 \to 0$. By making the area A_1 very large, the designer of a well type of manometer can create a condition where $z_2 \to h$. The difference in area ratios is usually taken care of by scale graduations.

Commercial manufacturers of the well type of manometer correct for the area ratios so that $(p_1 - p_2) = (\gamma_m - \gamma_f)S$, where S is the scale reading and is equal to $z_2(1 + A_2/A_1)$. For this reason, scales should not be interchanged between U-tube or well type, nor between well types without consulting the manufacturer.

The *inclined manometer,* as shown in Fig. 2-9 is a special form of the well type. It is designed to enhance the readability of small

FIGURE 2-9 Inclined manometer

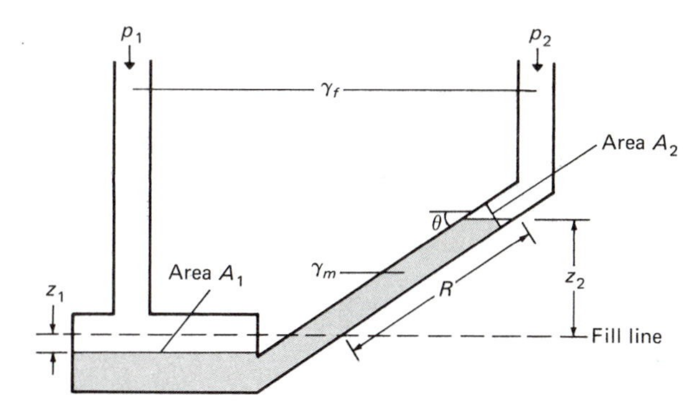

pressure differentials. From consideration of the geometry of this device, for displacement,

$$z_1 A_1 = R A_2$$

or

$$z_1 = \frac{R A_2}{A_1}$$

and for slope

$$z_2 = R \sin \theta$$

Substituting in Eq. (2-8),

$$(p_1 - p_2) = (\gamma_m - \gamma_f)\left(\frac{A_2}{A_1} + \sin \theta\right)R \tag{2-10}$$

where R is the distance along the inclined tube. Commercial inclined manometers also have special scales so that

$$(p_1 - p_2) = (\gamma_m - \gamma_f)S$$

where

$$S = \left(\frac{A_2}{A_1} + \sin \theta\right)R.$$

In actual practice, inclined manometers are used for measurement of small air pressures. Their scales are usually graduated to read in inches of water, but they use many other fluids. Care must be taken to "level" these instruments and to insure that the correct liquid is used as specified by the manufacturer. Scales should never be interchanged.

Application

The equations derived above are simple, but actual installations may require more complex ones. Since there are almost an infinite number of combinations and arrangements that can be used, it is better to derive an equation for each actual case, as will be shown in the examples that follow.

EXAMPLE OF A U-TUBE MANOMETER APPLICATION For the arrangement and conditions shown in Fig. 2-10, compute $p_A - p_B$.

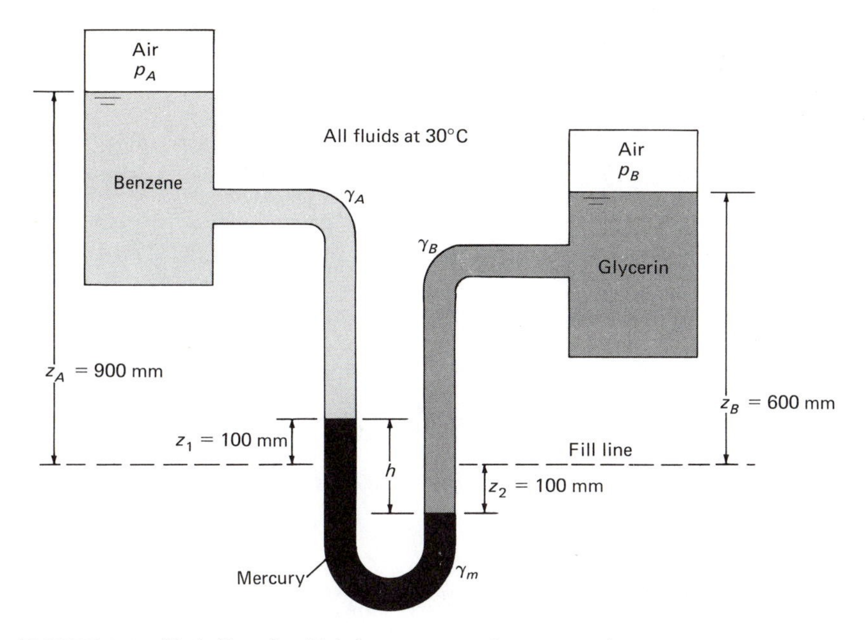

FIGURE 2-10 Notation for U-tube manometer example

Step A *Data Reduction*

(1) *assume standard gravity,*

(B-51) $g = 9.807 \text{ m/s}^2$

(2) *fluid properties,* 30°C

Table A-1 *benzene,* $\rho_A = 868.0 \text{ kg/m}^3$

(1-35) $\gamma_A = \rho_A g = 868 \times 9.807 = 8\,512 \text{ N/m}^3$

Table A-1 *glycerin,* $\rho_B = 1\,255 \text{ kg/m}^3$

(1-35) $\gamma_B = \rho_B g = 1\,255 \times 9.807 = 12\,308 \text{ N/m}^3$

Table A-1 *mercury,* $\rho_m = 13\,520 \text{ kg/m}^3$

(1-35) $\gamma_m = \rho_m g = 13\,520 \times 9.807 = 132\,591 \text{ N/m}^3$

Step B *Develop an Equation for This Application Based on Equilibrium*

(2-7) $p_A + \gamma_A(z_A - z_1) + \gamma_m h = p_B + \gamma_B(z_B + z_2)$

or

$$p_A - p_B = \gamma_B(z_B + z_2) - \gamma_A(z_A - z_1) - \gamma_m h$$

Step C *Solve Developed Equation*

$$p_A - p_B = (12\,308)(0.6 + 0.1) - (8\,512)(0.9 - 0.1) - (132\,591)(0.1 + 0.1)$$

$$p_A - p_B = -24\,712 \text{ N/m}^2 = -24.71 \text{ kPa}$$

$$p_B > p_A$$

EXAMPLE OF AN INVERTED U-TUBE MANOMETER APPLICATION

Compute $p_A - p_B$ for the arrangement and conditions shown in Fig. 2-11.

Step A *Data Reduction*

(1) *unit conversion*

(B-56) $z_A = {}^{34}\!/_{12} = {}^{17}\!/_{6}$, $z_B = {}^{24}\!/_{12} = 2$,

$$z_1 = z_2 = {}^{4}\!/_{12} = {}^{1}\!/_{3},\ h = \frac{4+4}{12} = {}^{2}\!/_{3}$$

(2) *assume standard gravity*

(B-50) $g = 32.17 \text{ ft/sec}^2$

(3) *fluid properties,* 68°F

Table A-1 *fresh water,* $\rho_A = 1.937 \text{ slugs/ft}^3$

(1-35) $\gamma_A = \rho_A g = 1.937 \times 32.17 = 62.31 \text{ lbf/ft}^3$

Table A-1 *salt water,* $\rho_B = 1.988 \text{ slugs/ft}^3$

(1-35) $\gamma_B = \rho_B g = 1.988 \times 32.17 = 63.95 \text{ lbf/ft}^3$

Table A-1 *linseed oil,* $\rho_m = 1.828 \text{ slugs/ft}^3$

(1-35) $\gamma_m = \rho_m g = 1.828 \times 32.17 = 58.81 \text{ lbf/ft}^3$

FIGURE 2-11 Notation for inverted U-tube manometer example

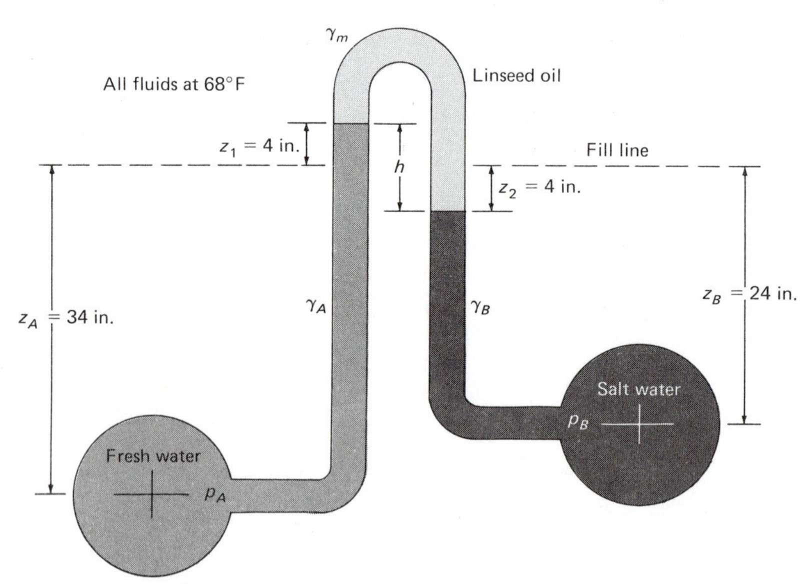

Step B *Develop an Equation for this Application Based on Equilibrium*

(2-7) $$p_A - \gamma_A(z_A + z_1) = p_B - \gamma_B(z_B - z_2) - \gamma_m h$$

or

$$p_A - p_B = \gamma_A(z_A + z_1) - \gamma_B(z_B - z_2) - \gamma_m h$$

Step C *Solve Developed Equation*

$$p_A - p_B = 62.31(17/6 + 1/3) - (63.95)(2 - 1/3) - (58.81)(2/3) = 51.52 \text{ lbf/ft}^2$$

(B-87) $$p_A = p_B = \frac{51.52}{144} = 0.3578 \text{ psi}$$

2-6 RELATIONS FOR STATIC COMPRESSIBLE FLUIDS

The equation for a static fluid in a gravitational field may be written as

(2-5) $$-\frac{dp}{\rho} = g\,dz$$

To integrate the left-hand term of this equation, the functional relationship between pressure and density must be established for a compressible fluid. The right-hand term requires that the relationship between the acceleration due to gravity and altitude be established.

We may proceed to establish these by noting from Eq. (1-37)

$$\frac{1}{\rho} = vg_c,$$

from Eq. (1-46):

$$v = v_1\left(\frac{p_1}{p}\right)^{1/n}$$

and from the equation of state $p_1v_1 = RT_1$, substituting these values in Eq. (2-5),

$$-\frac{dp}{\rho} = -dp\,vg_c = -g_cv_1p_1^{1/n}\,dp\,p^{-1/n} = -g_cRT_1p_1^{1/n-1}\,dp\,p^{-1/n} = g\,dz$$

or

$$-g_cRT_1p_1^{1/n-1}\,dp\,p^{-1/n} = g\,dz \tag{2-11}$$

For the right-hand term from Eq. (1-17)

$$\frac{g}{g_\phi} = \frac{r^2}{(r+z)^2}$$

Substituting this in Eq. (2-11),

$$-g_c RT_1 p_1^{1/n-1}\, dp\, p^{-1/n} = (g_\phi) r^2\, dz (r+z)^{-2}$$

or

$$-(g_c/g_\phi) RT_1 p_1^{1/n-1}\, dp\, p^{-1/n} = r^2\, dz(r+z)^{-2} = dH \tag{2-12}$$

where H is the *geopotential* altitude (the altitude if gravitational acceleration were constant) as defined by Eq. (2-12). The relationship between the geopotential altitude H and *geometric* altitude z may be established as follows:

$$\frac{g}{g_\phi}\int_0^z dz = r^2 \int_0^z (r+z)^{-2}\, dz = \frac{z}{1+z/r} = \int_0^H dH = H \tag{2-13}$$

Returning now to Eq. (2-12) we see that mathematically there are only two values of $1/n$ to be considered, $n = 1$ and $n \neq 1$. Since the value of $n = 1$ is for an isothermal process, we will have two equations, one for isothermal processes and another for nonisothermal processes.

Isothermal Process

For an isothermal process Eq. (2-12) may be integrated as follows:

$$\left(\frac{g_c}{g_\phi}\right) RT_1 p_1^{1/1-1} \int_2^1 dp\, p^{-1/1} = \left(\frac{g_c}{g_\phi}\right) RT \int_2^1 \frac{dp}{p} = \left(\frac{g_c}{g_\phi}\right) RT_1 \log_e\left(\frac{p_1}{p_2}\right)$$

$$= \int_1^2 dH = H_2 - H_1$$

which reduces to

$$\left(\frac{g_c}{g_\phi}\right) RT \log_e\left(\frac{p_1}{p_2}\right) = H_2 - H_1 \qquad \text{(ideal gas } T = C\text{)} \tag{2-14}$$

Nonisothermal Processes

For all other processes, Eq. (2-8) integrates as follows:

$$\left(\frac{g_c}{g_\phi}\right) R T_1 p_1^{1/n-1} \int_2^1 dp\, p^{-1/n} = \left(\frac{g_c}{g_\phi}\right) R T_1 p_1^{1/n-1} \left[\frac{p^{-1/n+1}}{1-1/n}\right]_2^1$$

$$= \int_1^2 dH = H_2 - H_1$$

which reduces to

$$\left(\frac{g_c}{g_\phi}\right)\left(\frac{n}{n-1}\right) R T_1 [1 - (p_2/p_1)^{(n-1)/n}] = H_2 - H_1 \qquad \text{(ideal gas } T \neq C\text{)} \qquad \textbf{(2-15)}$$

Temperature relations may be established from Eq. (1-47),

$$\left(\frac{p_2}{p_1}\right)^{(n-1)/n} = \frac{T_2}{T_1}.$$

Substituting Eq. (1-47) in Eq. (2-15),

$$\left(\frac{g_c}{g_\phi}\right)\left(\frac{n}{n-1}\right) R T_1 \left(1 - \frac{T_2}{T_1}\right) = \left(\frac{g_c}{g_\phi}\right)\left(\frac{nR}{n-1}\right)(T_1 - T_2) = H_2 - H_1$$

Solving for temperature gradient,

$$\frac{T_2 - T_1}{H_2 - H_1} = \frac{\Delta T}{\Delta H} = \frac{(g_\phi/g_c)(1-n)}{nR} \qquad \text{(ideal gas } T \neq C\text{)} \qquad \textbf{(2-16)}$$

2-7 ATMOSPHERE

The atmosphere is a gaseous envelope that surrounds the Earth, extending from sea level to an altitude of several hundred miles. The altitude for near space has been set arbitrarily at 50 miles (80 km).

The earth's atmosphere is divided into five levels based on temperature variation. The *troposphere* extends from sea level to 54,000 ft at the equator, decreasing to 28,000 ft at the poles, and is composed of approximately 79% nitrogen and 21% oxygen. With increasing altitude from sea level, the temperature decreases from 59°F to −69.7°F. Above the troposphere is the *stratosphere,* which extends to approximately 65,000 ft and exists at a relatively constant temperature of −69.70°F. The *mesosphere* extends from nearly 65,000 ft to 300,000 ft, and its temperature increases from −69.70°F to +28.67°F, then decreasing to −134°F. The mesosphere is characterized by an ozone layer which absorbs the ultraviolet radiation from the sun. Above the mesosphere is the *thermosphere,* also called

TABLE 2-1 *Standard U.S. Atmosphere 1962 to 100,000 Feet, 1966 Supplement over 100,000 Feet**

Altitude				Temperature		Pressure Ratio	Density Ratio
GEOPOTENTIAL H		GEOMETRIC z				p/p_o	ρ/ρ_o
m	*ft*	*m*	*ft*	°C	°F		
0	0	0	0	15	59.00	1.0000	1.0000
11 000	36,089	11 019	36,151	−56.5	−69.70	0.2233	0.2971
20 000	65,617	20 062	65,823	−56.5	−69.70	0.05403	0.07187
32 000	104,987	32 162	105,518	−44.5	−48.10	0.008567	0.010544
47 000	154,199	47 350	155,347	−2.5	−27.50		
52 000	170,603	52 429	172,011	−2.5	−27.50		
61 000	200,131	61 591	202,070	−20.5	−4.90		
69 000	226,378	69 757	228,862	−52.5	−62.50		
79 000	259,186	79 994	262,448	−82.5	−116.50		
90 000	295,276	91 293	299,518	−82.5	−116.50		
100 000	328,084	101 598	333,327	−62.5	−80.50		
110 000	360,892	111 937	367,247	−18.9	2.02		
117 776	386,404	120 000	393,700	109.09	228.36		

Notes:

(1) Temperature variation between given altitudes is linear.

(2) For over 100,000 ft:

$\log p_o/p = 191z \times 10^{-6} + 0.0140$

$\log \rho_o/\rho = 189z \times 10^{-6} - 0.1000$

*45° N spring/fall.

the *ionosphere,* which extends from approximately 300,000 ft to 1,000,000 ft. The temperature in this layer increases from −134°F to nearly 2200°F. The composition is primarily ionized atoms of the lighter gases. The last level is the *exosphere,* which extends to the space environment.

Because of wide variations in the atmosphere, a standard atmosphere is used for design purposes. The United States Standard Atmosphere formulated in 1962 extends from sea level to 100,000 ft. This was increased in 1966 to 393,700 ft (geometric height of 120,000 m). Table 2-1 gives the basic data for the U.S. Standard Atmosphere. Table A-9 is a condensed version of the 1962 U.S. Standard Atmosphere. Fig. 2-12 shows the temperature-altitude profile of this atmosphere.

Planetary atmospheres at this time are not well known. Venus is thought to have a surface temperature of 800°F, pressure 235 lbf/in^2, and an atmosphere composed of 95% carbon dioxide, possibly nitrogen, and a trace of water vapor. Mars has a surface temperature of −190 to 90°F, pressure of 0.15 lbf/in^2 and an atmosphere mostly made up of carbon dioxide. Mercury has a surface temperature of 750°F and a pressure 15 lbf/in^2. The composition of the gas is not known.

EXAMPLE Atmospheric pressure at sea level is 101.3 kPa, and temperature is 15°C. Using Table 2-1, estimate the absolute pressure, temperature, and density at 5000 m above sea level

FIGURE 2-12 Temperature-altitude profile

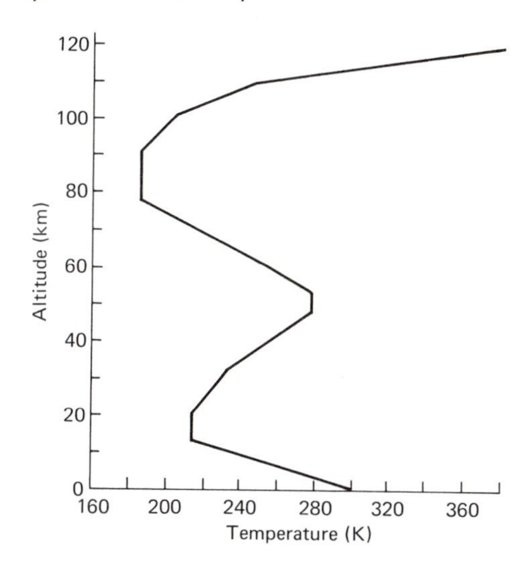

assuming the air: (a) to be an incompressible fluid, (b) to follow an isothermal process, and (c) to follow an isentropic process. Compare results with data given in Table A-9.

Step A — Data Reduction

(1) *unit conversion*

(B-27) $T_K = t_C + 273.15 = T_1 = 15 + 273.15 = 288.15 \text{ K}$

(2) *proportionality constant*

(1-9) $g_c = 1 \text{ kg} \cdot \text{m/N} \cdot \text{s}^2$

(3) *gravity*

(B-51) assume $g = g_o = g_\phi = 9.807 \text{ m/s}^2$

(4) *Earth's radius*

Table 1-3 — 6 371 km

(5) *geopotential altitude*

(2-13) $$H = z\left(1 + \frac{z}{r}\right)^{-1} = 5 \times 10^3\left(\frac{1 + 5\,000}{6\,371\,000}\right)^{-1}$$

$$H = 4\,996 \text{ m}$$

(6) *fluid properties* air

Table A-7 — $R = 287.1 \text{ J/kg} \cdot \text{K}$, $k = 1.400$

Step B — (a) *Assuming Air to Be an Incompressible Fluid*

(1-44) $$\rho_2 = \rho_1 = \frac{p_1}{g_c R T_1} = \frac{101\,300}{(1)(287.1)(288.15)}$$

$$= 1.225 \text{ kg/m}^3$$

(1-35) $\gamma = \rho g = (1.225)(9.807) = 12.01 \text{ N/m}^3$

(2-6) $p_2 = p_1 - \gamma H = 101\,300 - 12.01(4\,996) = 41.30 \times 10^3 \text{ N/m}^2$

If we assume that the air is incompressible, but still an ideal gas, then to maintain the same density for the same mass, the specific volume must be unchanged, and for a constant volume process $n = \infty$.

(1-47) $$T_2 = T_1(p_2/p_1)^{(n-1)/n} = (288.15)(41\,300/101\,300)^{(\infty-1)/\infty}$$

$$T_2 = 117.48 \text{ K}$$

Step C — (b) *Assuming Air to Follow an Isothermal Process*

$$n = 1$$

(2-14) $$\log_e p_2 = \log_e p_1 - \left(\frac{g_\phi}{g_c}\right)\left(\frac{H_2 - H_1}{RT_1}\right)$$

$$= \log_e (101\,300) - \left(\frac{9.807}{1}\right)\left(\frac{4\,996 - 0}{(287.1)(288.15)}\right)$$

$$= 10.933\,6.$$

$$p_2 = 56.03 \times 10^3 \text{ N/m}^2$$

$T_2 = T_1$ by definition of an isothermal process

(1-44)
$$\rho_2 = \frac{p_2}{g_c R T_2} = \frac{56\,030}{(1)(287.1)(288.15)}$$

$$= 0.677\,3 \text{ kg/m}^3$$

Step D (c) *Assuming Air to Follow an Isentropic Process*

$$n = k$$

(2-16)
$$T_2 = \frac{(g_\phi/g_c)(H_2 - H_1)(1 - n)}{nR} + T_1$$

$$= \left(\frac{9.807}{1}\right)\left(\frac{(4\,996 - 0)(1 - 1.400)}{(1.400)(287.1)}\right) + 288.15 = 239.39 \text{ K}$$

(1-47)
$$p_2 = p_1\left(\frac{T_2}{T_1}\right)^{n/(n-1)} = (101\,300)\left(\frac{239.39}{288.15}\right)^{1.400/(1.400-1)}$$

$$p_2 = 52.94 \times 10^3 \text{ N/m}^2$$

(1-44)
$$\rho_2 = \frac{p_2}{g_c R T_2} = \frac{52\,940}{(1)(287.1)(239.39)}$$

$$\rho_2 = 0.770\,3 \text{ kg/m}^3$$

Step E (d) *Using Table 2-1*

Table 2-1

at $H = 0$, $t = 15°\text{C}$

at $H = 11\,000$ m, $t = -56.5°\text{C}$

$$\frac{\Delta T}{\Delta H} = \frac{-56.5 - 15}{11\,000 - 0} = -6.5 \times 10^{-3} \text{ °C/m}$$

(2-16)
$$\frac{\Delta T}{\Delta H} = \frac{(g_\phi/g_c)(1 - n)}{nR} = -6.5 \times 10^{-3}$$

$$= \frac{(9.807/1)(1 - n)}{(n)(287.1)}$$

Solving for n, $n = 1.235$,

$$\frac{\Delta T}{\Delta H} = \frac{T_2 - T_1}{H_2 - H_1} = -6.5 \times 10^{-3} = \frac{T_2 - 288.15}{5000 - 0}$$

Solving for T_2, $T_2 = 255.65$ K,

(1-47)
$$p_2 = p_1\left(\frac{T_2}{T_1}\right)^{n/(n-1)}$$

$$= (101\,300)\left(\frac{255.65}{288.15}\right)^{1.235/(1.235-1)}$$

$$= 54.01 \times 10^3 \text{ N/m}^2$$

(1-44) $$\rho_2 = \frac{p_2}{g_c R T_2} = \frac{54\,010}{(1)(287.1)(255.65)}$$
$$= 0.7359 \text{ kg/m}^3$$

Step F (e) *Compare Results with Table A-9 Data*

Summary of Methods

Method	n	T_2 K	p_2 kPa	ρ_2 kg/m³
(a) Incompressible	∞	117.48	41.30	1.225
(b) Isothermal	1	288.15	56.03	0.6773
(c) Isentropic	1.400	239.39	52.94	0.7703
(d) Table 2-1	1.235	255.65	54.01	0.7359
(e) Table A-9	—	255.68	54.05	—

2-8 LIQUID FORCE ON PLANE SURFACES

Pressure-Height Relations

The total or absolute pressure on the vertical side of the tank shown in Fig. 2-13 at a depth h below the surface is p_t. Let the liquid pressure be denoted as p, then

(2-6) $$\Delta p = p_t - p_b = p = \gamma h$$

Note that the pressure exerted by the liquid is *gage pressure.*

FIGURE 2-13 Notation for liquid pressure study

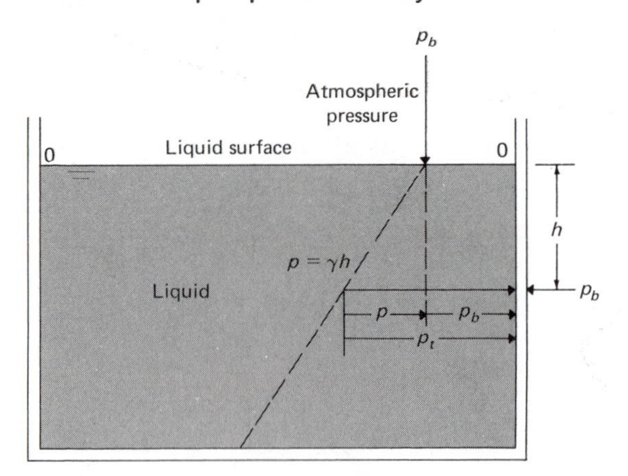

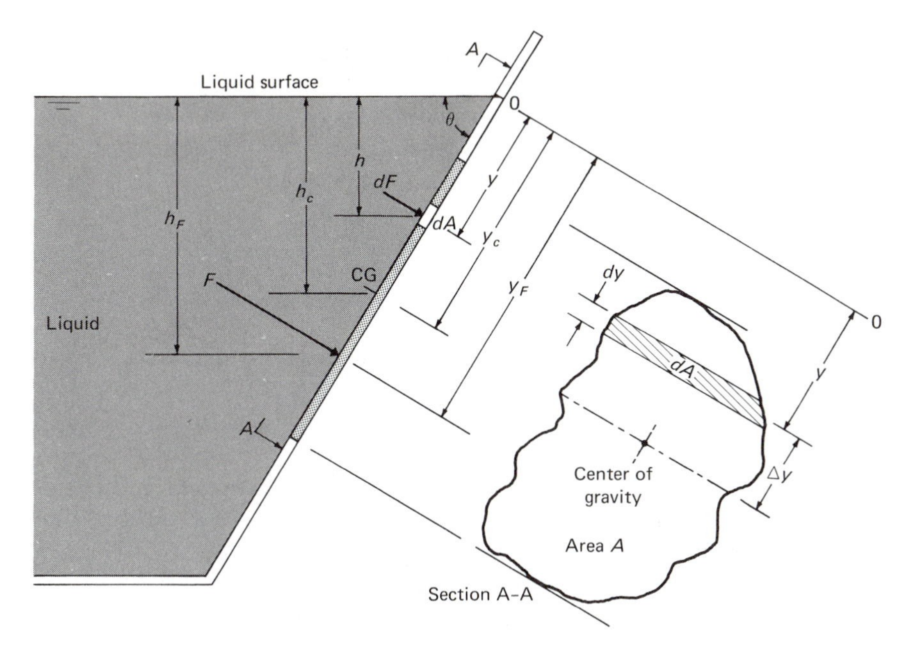

FIGURE 2-14 Notation for liquid force on plane submerged surfaces

Centroids of Plane Areas

Figure 2-14 shows an inclined end of an open tank partly filled with a liquid. Area A is a surface of this end in contact with the liquid. The first area moment about the axis 0-0 (liquid surface) is

$$M_{0A} = \int y\,dA \tag{2-17}$$

The *centroid* of an area is the point at which the area might be concentrated and still leave unchanged the first moment of the area around any axis. The centroid of an area is also its *center of gravity,* thus

$$M_{0A} = \int y\,dA = y_c A \tag{2-18}$$

where y_c is the distance from the liquid surface 0-0 to the center of gravity of the area.

Force Exerted

The force F exerted at a depth h from the liquid surface is

$$F = \int dF = \int p\,dA = \gamma \int h\,dA \qquad \textbf{(2-19)}$$

From geometry $h = y \sin \theta$, so that Eq. (2-19) becomes

$$F = \gamma \int h\,dA = \gamma \sin \theta \int y\,dA \qquad \textbf{(2-20)}$$

From Eq. (2-18)

$$\int y\,dA = y_c A$$

and again from geometry $h_c = y_c \sin \theta$, so that

$$F = (\gamma \sin \theta)(y_c A) = \gamma h_c A \qquad \textbf{(2-21)}$$

where h_c is the vertical distance from the liquid surface to the center of gravity. Equation (2-21) is a very important statement of fluid statics.

EXAMPLE OF LIQUID FORCE CALCULATION A cylindrical tank 3 ft in diameter has its axis horizontal. At the middle of the tank, on top, is a pipe 4 in. in diameter, which extends vertically. The tank and pipe are filled with castor oil at 68°F. The tank ends are designed for a maximum force of 9000 lbf. What is the safe maximum level of the free surface of the castor oil in the pipe above the tank top? (See Fig. 2-15.)

FIGURE 2-15 Notation for liquid force calculation example

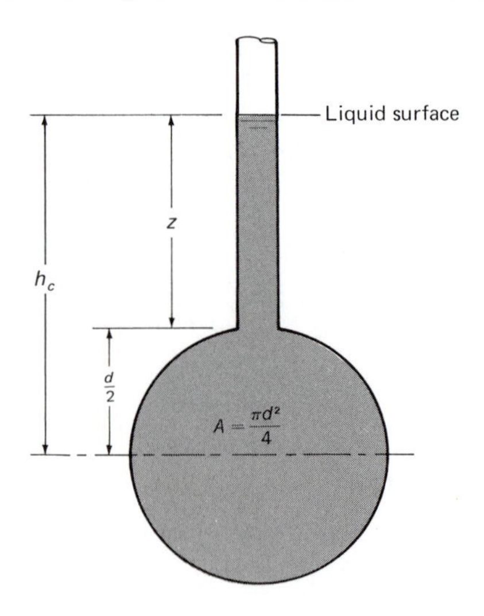

Step A *Data Reduction*

(1) *gravity*

(B-50) Assume $g = g_o = 32.17$ ft/sec^2

(2) *fluid properties,*

castor oil at 68°F

Table A-1 $\rho = 1.863$ slugs/ft^3

(1-35) $\gamma = \rho g = 1.863 \times 32.17 = 59.93$ lbf/ft^3

Step B *Derive an Equation for this Application*

Fig. 2-15 $h_c = z + \frac{d}{2}$, substituting in Eq. (2-21),

(2-21) $$F = \gamma h_c A = \gamma\left(\frac{z+d}{2}\right)\left(\frac{\pi d^2}{4}\right)$$

Solving for z,

$$z = \frac{4F}{\gamma \pi d^2} - \frac{d}{2} \qquad \textbf{(a)}$$

Step C *Calculate Maximum Safe Level*

$$z = \frac{4 \times 9000}{59.93 x \pi x (3)^2} - \frac{3}{2} = 19.75 \text{ ft} = 19 \text{ ft } 9 \text{ in.} \qquad \textbf{(a)}$$

Location of Liquid Force

If the moment of forces around the axis 0-0 (liquid surface of Fig. 2-14) is taken, then

$$\int dM_0 = \int y\, dF = y_F F$$

or

$$y_F = \left(\frac{1}{F}\right) \int y\, dF \qquad \textbf{(2-22)}$$

where y_F is the distance from the liquid surface to the point where F would act if it were concentrated in one location (center of force). From Eq. (2-20) $dF = \gamma \sin\theta\, y\, dA$ and from Eq. (2-21) $F = \gamma y_c A \sin\theta$. Substituting in Eq. (2-22),

$$y_F = \frac{\int y(\gamma \sin\theta y\, dA)}{y_c A \sin\theta} = \frac{\gamma \sin\theta \int y^2\, dA}{\gamma \sin\theta y_c A} = \frac{\int y^2\, dA}{y_c A} \qquad \textbf{(2-23)}$$

Noting that $\int y^2 \, dA$ is the second moment or moment of inertia I_0, and substituting in Eq. (2-23),

$$y_F = \frac{I_0}{y_c A} \tag{2-24}$$

Because Eq. (2-24) requires that the moment of inertia around the liquid surface be known, it is not always convenient to apply. To transfer the moment of inertia to the center of gravity of the area, we may proceed as follows, using the parallel axis theorem:

$$dI_0 = y^2 \, dA = (y_c - \Delta y)^2 \, dA \tag{2-25}$$

Integrating Eq. (2-25),

$$I_0 = \int dI_0 = \int (y_c - \Delta y)^2 \, dA$$

$$= y_c^2 \int dA - 2y_c \int \Delta y \, dA + \int \Delta y^2 \, dA \tag{2-26}$$

By definition of a centroid, the first moment around the center of gravity $\int \Delta y \, dA = 0$, and the second moment or moment of inertia around the center of gravity $I_G = \int \Delta y^2 \, dA$, substituting in Eq. (2-26),

$$I_0 = y_c^2 \int dA - 2y_c(0) + I_G = y_c^2 A + I_G \tag{2-27}$$

Substituting for I_0 in Eq. (2-24),

$$y_F = \frac{I_0}{y_c A} = \frac{y_c^2 A + I_G}{y_c A}$$

or

$$y_F - y_c = \frac{I_G}{y_c A} \tag{2-28}$$

Properties of areas and volumes for selected shapes are given in Table A-10.

EXAMPLE OF LOCATION OF LIQUID FORCE The rectangular gate shown in Fig. 2-16 is 6 m high and 5 m wide and placed vertically on the side of an open rectangular container of linseed oil at 50°C. The free oil surface is 3 m above the upper edge of the gate. What force must be applied at the upper edge of the gate to keep it closed if the gate is hinged at the lower edge?

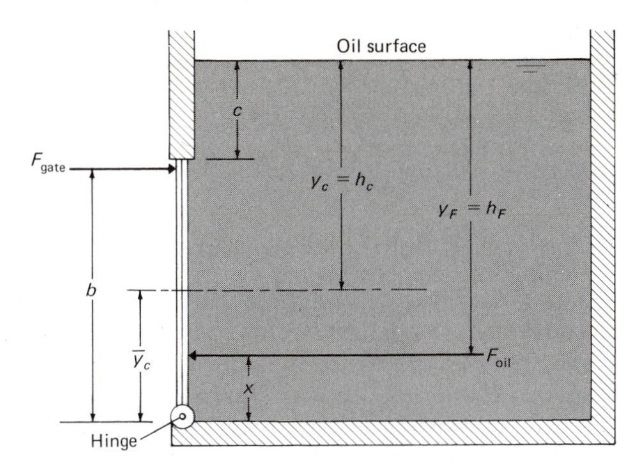

FIGURE 2-16 Notation for liquid force location example

Step A	*Data Reduction*
	(1) *gravity*
(B-51)	assume $g = g_o = 9.807 \text{ m/s}^2$
	(2) *fluid properties,* linseed oil at 50°C
Table A-1	$\rho = 922.6 \text{ kg/m}^3$
(1-35)	$\gamma = \rho g = (922.6 \times 9.807) = 9\,048 \text{ N/m}^3$
	(3) *geometric,* rectangle
Table A-10	$A = ab = 5 \times 6 = 30 \text{ m}^2,\ \bar{y}_c = \dfrac{b}{2} = \dfrac{6}{2} = 3 \text{ m}$
Table A-10	$\dfrac{I_G}{A} = \dfrac{b^2}{12} = \dfrac{(6)^2}{12} = 3 \text{ m}^2$
Fig. 2-16	$h_c = y_c = b - \bar{y}_c + c = 6 - 3 + 3 = 6 \text{ m}$
Step B	*Calculate the Force Exerted by the Oil and Its Location*
(2-21)	$F_{\text{oil}} = \gamma h_c A = (9\,048)(6)(30) = 1\,628\,640 \text{ N}$
(2-28)	$y_F = \dfrac{I_G/A}{y_c} + y_c = \dfrac{3}{6} + 6 = 6.5 \text{ m}$
Fig. 2-16	$x = b + c - y_F = 6 + 3 - 6.5 = 2.5 \text{ m}$
Step C	*Compute Gate Force*
Fig. 2-16	Taking moments around the hinge, $(F_{\text{gate}})(b) = (F_{\text{oil}})(x)$

or

$$F_{\text{gate}} = (x/b)F_{\text{oil}} \tag{a}$$

$$F_{\text{gate}} = (2.5/6)\ (1\,628\,640) = 678\,600\ \text{N} = 678.6\ \text{kN} \tag{a}$$

2-9 LIQUID FORCE ON CURVED SURFACES

Liquid Forces on curved surfaces may be readily calculated by considering their horizontal and vertical components separately and resolving them. Consider the curved surface AE shown in Fig. 2-17, whose width is w, submerged in a liquid so that its upper edge EE' is a distance of c below the liquid surface.

There are two *vertical forces* acting on the surface AEw. The first is the weight of the liquid above line BE. The volume of the liquid above line BE is acw, so that from the definition of specific weight

$$F_{z1} = \gamma \text{Vol} = \gamma acw \tag{2-29}$$

The second vertical force is the weight of the liquid below line BE and above the curved line AE. Again from the definition of specific weight

$$F_{z2} = \gamma \text{Vol} = \gamma \text{area}_{ABE} w = \gamma A w \tag{2-30}$$

The total vertical force acting on the surface AEw is the sum of these two, or

$$F_z = F_{z1} + F_{z2} \tag{2-31}$$

FIGURE 2-17 Notation for liquid force on curved surfaces

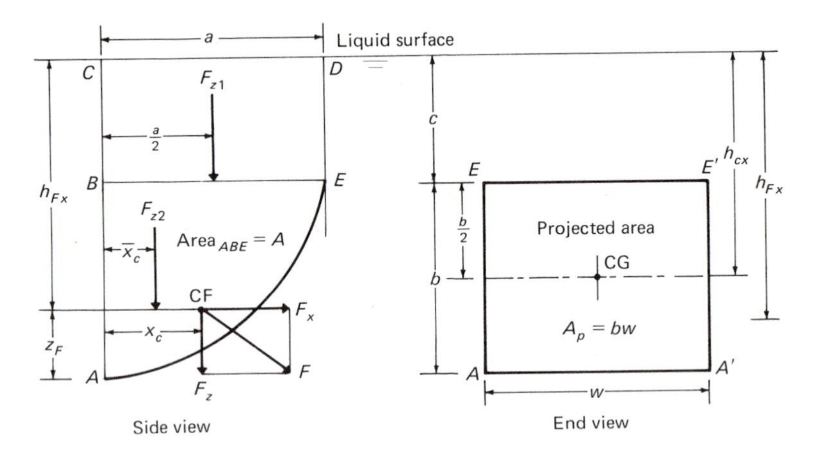

The upper vertical force F_{z1} acts through the center of gravity of area$_{BCDE}$, or from line AC, a distance of $a/2$. The lower vertical force F_{z2} acts through the center of gravity of area$_{ABE}$ or a distance of $\bar{x}_c$ from line AC. The *location* x_c where the combined force acts may be determined by taking moments around line AC:

$$x_c F_z = (a/2)F_{z1} + \bar{x}_c F_{z2}$$

or

$$x_c = \frac{aF_{z1}/2 + \bar{x}_c F_{z2}}{F_z} \quad \text{(2-32)}$$

Substituting

$$F_{z1} = \gamma acw, \; F_{z2} = \gamma Aw \qquad \text{and} \qquad F_z = F_{z1} + F_{z2}$$

from Eqs. (2-29), (2-30), and (2-31) respectively in Eq. (2-32),

$$x_c = \frac{a\gamma acw/2 + \bar{x}_c \gamma Aw}{\gamma acw + \gamma w}$$

which reduces to

$$x_c = \frac{a^2c/2 + \bar{x}_c A}{ac + A} \quad \text{(2-33)}$$

The horizontal force F_x on the curved surface AEw may be obtained by application of Eq. (2-21):

$$F_x = \gamma h_{cx} A_p \quad \text{(2-34)}$$

where h_{cx} is the distance from the liquid surface to the center of gravity of the projected area $AEE'A'$. From Fig. 2-17, $h_{cx} = c + b/2$, and the projected area $A_p = bw$. Substituting in Eq. (2-34),

$$F_x = \gamma h_{cx} A_p = \gamma(c + b/2)(bw) \quad \text{(2-35)}$$

Location of horizontal force may be determined by application of Eq. (2-28), noting that for a vertical distance $y_c = h_{cx}$ and $y_F = h_{Fx}$,

$$h_{Fx} = h_{cx} + \frac{I_{Gx}}{A_p h_{cx}} \quad \text{(2-36)}$$

From Table A-10, I_G/A for a rectangle is $b^2/12$; again from Fig. 2-17,

$$h_{cx} = c + \frac{b}{2}$$

Substituting in Eq. (2-36),

$$h_{Fx} = \left(c + \frac{b}{2}\right) + \frac{b^2/12}{c + b/2} \tag{2-37}$$

The magnitude of the *resultant force* F may be determined by noting that F is the hypotenuse of the right triangle formed by F_x and F_z in Fig. 2-17. From trigonometry,

$$F = \sqrt{F_x^2 + F_z^2} \tag{2-38}$$

This force F will act at the intersection of h_{Fx} and x_c, shown in Fig. 2-17 as CF (center of force).

EXAMPLE The curved surface shown in Fig. 2-17 is immersed in a tank filled with ethyl alcohol at 50°F. The edge EE' is horizontal and is 30 ft below the liquid surface. The curved surface is a parabola whose vertex is at A. The horizontal distance a is 20 ft, and the vertical distance b is 24 ft. Calculate the magnitude and the location of the total liquid force on the surface AEw per foot of width.

Step A *Data Reduction*

(1) *gravity*

(B-50) Assume $g = g_o = 32.17$ ft/sec²

(2) *fluid properties,* ethyl alcohol at 50°F

Table A-1 $\rho = 1.548$ slugs/ft³

(1-35) $\gamma = \rho g = 1.548 \times 32.17 = 49.80$ lbf/ft³

(3) *geometric* half parabola

Given $a = 20$ ft, $b = 24$ ft, $c = 30$ ft, and $w = 1$ ft

Table A-10 $$A = \frac{2ab}{3} = \frac{2 \times 20 \times 24}{3} = 320 \text{ ft}^2$$

Table A-10 $$\bar{x}_c = \frac{3a}{8} = \frac{3 \times 20}{8} = 7.5 \text{ ft}$$

Step B *Compute Total Vertical Force*

(2-29) $F_{z1} = \gamma acw = 49.80 \times 20 \times 30 \times 1 = 29{,}880$ lbf/ft

(2-30) $F_{z2} = \gamma Aw = 49.80 \times 320 \times 1 = 15{,}936$ lbf/ft

(2-31) $F_z = F_{z1} + F_{z2} = 29{,}880 + 15{,}936 = 45{,}816$ lbf/ft

Step C *Locate Vertical Force*

(2-33) $$x_c = \frac{a^2c/2 + \bar{x}_c A}{ac + A} = \frac{20^2 \times 30/2 + 7.5 \times 320}{20 \times 30 + 320} = 9.130 \text{ ft}$$

Step D *Compute Horizontal Force*

(2-35)
$$F_x = \gamma\left(c + \frac{b}{2}\right)(bw) = 49.80\left(30 + \frac{24}{2}\right)(24 \times 1)$$

$$F_x = 50{,}198 \text{ lbf/ft}$$

Step E *Locate Horizontal Force*

(2-37)
$$h_{Fx} = \left(c + \frac{b}{2}\right) + \frac{b^2/12}{c + b/2}$$

$$h_{Fx} = \left(30 + \frac{24}{2}\right) + \frac{24^2/12}{30 + 24/2} = 43.14 \text{ ft}$$

Fig. 2-17
$$z_F = c + b - h_{Fx} = 30 + 24 - 43.14 = 10.86 \text{ ft}$$

Step F *Compute Resultant Force*

(2-38)
$$F = \sqrt{F_x^2 + F_z^2} = \sqrt{(50{,}198)^2 + (45{,}816)^2}$$

$$F = 67{,}963 \text{ lbf/ft}$$

2-10 STRESS IN PIPES DUE TO INTERNAL PRESSURE

Stress

When a fluid is contained the forces due to fluid pressure produce equal but opposite resisting forces in the container. These resisting forces produce stress in the material of the container.

Definition / Force per unit area
Symbol / S
Dimensions / FL^{-2} or $ML^{-1}T^{-2}$
**Units* / U.S.: lbf/in^2 SI: N/m^2 or Pa

Tensile Stress

When the internal pressure exceeds the external the size of the container is increased because of the elasticity of the container material. This increase in size produces tension in the structure, and hence the material is subjected to tensile stress.

* More trouble will be encountered with U.S. customary units as we invade the field of strength of materials, which uses inch as the unit for length.

Thin Wall

All stress relations developed in this article are based on the assumption that stress in a given cross section is uniform. For this to be valid it is necessary that the thickness of the walls with respect to the size of the container be small. For the purpose of defining "thin," pipes whose wall thickness is less than one-tenth of their internal diameters will be considered thin-walled.

Pipes

Figure 2-18 shows a cylinder subjected to an internal pressure. The stress produced may be reduced to longitudinal (S_L) and circumferential (S_c) components. Figure 2-18*a* shows the circumferential areas and 2-18*b* shows the stresses. The fluid force $F_p = p'A_p$, where p' is the difference between the internal and ex-

FIGURE 2-18 Stress in pipes

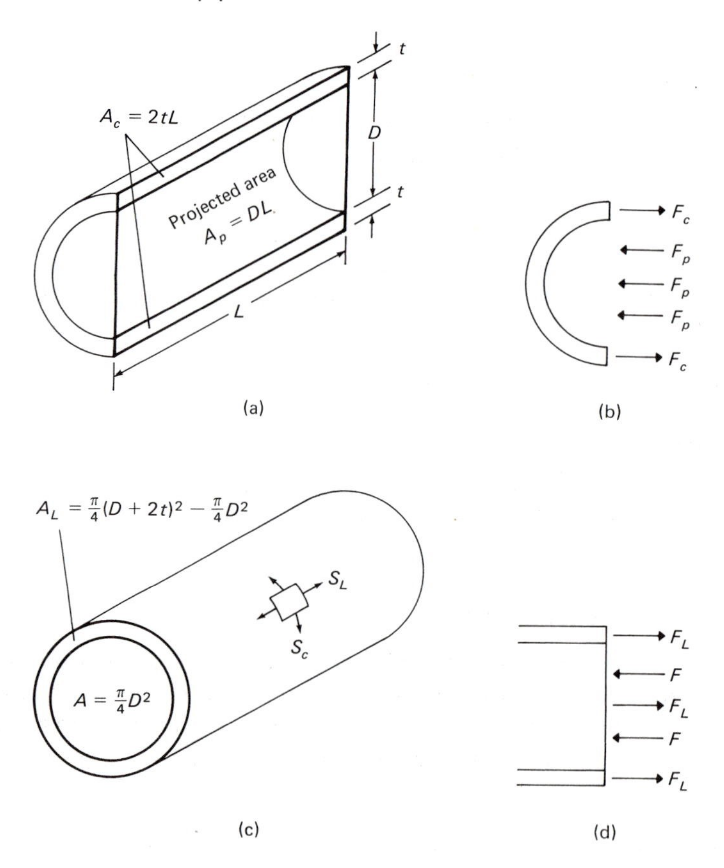

ternal pressures and A_p is the projected area, and equal to DL. The resisting force $F_c = S_c A_c$, where A_c is the circumferential stress area and is equal to $2tL$. From the free-body diagram of Fig. 2-18b the resisting forces must equal the fluid forces, or

$$F_p = p' A_p = p' DL = F_c = S_c A_c = S_c 2tL,$$

which reduces to

$$\frac{p'}{S_c} = \frac{2t}{D} \tag{2-39}$$

For the longitudinal component Fig. 2-18c shows the areas and 2-18d shows the stresses. The fluid force $F = p'A$, where A is the cross-sectional area and equal to $\pi D^2/4$. The resisting force $F_L = S_L A_L$ where A_L is the area of the annulus and is equal to the difference between the cross-sectional areas $\pi(D + 2t)^2/4$ and $\pi D^2/4$, which reduces to $\pi t(D + t)$. Since t is small with respect to D, $A_L \approx \pi t D$. From the free-body diagram of Fig. 2-18d the resisting forces must equal the fluid forces, or

$$F = p'A = p'\pi D^2/4 = F_L = S_L A_L = S_L \pi t(D + t) \approx S_L \pi t D$$

which reduces to

$$\frac{p'}{S_L} = 4\left(\frac{t}{D}\right) + 4\left(\frac{t}{D}\right)^2 \approx \frac{4t}{D} \tag{2-40}$$

Dividing Eq. (2-39) by Eq. (2-40),

$$\frac{p'/S_c}{p'/S_L} \approx \frac{2t/D}{4t/D}$$

which reduces to

$$S_L \approx \frac{S_c}{2}$$

Because the longitudinal stress is only half the circumferential stress, the circumferential stress is the determining one for thickness calculations. Equations (2-39) and (2-40) were derived to show only theoretical relations and should not be used for design.

Design Equations

The American National Standard ANSI B.31 Code for Pressure Piping recommends an equation which may be derived from Eq. (2-39) as follows: replace t by $t_m - A$, where t_m is the minimum wall thickness and A is additional wall thickness required to compensate for material removed in threading, grooving, etc., and to provide for mechanical strength, replace

$$D \text{ by } D_o - 2y(t_m - A)$$

where D_o is the outside diameter and y is a correction factor for material and temperature; S_c by S, where S is the allowable stress. Substituting these values in Eq. (2-39),

$$\frac{p'}{S_c} = \frac{2t}{D} = \frac{p'}{S} = \frac{2(t_m - A)}{D_o - 2y(t_m - A)} \tag{2-41}$$

For service below 900°F, $y = 0.4$ and Eq. (2-41) becomes

$$\frac{p'}{S} = \frac{2(t_m - A)}{D_o - 0.8(t_m - A)} \qquad (t_F < 900°\text{F}) \tag{2-42}$$

Piping Schedules

Table A-11 shows some properties of wrought steel and wrought iron pipe from American National Standard ANSI B36.10-1970. In 1939 the B36.10 committee surveyed the pipe sizes then in use and assigned schedule numbers to them. These numbers were based on an allowance of 0.1 for A, $y = 0$, and $t_m = 7t_s/8$, where t_s is schedule thickness and the factor $7/8$ to allow for a $12\frac{1}{2}\%$ variation in wall thickness. Substituting in Eq. (2-41),

$$\frac{p'}{S} = \frac{2(7t_s/8 - 0.1)}{D_o - 2(0)(7t_s/8 - 0.1)} \approx \frac{N}{1000}$$

or

$$N \approx \frac{1000p'}{S} \approx \frac{1750(t_s - 0.1)}{D_o}. \tag{2-43}$$

where N is the schedule number.

The relationship $N \approx 1000p'/S$ is very approximate owing to the rounding off of values of existing sizes and the variation between Eqs. (2-42) and (2-43) and should not be used for design. Schedule numbers always give conservative values. In using piping schedules, values of t_m must be increased by wall thickness tolerance to obtain t_s, and the values of t_s selected must always be equal to or greater than the calculated value of t_s. *For design the American National Standards Codes should be used.*

EXAMPLE A 12-in. wrought steel pipe is required for 1800 psig and 300°F service. The B-31.10 code gives a value of $A = 0$, and an allowable stress of 10,200 psi for this service. For a mill tolerance of $12\frac{1}{2}\%$ for wall thickness, what pipe schedule should be used?

Step A *Data Reduction*

Table A-11 $D_o = 12.750$ in.

Step B *Compute Minimum Wall Thickness*

$t_F = 300°\text{F} < 900°\text{F}$

(2-42)
$$p'/S = \frac{2(t_m - A)}{D_o - 0.8(t_m - A)}$$

Solving for t_m,

$$t_m = \frac{p'D_o}{2S + 0.8p'} + A \qquad \text{(a)}$$

$$t_m = \frac{(1800)\,(12.750)}{2 \times 10{,}200 + 0.8 \times 1800} + 0 = 1.051 \text{ in.} \qquad \text{(a)}$$

Step C *Select Schedule*

$$t_s = \frac{t_m}{1 - .125/100} = \frac{8t_m}{7} = 8(1.051/7) = 1.201 \text{ in.}$$

Table A-11 Schedule 140, $t_s = 1.125$

Schedule 160, $t_s = 1.312$

Select schedule 160, $1.125 < 1.201 < 1.312$

2-11 ACCELERATION OF FLUID MASSES

Static Acceleration

Fluid masses may be subject to various types of uniform acceleration without relative motion occurring between the fluid particles or between fluid particles and their boundaries. As was discussed in Sec. 2-1, shear stress must be absent, thus permitting the accelerated fluid mass to be treated as a static fluid. Under these conditions the basic equation for fluid statics, Eq. (2-4), applies. Integrating Eq. (2-4) for an incompressible fluid (ρ constant) in a field of constant gravity (g constant), for uniform acceleration (a_x, a_y, and a_z constant), results in

$$\int_1^2 dp = -\rho a_x \int_1^2 dx - \rho a_y \int_1^2 dy - \rho(a_z + g) \int_1^2 dz = p_2 - p_1$$

$$= -\rho[a_x(x_2 - x_1) + a_y(y_2 - y_1) + (a_z + g)(z_2 - z_1)] \tag{2-44}$$

D'Alembert's Principle

Jean le Rond d'Alembert (1717–1783), a French scientist, noted that Newton's second law could be written as

$$F - ma = 0 \tag{1-20}$$

where $-ma$ is a fictitious force and is sometimes called the reversed effective force or the *inertia* force. This principle may be used to reduce a problem of dynamics to one of statics. In the derivation of the basic equation of fluid statics, Eq. (2-4), the body force accelerations a_x, a_y, and a_z were assumed to act *opposite* the directions x, y, and z respectively. With the employment of the inertia force concept then the accelerations may be assumed to act *in* the directions of x, y, and z respectively in Eq. (2-44).

Translation

Consider the liquid mass shown in Figs. 2-19 and 2-20 being uniformly accelerated upward at an angle of α and a rate of a. The acceleration in the y direction, a_y, is zero. Letting $p_2 - p_1 = p_s - p$, $x_2 - x_1 = -L$, $z_2 - z_1 = h$ and $a_y = 0$, Eq. (2-44) becomes

FIGURE 2-19 Notation for translation

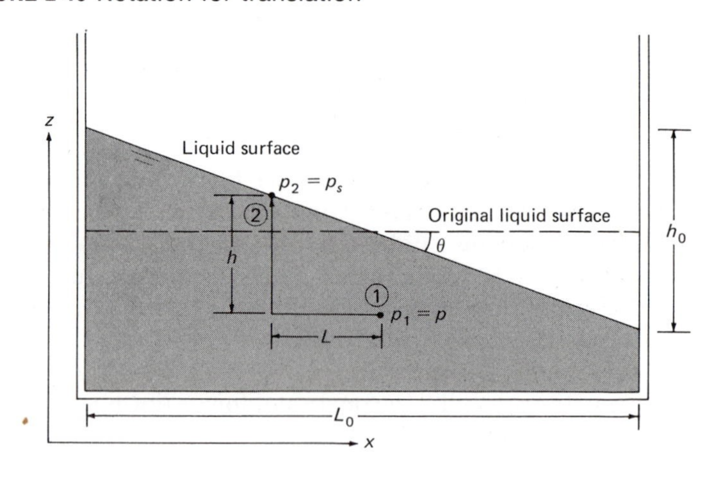

$$p_s - p = -\rho[a_x(-L) + (0)(y_2 - y_1) + (a_z + g)(h)]$$

which reduces to

$$\Delta p = p - p_s = \rho[(a_z + g)h - a_x L] \tag{2-45}$$

At the *liquid surface*

$$p = p_s \quad \text{or} \quad \Delta p = 0,\ h = h_o,\ L = L_o,$$

which when substituted in Eq. (2-45) becomes

$$0 = \rho[(a_z + g)h_o - a_x L_o]$$

which reduces to

$$\frac{h_o}{L_o} = \frac{a_x}{a_z + g} = \tan\theta \tag{2-46}$$

Pressure-height relations for the accelerated fluid may be established by letting $L = 0$ in Eq. (2-45) and noting from Eq. (1-35) that $\rho = \gamma/g$:

$$\Delta p = \frac{\gamma}{g}[(a_z + g)h - a_x(0)] = \gamma\left(1 + \frac{a_z}{g}\right)h \qquad (L = 0) \tag{2-47}$$

Comparing Eq. (2-47) with Eq. (2-6) $\Delta p = \gamma h$ for an unaccelerated liquid, it becomes evident that the ratio of the two is $1 + a_z/g$. The

FIGURE 2-20 Notation for inclined plane example

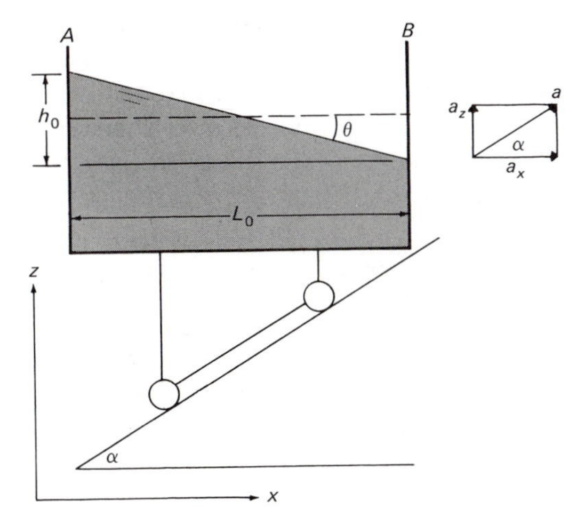

horizontal force may be obtained by multiplying Eq. (2-34) by this ratio, or

$$F_x = \gamma(1 + a_z/g)\,h_{cx}A_p \tag{2-48}$$

In a like manner the *vertical force* is the effective weight of the liquid above the bottom. From Eq. (2-29) multiplied by the ratio $(1 + a_z/g)$ the vertical force becomes

$$F_z = \gamma(1 + a_z/g)\text{Vol} \tag{2-49}$$

EXAMPLE OF ACCELERATION OF A FLUID UP AN INCLINED PLANE

The open tank shown in Fig. 2-20 contains 200 m³ of water at 20°C. The tank is 6 m high, 5 m wide and 10 m long. The angle of the incline is 30°. Determine (a) the maximum acceleration to which the tank may be subjected without spilling any water, (b) the force on each end during the acceleration, and (c) the total force required to accelerate the fluid mass.

Step A — *Data Reduction*

(1) *gravity*

(B-51) Assume $g = g_o = 9.807\ \text{m/s}^2$

(2) *fluid properties,* water at 20°C

Table A-1 $\rho = 998.3\ \text{kg/m}^3$

(1-35) $\gamma = \rho g = 998.3 \times 9.807 = 9\,790\ \text{N/m}^3$

Step B — *Maximum Acceleration Without Spilling*

Fig. 2-20 — The bottom area of this tank is $5 \times 10 = 50\ \text{m}^2$. When the tank was at rest the height of liquid was 200/50, or 4 m. During acceleration the height of the liquid at end A may not exceed the tank height, or 6 m. The rise of liquid on end A will be $6 - 4$, or 2 m. The drop in level at end B must equal this rise, so that $h_o = 2 + 2 = 4$ m. From geometry $a_x = a\cos\alpha$ and $a_z = a\sin\alpha$.

(2-46)
$$\frac{h_o}{L_o} = \frac{a_x}{a_z + g} = \frac{a\cos\alpha}{a\sin\alpha + g}$$

Solving for a,

$$a = \frac{g}{(L_o/h_o)\cos\alpha - \sin\alpha} = \frac{9.807}{(10/4)\cos 30° - \sin 30°} \tag{a}$$

$$a = 5.890\ \text{m/s}^2,\ a_z = 5.890 \sin 30° = 2.945\ \text{m/s}^2$$

Step C — *Force on Each End*

(2-48)
$$F_{xA} = \gamma\left(1 + \frac{a_z}{g}\right)h_{cxA}A_{pA}$$

$$F_{xA} = (9\,790)\left(1 + \frac{2.945}{9.807}\right)\left(\frac{6}{2}\right)(6 \times 5) = 1\,145\,691 \text{ N}$$

$$F_{xA} = 1\,146 \text{ kN}$$

(2-48) $$F_{xB} = \gamma\left(1 + \frac{a_z}{g}\right)h_{cxB}A_{pB}$$

$$F_{xB} = (9\,790)\left(1 + \frac{2.945}{9.807}\right)\left(\frac{2}{2}\right)(2 \times 5) = 127\,299 \text{ N}$$

$$F_{xB} = 127.3 \text{ kN}$$

Step D *Total Force*

(1-20) $$F = ma = \rho \text{Vol}\, a = 998.3 \times 200 \times 5.890 = 1\,175\,997 \text{ N}$$

$$F = 1.176 \text{ MN}$$

EXAMPLE OF A HORIZONTAL ACCELERATION The U-tube manometer shown in Fig. 2-21 with vertical legs 20 in. apart is partly filled with water to be used as an accelerometer. It is installed in an automobile which accelerates uniformly from 15 mph to 50 mph in 15 sec. What is the difference in level of the two legs during the acceleration?

Step A *Data Reduction*

(1) *unit conversions*

(B-65/54) $$V_{x_1} = 15\left(\frac{5{,}280}{3{,}600}\right) = 22.00 \text{ ft/sec}$$

(B-65/54) $$V_{x_2} = 50\left(\frac{5{,}280}{3{,}600}\right) = 73.33 \text{ ft/sec}$$

(B-55) $$L_o = \frac{20}{12} = \frac{5}{3} \text{ ft}$$

(2) *gravity*

(B-50) Assume $g = g_o = 32.17 \text{ ft/sec}^2$

FIGURE 2-21 Notation for horizontal acceleration example

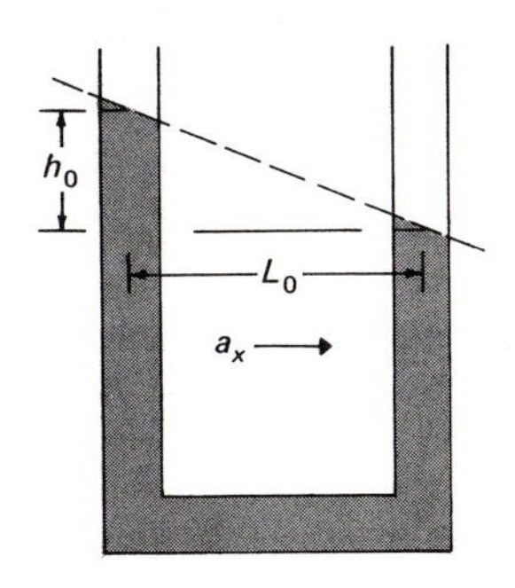

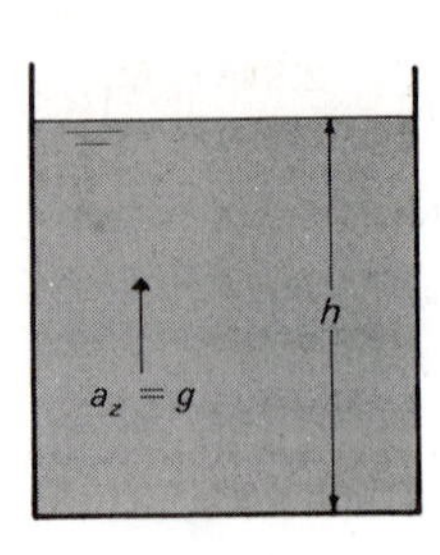

FIGURE 2-22 Notation for vertical acceleration example

(3) Acceleration

(1-26) $$a_x = \frac{dV_x}{dt} = \frac{\Delta V_x}{\Delta t} = \frac{73.33 - 22.00}{15}$$

$$a_x = 3.422 \text{ ft/sec}^2$$

$$a_z = 0 \text{ (assume level road)}$$

Step B *Compute Level Difference*

(2-46) $$h_o = \frac{L_o a_x}{a_z + g}$$

$$h_o = \frac{(5/3)(3.422)}{(0 + 32.17)} = 0.17729 \text{ ft}$$

(B-55) $$h_o = 0.17729 \times 12 = 2.127 \text{ in.}$$

EXAMPLE OF A VERTICAL ACCELERATION A glass of water (Fig. 2-22) at 30°C is on an elevator which is accelerating upward at 1 g. The water in the glass is 100 mm deep. Find the pressure exerted by the water on the bottom of the glass.

Step A *Data Reduction*

(1) *gravity*

(B-51) Assume $g = g_o = 9.807 \text{ m/s}^2$

(2) *fluid properties,* water at 30°C

Table A-1 $\rho = 995.7 \text{ kg/m}^3$

(1-35) $\gamma = \rho g = 995.7 \times 9.807 = 9\,765 \text{ N/m}^3$

Step B *Compute Pressure on Bottom*

(2-47) $$\Delta p = \gamma(1 + a_z/g)h = (9\,765)(1 + 1/1)(0.100)$$
$$\Delta p = 1\,953 \text{ N/m}^2$$

Rotation

Consider the fluid mass shown in Fig. 2-23 being rotated around the z axis at a constant angular velocity of ω radians per second. The acceleration of the fluid mass $\rho\, dy\, dx\, dz$ is $-\omega^2 x$ (radially inward). The acceleration in the y direction a_y is zero and gravity is the only force in the z direction, so that a_z is also zero. Using the inertia force concept discussed in conjunction with translation, $a_x = -\omega^2 x$ and Eq. (2-4) for rotation becomes

$$dp = -\rho(-\omega^2 x\, dx + g\, dz) \tag{2-50}$$

Integrating Eq. (2-50) and using the relation $\rho = \gamma/g$ of Eq. (1-35),

$$\int_1^2 dp = \frac{\gamma\omega^2}{g}\int_1^2 x\, dx - \gamma\int_1^2 dz = (p_2 - p_1) = \frac{\gamma\omega^2}{2g}(x_2^2 - x_1^2) - \gamma(z_2 - z_1) \tag{2-51}$$

Lines of constant pressure occur when $p_2 = p_1$, reducing Eq. (2-51) to

FIGURE 2-23 Notation for rotation

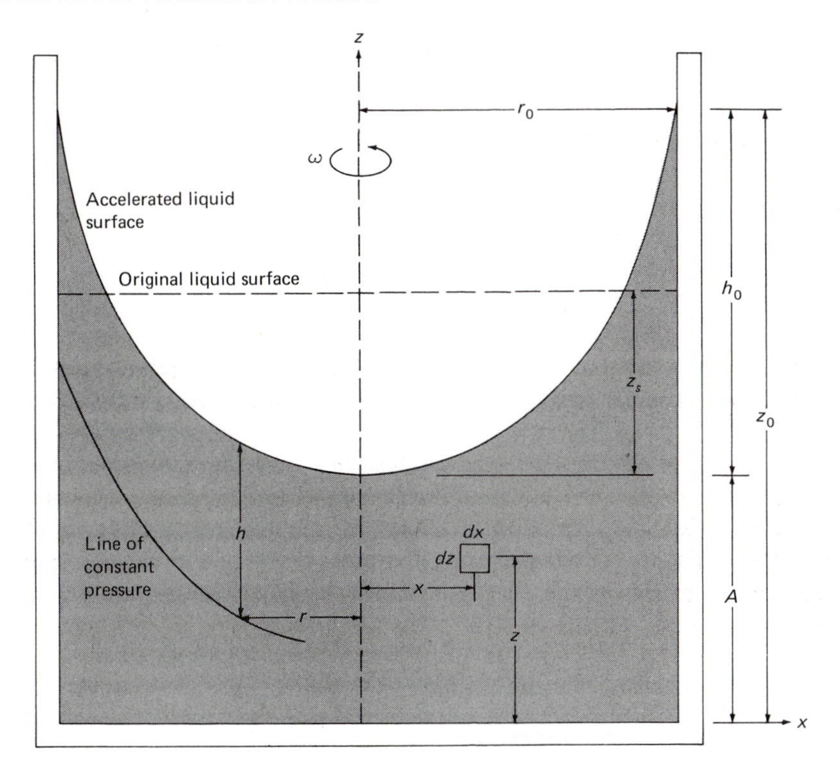

$$(z_2 - z_1) = \frac{\omega^2}{2g}(x_2^2 - x_1^2) \qquad \text{(2-52)}$$

The liquid surface is a special case of constant pressure. From Fig. 2-23, when $x_1 = 0$, $z_1 = A$. If we let $x_2 = r$, then $z_2 - z_1 = z - A$ and Eq. (2-52) becomes

$$(z - A) = \frac{\omega^2 r^2}{2g} \qquad \text{(2-53)}$$

At the wall of the container $r = r_o$ and $h_o = z_o - A$:

$$h_o = \frac{\omega^2 r_o^2}{2g} \qquad \text{(2-54)}$$

From Eqs. (2-52) through (2-54), it may be seen that *all constant pressure* lines including the surface are *parabolic* (Table A-10). The volume of a paraboloid of revolution is one-half that of its circumscribed cyclinder. If no liquid is spilled, then

$$z_s = {}^1\!/_2\, h_o \qquad \text{(2-55)}$$

If some liquid is spilled then Eq. (2-55) represents the surface distance *after* rotation.

Pressure-height relations for the rotating fluid may be established by letting $x_1 = x_2$. Then Eq. (2-51) becomes

(2-6)
$$(p_2 - p_1) = -\gamma(z_2 - z_1)$$

or

$$\Delta p = \gamma h \qquad (x \text{ constant})$$

This is, of course, the same as for unaccelerated fluids, and Eq. (2-21) may also be used along any $z = a$ constant or vertical line.

EXAMPLE 1 Rotation of an Open Tank An open cylindrical tank (Fig. 2-24) 3 ft in diameter and 20 ft high is filled to its brim with water at 86°F and rotated about its vertical center line at 200 rpm. Determine (a) the volume of water spilled, and (b) the gage pressure exerted by the liquid on the bottom of the tank 1 ft from the center line.

Step A *Data Reduction*

(1) *unit conversion*

(B-24)
$$\frac{\pi}{180^\circ} = 1 \text{ radian, } 1 \text{ revolution} = \left(\frac{\pi}{180}\right)(360) = 2\pi$$

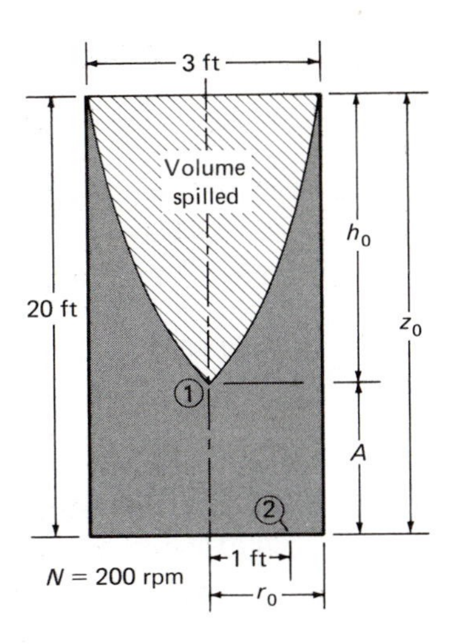

FIGURE 2-24 Notation for open-tank problem

(B-24/68) $$\omega = \frac{2\pi N}{60} = (2\pi)\left(\frac{200}{60}\right) = 20.94 \text{ rad/sec}$$

(2) *gravity*

(B-50) Assume $g = g_o = 32.17$ ft/sec^2

(3) *fluid properties,* water at 86°F

Table A-1 $$\rho = 1.932 \text{ slugs/ft}^3$$

(1-35) $$\gamma = \rho g = 1.932 \times 32.17 = 62.15 \text{ lbf/ft}^3$$

(4) *geometric,* paraboloid of revolution

Table A-10 $$\text{vol} = \frac{\pi b r^2}{2} = \frac{\pi h_o r_o^2}{2} \quad \text{(a)}$$

Step B *Compute Volume Spilled*

Since the tank was filled before rotation, the cross-hatched area in Fig. 2-24 is a paraboloid of revolution representing the amount spilled.

(2-54) $$h_o = \frac{\omega_o^2 r_o^2}{2g} = \frac{(20.94)^2 (3/2)^2}{2(32.17)} = 15.33 \text{ ft.}$$

$$\text{vol}_{SP} = \frac{\pi h_o r_o^2}{2} = \frac{\pi (15.33)(1.5)^2}{2} \quad \text{(a)}$$

$$\text{vol}_{SP} = 54.18 \text{ ft}^3$$

Step C *Calculate the Pressure on the Tank Bottom*

Fig. 2-24 For an open tank $p_1 = 0$ psig, at center line $x_1 = 0$. Taking datum at bottom of tank,

$$z_2 = 0,\ z_1 = z_o - h_o = 20 - 15.33 = 4.67 \text{ ft}$$

(2-51) $$(p_2 - p_1) = \frac{\gamma\omega^2}{2g}(x_2^2 - x_1^2) - \gamma(z_2 - z_1)$$

$$(p_2 - 0) = \frac{62.15(20.94)^2(1^2 - 0^2)}{(2 \times 32.17)}$$

$$-62.15(0 - 4.67) = 713.8 \text{ lbf/ft}^2 \text{ gage}$$

(B-87) $$p_2 = \frac{713.8}{144} = 4.96 \text{ psig}$$

EXAMPLE 2 Rotation of a Closed Tank A closed cylindrical tank (Fig. 2-25) 4 m in diameter and 10 m high is filled with linseed oil at 20°C. It is rotated at 4 rps around a vertical axis 3 m from its center line. Compute the maximum pressure differential in the tank.

Step A *Data Reduction*

(1) *unit conversion*

(B-24) $\pi/180° = 1$ radian, 1 revolution $= (\pi/180)(360) = 2\pi$

$$\omega = 2\pi N = 2\pi(4) = 25.13 \text{ rad/s}$$

(2) *gravity*

(B-51) Assume $g = g_o = 9.807 \text{ m/s}^2$

FIGURE 2-25 Notation for closed-tank example

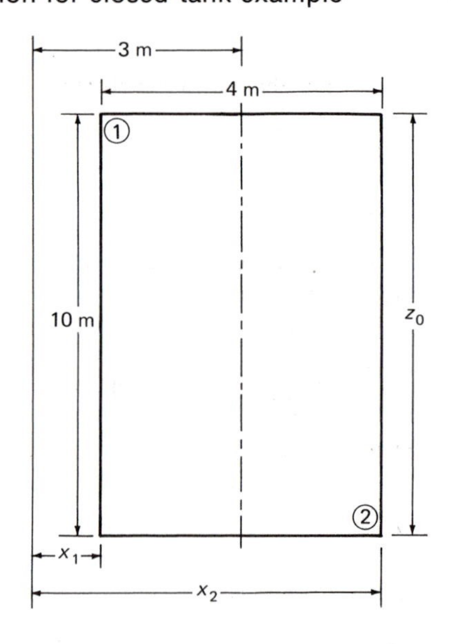

	(3) *fluid properties,* linseed oil at 20°C
Table A-1	$\rho = 942.0 \text{ kg/m}^3$
(1-35)	$\gamma = \rho g = 942.0 \times 9.807 = 9\,238 \text{ N/m}^3$

Step B — *Compute Maximum Pressure Differential*

(2-51) Analysis of equation indicates that the minimum pressure will occur at minimum radius and maximum elevation. This is point 1 of Fig. 2-25. In a like manner, it is found that the maximum pressure will occur at point 2.

Fig. 2-25 $x_1 = 3 - 2 = 1$ m, $x_2 = 4 + 1 = 5$ m.

Taking tank bottom as the datum,

$z_1 = 10$ m, $z_2 = 0$ m

(2-51)
$$p_2 - p_1 = \frac{\gamma\omega^2}{2g}(x_2^2 - x_1^2) - \gamma(z_2 - z_1)$$

$$p_2 - p_1 = \frac{(9\,238)\,(25.13)^2}{2 \times 9.807}(5^2 - 1^2) - (9\,238)\,(0 - 10)$$

$$p_2 - p_1 = 7\,230\,897 = 7.230 \text{ M Pa}$$

2-12 BUOYANCY AND FLOTATION

Principles

The elementary principles of buoyancy and flotation were established by Archimedes (287–212 B.C.). These principles are usually stated as follows: (1) a body immersed in a fluid is buoyed up by a force equal to the weight of fluid displaced by the body, and (2) a floating body displaces its own weight of the fluid in which it floats. These principles are readily proved by the methods of Sec. 2-9.

Buoyant Force

Consider the body *ABCD* shown in Fig. 2-26. Dashed lines *AE* and *BF* are vertical projections. The force F_z exerted by the fluid vertically on the body is the weight of the fluid above *ABC*. This weight is

$$F_z = \gamma \text{Vol}_{EACBF} \tag{2-56}$$

In a like manner, the upward vertical force is the weight above *ADB*, or

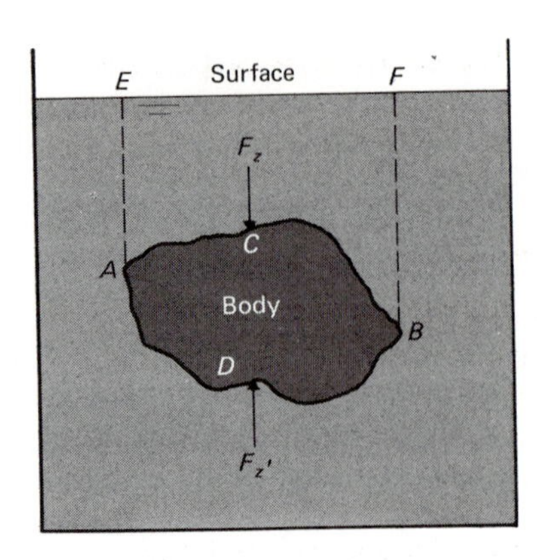

FIGURE 2-26 Notation for submerged bodies

$$F_{z'} = \gamma \text{Vol}_{EADBF} \tag{2-57}$$

The net upward force is the buoyant force F_B defined as follows:

$$F_B = F_{z'} - F_z = \gamma \text{Vol}_{EADBF} - \gamma \text{Vol}_{EACBF} = \gamma \text{Vol}_{ABCD} \tag{2-58}$$

Thus the buoyant force is the weight of the fluid displaced and always acts upward.

When an object floats as shown in Fig. 2-27, the buoyant force F_B then becomes

$$F_B = \gamma \text{Vol}_{ABD} \tag{2-59}$$

The weight of the body F_g acts downward, so that for vertical equilibrium,

$$\Sigma F_z = 0 = F_g - F_B$$

or

$$F_B = F_g \tag{2-60}$$

FIGURE 2-27 Notation for floating bodies

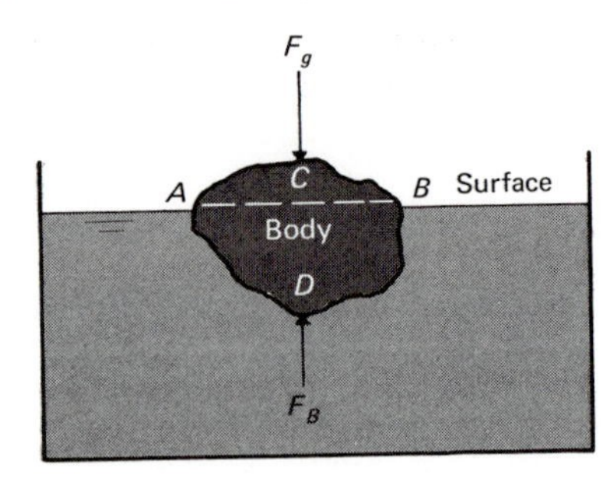

Free Body Analysis

The equation developed for flotation is a special case where the body is lighter than the fluid it can displace. A more general approach is that of the free body diagram. If an object immersed in a liquid is heavier than the fluid it can displace, it will sink to the bottom unless an upward force is applied to prevent it. A lighter-than-air ship or balloon will continue to rise unless a downward force is applied or it reaches an altitude where its density is the same as the atmosphere.

Figure 2-28 is a free-body diagram of an object immersed in a fluid. For vertical equilibrium,

$$\Sigma F_z = 0 = F_B - F_g - F_L \tag{2-61}$$

where

F_B = buoyant force

F_g = weight of body

F_L = force required to prevent body from rising.

$(-F_L)$ = force required to raise object.

EXAMPLE OF FLOTATION Ten percent of an iceberg floats above the surface of sea water at 0°C. What is the specific weight of the ice?

Step A *Data Reduction*

(B-51) (1) *gravity*

Assume $g = g_o = 9.807\ \text{m/s}^2$

(2) *fluid properties,* sea water at 0°C

Table A-1 $\rho = 1\ 028\ \text{kg/m}^3$

(1-35) $\gamma_{\text{sea}} = \rho g = (1\ 028)(9.807) = 10\ 082\ \text{N/m}^3$

FIGURE 2-28 Free-body diagram

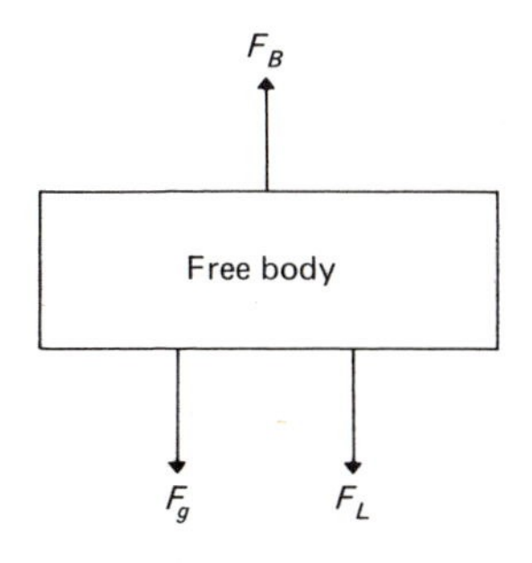

Step B *Compute Specific Weight of Ice*

(2-61) $0 = F_B - F_g - F_L$

for flotation $F_L = 0$ or

$F_B = F_g$, if 10% of the iceberg is above the surface, then 90% is below, or

$$F_B = \gamma_{sea}\ 0.9\ \text{Vol} = F_g = \gamma_{ice}\ \text{Vol} \quad \textbf{(a)}$$

or

$$\gamma_{ice} = 0.9\gamma_{sea} = 0.9 \times 10\,082 = 9\,074\ \text{N/m}^3 \quad \textbf{(b)}$$

EXAMPLE OF SUBMERGENCE A sphere 4 ft in diameter weighs 2,200 lbf and lies at the bottom of the sea. If the ocean water is 86°F, what is the minimum force required to lift this sphere from the ocean floor?

Step A *Data Reduction*

B-50 (1) *gravity*

Assume $g = g_o = 32.17$ ft/sec^2

(2) *fluid properties,* sea water at 86°F

Table A-1 $\rho = 1.982$ slugs/ft^3

(1-35) $\gamma = \rho g = 1.982 \times 32.17 = 63.76$ lbf/ft^3

(3) *geometric,* volume of a sphere

Table A-10 $$\text{Vol} = \frac{4\pi r^3}{3} = \frac{\pi D^3}{6} = \frac{\pi (4)^3}{6} = 33.51\ \text{ft}^3$$

Step B *Compute Required Force*

(2-59) $F_B = \gamma \text{Vol} = 63.76 \times 33.51 = 2137$ lbf

(2-61) $\Sigma F_z = 0 = F_B - F_g - F_L$

or

$$F_L = F_B - F_g = 2137 - 2200$$
$$= -63 \text{ lbf or 63 lbf upward to lift sphere}$$

EXAMPLE OF AN AIRSHIP An airship has a volume of 56 600 m^3. Its total weight including the gas is 45 kN. What force is required to anchor this airship under atmospheric conditions of 15°C and 101.3 kPa?

Step A *Data Reduction*

(1) *unit conversion*

(B-27) $T_K = t_c + 273.15 = 15 + 273.15$

$T = 288.15$ K

(2) *gravity*

(B-51) Assume $g = g_o = 9.807$ m/s^2

(3) *proportionality constant*

(1-9) $g_c = 1$ kg · m/N · s^2

(4) *fluid properties,* air

Table A-7 $R = 287.1$ J/kg · K

(1-44) $\rho = p/g_cRT = (101.3 \times 10^3)/(1)(287.1)(288.15)$

$\rho = 1.224$ kg/m^3

(1-35) $\gamma = \rho g = 1.224 \times 9.807 = 12.01$ N/m^3

Step B *Compute Anchor Force*

(2-59) $F_B = \gamma\text{Vol} = (12.01)(56\ 600) = 679\ 766$ N

(2-61) $F_L = F_B - F_g = 679\ 766 - 45\ 000$

$F_L = 634\ 766 = 634.8$ kN (anchor force)

2-13 STATIC STABILITY

Definition

A body is said to be in static equilibrium when the imposition of a small displacement brings into action forces that tend to restore the body to its original position.

Stability of Submerged Bodies

Consider the balloon and basket shown in Fig. 2-29. The center of gravity is indicated by the letter G, the center of buoyancy by the letter B. For vertical equilibrium $F_B = F_g$. The balloon in its normal vertical position is shown on the left and in a displaced position on the right. In this second position, there is a couple F_gd tending to restore the system to its original position. This illustration shows that for completely submerged bodies there are two requirements for stability:

(1) The center of buoyancy and the center of gravity must lie in the same vertical line.
(2) The center of buoyancy must be located above the center of gravity.

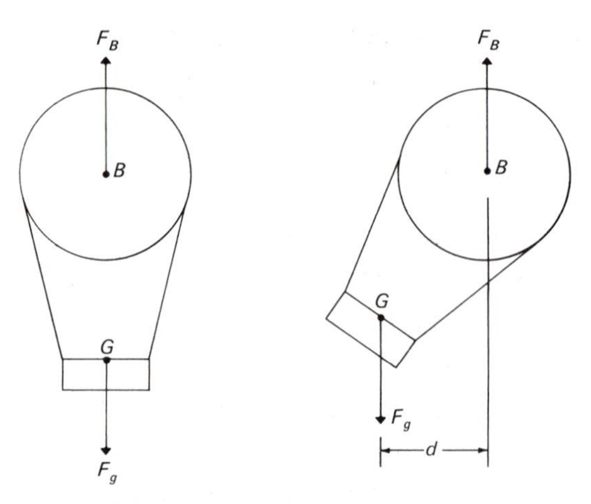

FIGURE 2-29 Notation for submerged bodies

Stability of Floating Bodies

The second requirement that the center of buoyancy must be above the center of gravity is not necessary for floating objects. Consider the simple object shown in Fig. 2-30 floating in an upright position on the left and in a displaced position on the right. The center of gravity G is located above the center of buoyancy B. The center of gravity is usually located above the center of buoyancy for surface ships except for some racing sailboats. Displacement sets up a restoring force so that the body is in stable equilibrium. Where the line of action of the buoyant force intersects the center line of the body is the point M. This point is called the metacenter. This is

FIGURE 2-30 Notation for floating bodies

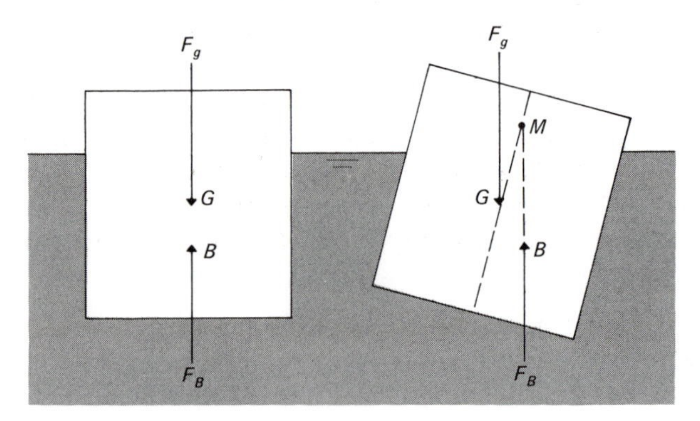

an important parameter in ship design but is beyond the scope of this text.

This illustration shows that for floating bodies there are two requirements for stability:

(1) The center of buoyancy and the center of gravity must lie on the same vertical line.
(2) When the body is displaced, the line of action of the buoyant force must intersect the center line (M, metacenter) above the center of gravity to set up a restoring couple.

PROBLEMS

1. A large closed tank located on the surface of the moon is filled with a liquid. The pressure on the liquid surface is 69 kPa, and 3.35 m below the surface it is 141 kPa. Estimate (a) the specific weight and (b) the density of the liquid.
2. Calculate the pressure at the bottom of the closed tank shown in Fig. 2-31 if the fluids are at 68°F.
3. A Bourdon tube gage is installed to sense the pressure of oxygen in a large closed tank. When the temperature of the oxygen is 40°C, the gage indicates 700 kPa. When the oxygen temperature is increased to 95°C, the gage indicates 840.5 kPa. What is the barometric pressure?
4. A precision Bourdon gage is used to measure the pressure of steam in a pipe whose center line is 20 ft above floor level. For accurate reading, the gage is mounted with its center line 16 ft below the center line of the pipe. The tubing connecting the gage

FIGURE 2-31 Notation for Problem 2

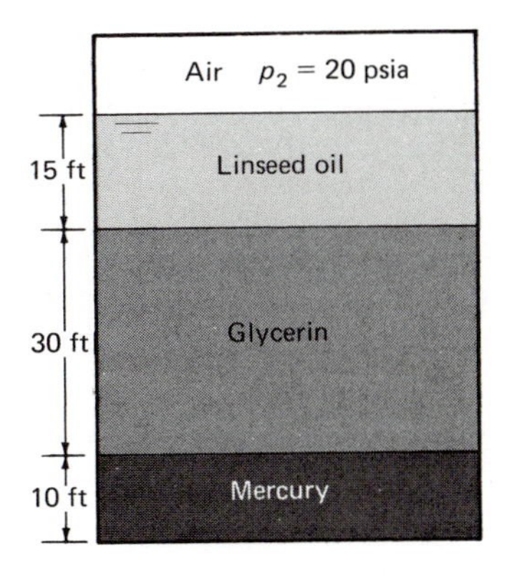

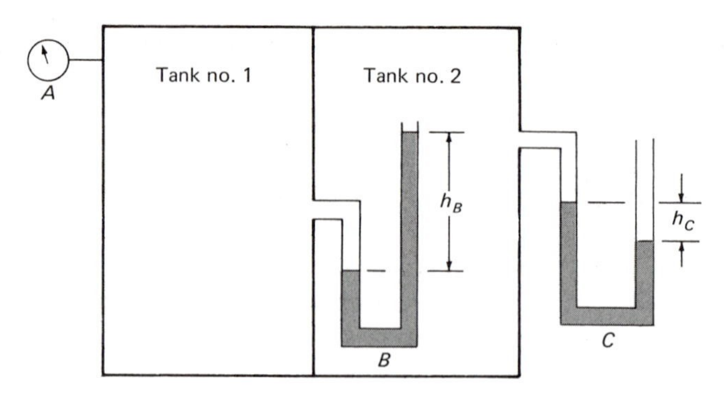

FIGURE 2-32 Notation for Problem 7

to the pipe is not insulated, so that the steam condenses in the tubing. The average temperature of the water is 86°F. The barometric reading is 30.00 in. Hg at 86°F. The reading of the gage is 100.22 lbf/in². What is the absolute pressure of the steam in the pipe?

5. A mercury barometer indicates 736 mm in a room whose temperature is 30°C. What is the atmospheric pressure in millimeters of mercury at 0°?
6. A barometer of the type shown in Fig. 2-6 is filled with 86°F carbon tetrachloride. How high will the carbon tetrachloride rise in the tube when the atmospheric pressure is 14.70 psia?
7. Tanks 1 and 2 of Fig. 2-32 are closed and filled with air. The barometric pressure is 100 kPa. Gage A indicates 206.9 kPa and $h_B = 1.816$ m. If both manometers contain mercury at 20°C, compute the value of h_C.
8. The liquid in the arrangement shown in Fig. 2-33 is ethyl alcohol at 30°C. The height of the liquid z_A is 1.2 m above the mercury. If the barometric pressure is 101.3 kPa, what is the manometer reading h?
9. An inclined manometer of the type shown in Fig. 2-9 has a cistern diameter of 1 in. and a reading tube diameter of 1/8 in. What angle θ is required for a 10 to 1 reading advantage over a U-tube manometer?
10. Assume that Venus has an isentropic atmosphere of carbon dioxide. If the surface temperature is 800°F and the pressure is 235 psia on the surface, estimate the pressure and temperature at an altitude of 50,000 ft.
11. Compute the pressure and temperature of the U.S. standard atmosphere at 20,000 ft using data from Table 2-1 and the ideal gas laws.

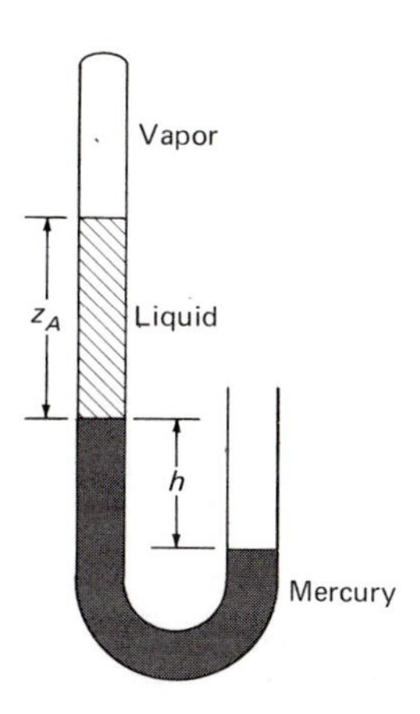

FIGURE 2-33 Notation for Problem 8

12. Check the relation p/p_o of the U.S. standard atmosphere at $H = 32{,}000$ m. Use the standard temperature profile and ideal gas laws.
13. A cylindrical container 300 mm in diameter, with axis horizontal, is half filled with mercury whose temperature is 20°C. Compute the total force acting on one end.
14. A circular disk 4 ft in diameter is in a plane sloping 60° from the horizontal. Five feet of turpentine at 122°F stands above the disk center. Calculate the magnitude and location of the total force of the turpentine on the disk.
15. Calculate the magnitude and location of the total force on a gate 4 ft square, located in a 60° plane, and having its upper edge 10 ft below the surface of water whose temperature is 68°F.
16. A cylindrical tank 1 m in diameter is mounted with its axis horizontal. The tank is completely filled with glycerin at 20°C. Atmospheric pressure acts on the topmost point of the glycerin. For a 1 m of length along the axis, what is the resultant force of the liquid on the curved surface of a vertical half of the tank?
17. Consider *ABCDE* of Fig. 2-17 to be a tank whose width is 10 ft. The curved portion *AE* is a quarter-circle whose radius is 20 ft. This tank is filled to a depth of 40 ft with water at 68°F. Calculate the magnitude and location of the total liquid force on the curved portion.
18. A cylindrical steel tank 10 ft long is to contain 2000 lbm of air at 1500 psig and 80°F. For an allowable stress of 10,000 psi, estimate the diameter and the theoretical minimum tank thickness.
19. In a special application, a 24-in. schedule 160 steel pipe is to be used to contain a gas at 100°F. What is the maximum pressure

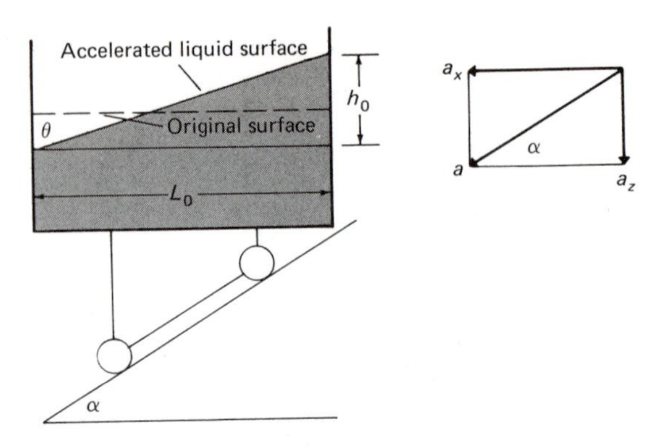

FIGURE 2-34 Notation for Problem 22

of the gas if the stress is not to exceed 10,000 psi? Use code value of $A = 0$ and a $12\frac{1}{2}\%$ tolerance for wall thickness.

20. It is proposed to set up a metric series of pipe size. What should be the wall thickness of a schedule 20 pipe with outside diameter of 600 mm?
21. What would be the equivalent schedule number for an 8-in. 250 psi cast iron pipe?
22. The open container of liquid shown in Fig. 2-34 accelerates down a 30° incline at 4.5 m/s². What is the slope of its free surface?
23. The rectangular tank shown in Fig. 2-35 is 5 ft wide, 6 ft deep, and 10 ft long and is filled to a depth of 4 ft with 50°F carbon tetrachloride. When this open tank is accelerated horizontally at $\frac{1}{2}$ g, calculate: (a) The volume of liquid spilled, (b) the force on each end, and (c) the inertia force of the accelerated mass.
24. An open conical container 6 m high and of 3 m maximum diameter is filled with water at 20°C and moves vertically downward with a deceleration of 10 m/sec². Calculate the liquid pressure on the bottom of the container.

FIGURE 2-35 Notation for Problem 23

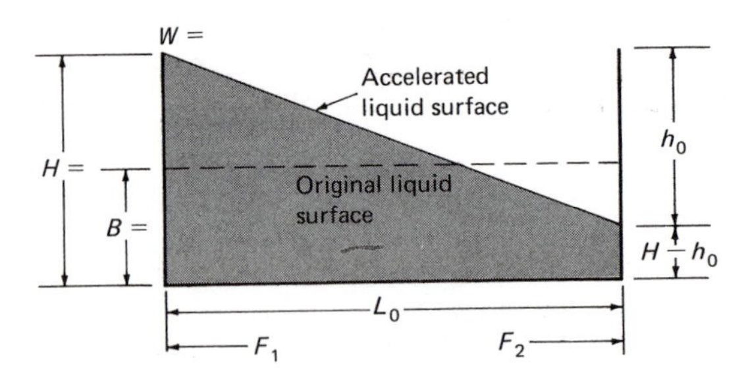

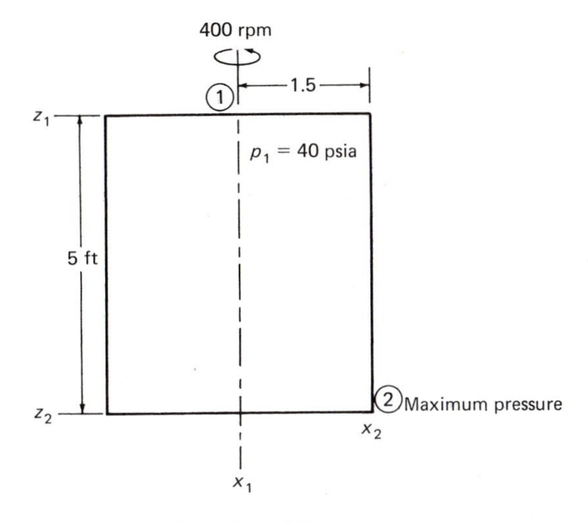

FIGURE 2-36 Notation for Problem 25

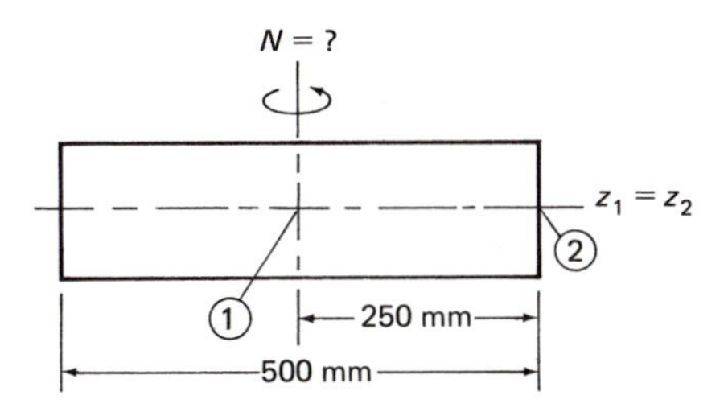

FIGURE 2-37 Notation for Problem 26

25. The open cylindrical tank shown in Fig. 2-36 is 3 ft in diameter and 5 ft deep and is filled with water at 68°F and closed. The pressure on the top is raised to 40 psia. Calculate the maximum pressure in the tank during rotation at 400 rpm about its vertical center line.
26. The centrifugal water pump impeller shown in Fig. 2-37 has a diameter of 500 mm. Assuming no flow through the pump, at what speed must the pump be operated to develop a difference of 690 kPa with water at 20°C?
27. The barge shown in Fig. 2-38 is 8 ft deep, has vertical sides, and is 20 ft wide, with a trapezoidal cross-section of 50 ft top length and 40 ft bottom length. The barge goes downstream in fresh river water at 68°F where it has a draft of 3 ft. The barge is docked and takes on its own weight in cargo and heads out to

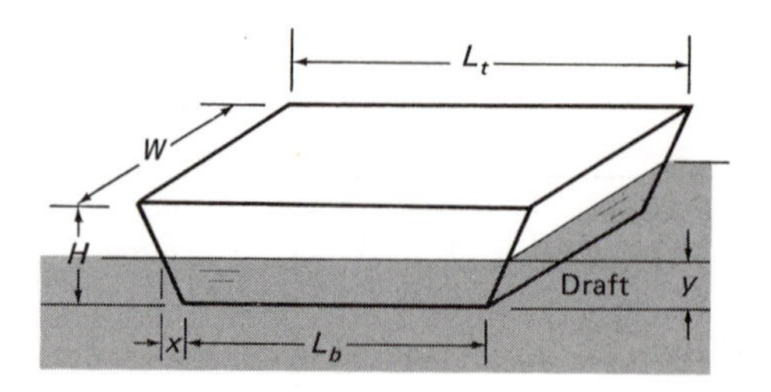

FIGURE 2-38 Notation for Problem 27

sea. If the sea water temperature is also 68°F what will be the draft at sea?

28. A cube of aluminum 1 m on a side requires a force of 17.16 kN to lift it when it is submerged in water at 3.98°C. What is its specific gravity?
29. It has been proposed (reference *) that natural gas could be transported by giant airships cruising at 48 mph at an altitude

FIGURE 2-39 Notation for Problem 30

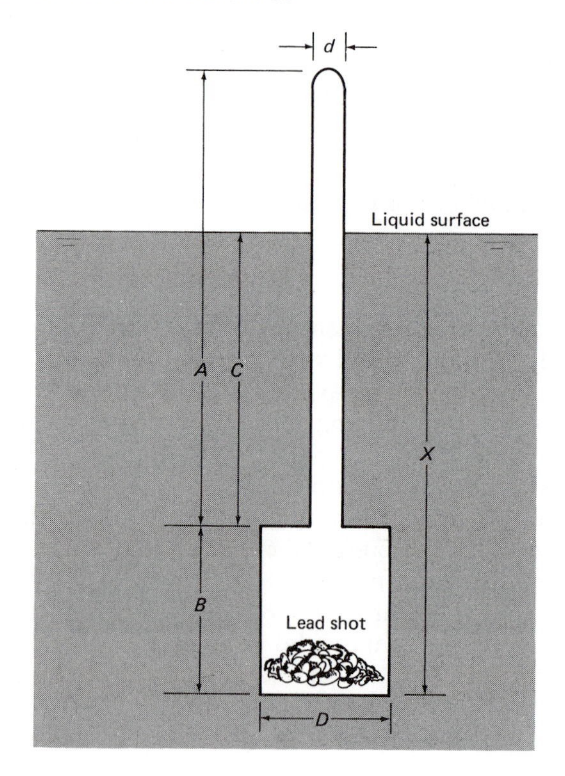

of 10,000 ft. For this purpose a hypothetical airship was assumed to be a prolate spheroid whose length is 5400 ft and width is 1400 ft with 80% of its volume devoted to natural gas. Assuming that natural gas has the same characteristics as methane, and is under a gage pressure of 42.3 lbf/ft^2, compute the gross lift of the airship in flight.

30. The crude hydrometer shown in Fig. 2-39 consists of a cylinder 13 mm in diameter and 50 mm in length surmounted by a cylinder 3 mm in diameter and 250 mm long. Lead shot is added until its total weight is 90 mN. To what depth will this hydrometer float in glycerin at 40°C?

REFERENCES

Hughes, W. F. and J. A. Brighton, *Theory and Problems of Fluid Dynamics.* Schaum Publishing Company, New York, 1967.

*Sonstegaard, Miles H. "Transporting Gas by Airship." *Mechanical Engineering,* June 1973, pp. 19–25.

Vennard, John K., *Elementary Fluid Mechanics,* 3d ed. John Wiley and Sons, 1954.

American National Standard Code for Power Piping. American Society of Mechanical Engineers, New York, B31.1.0-1967.

American National Standard for Wrought Steel and Wrought Iron Pipe. American Society of Mechanical Engineers, New York, B36.10-1970.

U.S. Standard Atmosphere 1962. Prepared under sponsorship of U.S. Air Force, National Aeronautics and Space Administration, and U.S. Weather Bureau, U.S. Government Printing Office, Washington, D.C., 1962.

U.S. Standard Atmosphere Supplements—1966. Prepared under sponsorship of Environmental Services Administration, National Aeronautics and Space Administration, and U.S. Air Force. Government Printing Office, Washington, D.C., 1967.

3 Fluid Kinematics

3-1 INTRODUCTION

Fluid kinematics is that branch of fluid mechanics that deals with the geometry of fluid motion without consideration of forces causing motion. It will be assumed that any fluid particle is very large in size with respect to a molecule and is hence continuous, so that we are concerned with a continuum.

A quantity such as velocity or fluid particle displacement must be measured relative to some convenient coordinate system. Two methods have been devised for representing fluid motion. One describes the behavior of a single fluid particle, the other is concerned with several fluid particles passing by certain points or sections of a fluid.

The description of the behavior of individual fluid particles is called the Lagrangian method after Joseph Louis Lagrange (1736–1813). This method of analysis involves establishing a coordinate system relative to a moving fluid particle as it moves through the continuum and measuring all quantities relative to the moving particle. The behavior of the individual fluid particle is of no practical importance in fluid mechanics, and this method is seldom used.

The establishment of a fixed coordinate system and the observation of the fluid passing through this system is called the Euler method after Leonard Euler (1707–1783). The Eulerian method will be used for the most part throughout this text.

3-2 STEADY AND UNSTEADY FLOW

If at every point in the continuum, the local velocity U, and other fluid property, remains unchanged with time, the flow is said to be *steady flow*. While flow is generally unsteady by nature, many real cases of unsteady flow may be reduced to the case of steady flow, a case that is far easier to analyze mathematically. One technique for doing this is to use a temporal mean or average.

Consider the velocity U' at a point in space and time shown in Fig. 3-1. The temporal mean average of U' is U as defined by

$$U = \frac{1}{t}\int U'\,dt \tag{3-1}$$

And more generally

$$(\text{temporal fluid property}) = 1/t \int (\text{instantaneous fluid property})\,dt \tag{3-2}$$

This technique may be used for small cyclic variations of fluid properties such as in turbulent flow or for large but rapidly changing cycles such as those produced by high-speed reciprocating machinery. The amount of error produced will, of course, vary with the application.

Another technique that may be used that is free of error is to change the space reference. Consider the boat shown in Fig. 3-2

FIGURE 3-1 Notation for unsteady flow

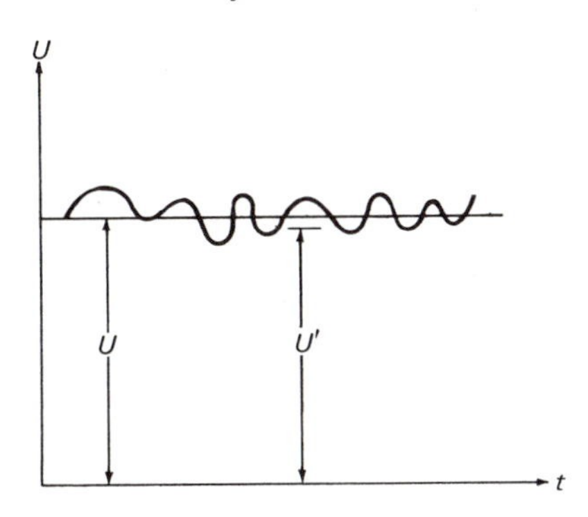

FIGURE 3-2 Boat moving through still water

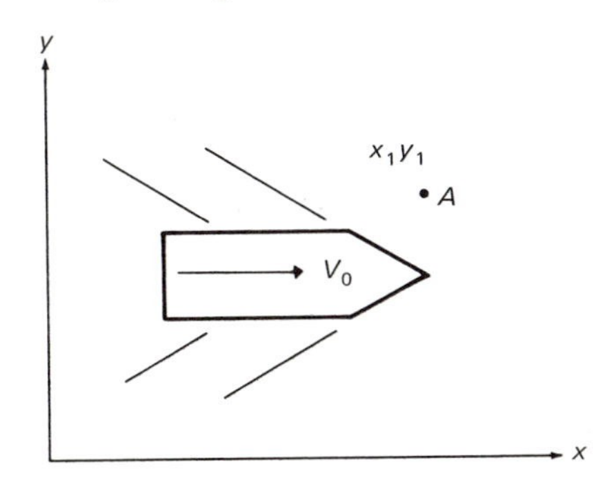

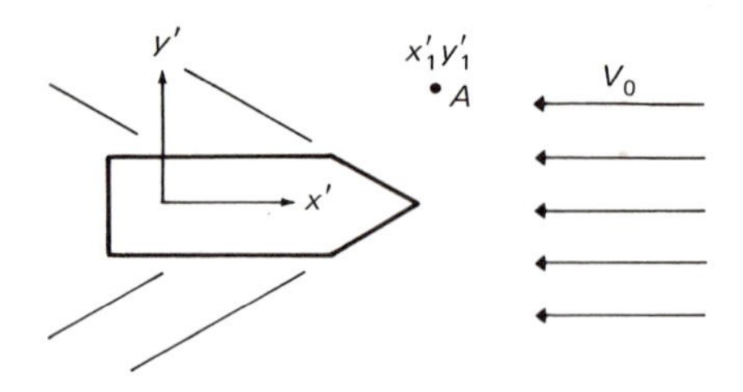

FIGURE 3-3 Water flowing around a boat

moving in still water with a speed of V_o. As the boat passes point A located at x_1, y_1 the wave produced will cause the fluid at point A to change from a velocity of zero to a complicated variation with time until long after the boat has passed before it returns to zero again. If the point of reference is switched to the boat, then point A has a velocity of $-V_o$ at all times, as shown in Fig. 3-3. This method may be used any time a body is moved at constant speed in an undisturbed fluid. Note that all that was actually done was to reverse the direction of the velocity.

3-3 STREAMLINES AND STREAMTUBES

Velocity is a vector and hence has both magnitude and direction. A *streamline* is a line which gives the direction of the velocity of the fluid at each point. If an almost instantaneous photograph were made of a flowing fluid, the movements of a given particle would appear as a short streak on the photograph. The direction of the streak would be tangent to the flow path at that point and at that instant, and the length of the streak would be proportional to the instantaneous velocity of that particle. Fig. 3-4 shows the construction of the streamline from the particle streaks. This streamline is also tangent everywhere to the velocity vectors.

When streamlines are drawn through a closed curve in a steady flow, they form a boundary which the fluid particles cannot pass. The space between streamlines becomes a tube or passage and is

FIGURE 3-4 Streamlines

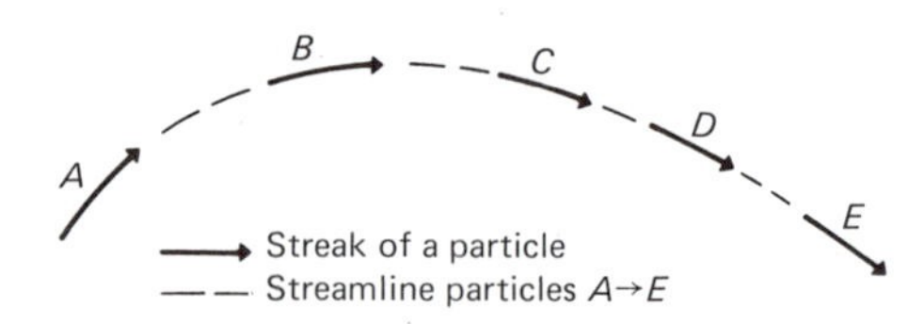

called a *streamtube*. The streamtube may be isolated from the rest of the fluid for analysis. The use of the streamtube concept broadens the application of fluid flow principles; for example, it allows treating the flow inside a pipe and the flow around an object with the same laws. A streamtube of small size approaches its own axis, a central streamline; thus equations developed for a streamtube may also be applied to a streamline.

3-4 VELOCITY PROFILE

Volumetric Flow Rate

In the flow of real fluids, the individual streamlines will have different velocities past a section. Fig. 3-5 shows the steady flow of a fluid in a circular pipe. The velocity profile is obtained by plotting the velocity U of each streamline as it passes section A-A. The steamtube that is formed by the space between the streamlines is an annulus whose area normal to the flow is dA as shown in Fig. 3-5 for the streamtube whose velocity is U. The volume rate of flow Q past section A-A is given by

$$Q = \int U \, dA \tag{3-3}$$

Average Velocity

In many engineering applications, the velocity profile is nearly a straight line or can be reduced to one so that the average velocity V may be used. The average velocity V is defined as follows:

$$V = \frac{Q}{A} = \frac{1}{A}\int U \, dA \tag{3-4}$$

FIGURE 3-5 Velocity profile

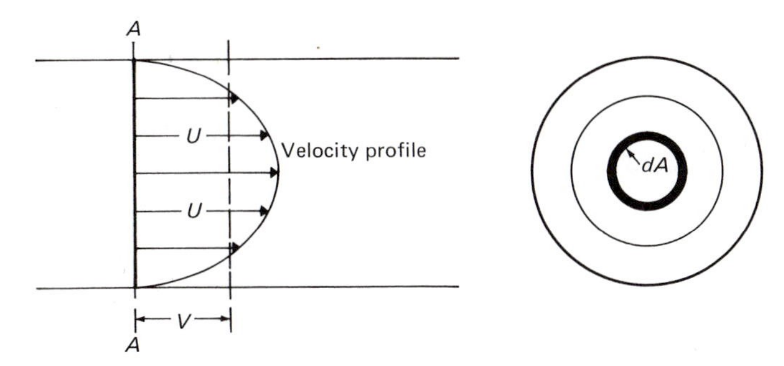

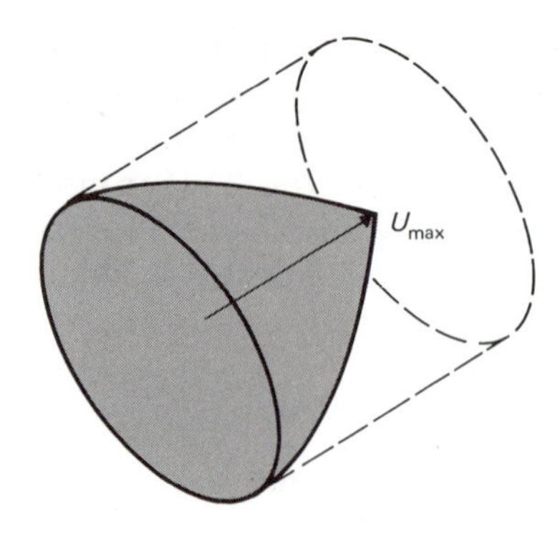

FIGURE 3-6 Three-dimensional flow

Methods of Flow Analysis

All flows take place between boundaries that are three-dimensional. The terms *one-dimensional, two-dimensional,* and *three-dimensional flow* refer to the number of dimensions required to describe the velocity profile of the streamtubes at a given section.

THREE-DIMENSIONAL FLOW A *volume* (L^3) is necessary to describe the velocity profile as shown in Figs. 3-5 and 3-6. The usual example is flow in a pipe, but the conduit or duct need not be circular for a three-dimensional velocity profile. Three-dimensional flow, of course, is the general case.

TWO-DIMENSIONAL FLOW An *area* (L^2) is necessary to describe the velocity profile as shown in Fig. 3-7; for example, the flow of a fluid between two parallel plates.

ONE-DIMENSIONAL FLOW A *line* (L) is necessary to describe the velocity profile as shown in Figs. 3-3 and 3-8.

EXAMPLE OF A ONE-DIMENSIONAL FLOW A fluid flows steadily through a 6 m by 12 m conical enlarger, as shown in Fig. 3-9. The average velocity at a section where the diameter is 9 m is 4 m/s. Calculate the volume flow rate past that section.

FIGURE 3-7 Two-dimensional flow

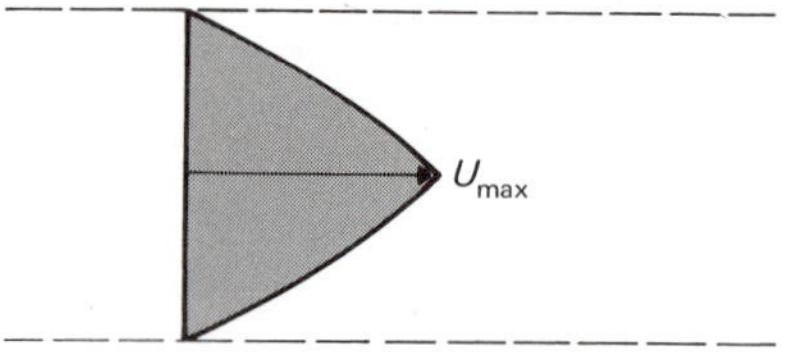

U_{max}

FIGURE 3-8 One-dimensional flow

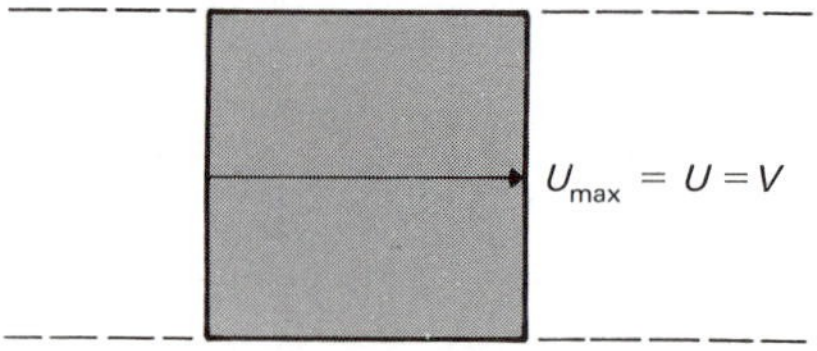

$U_{max} = U = V$

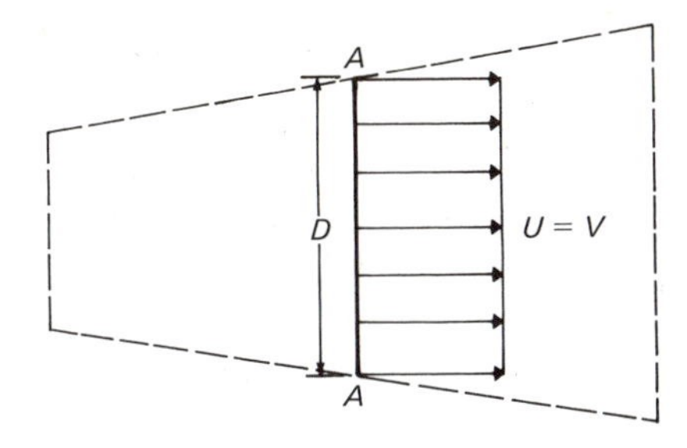

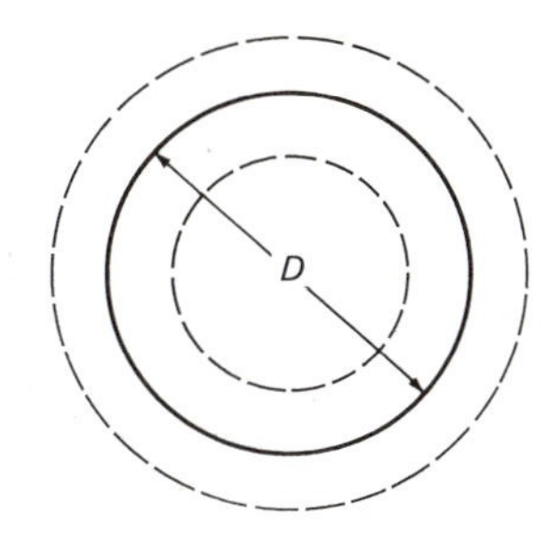

FIGURE 3-9 One-dimensional flow example

Step A	*Data Reduction*
	Geometric, circle
Table A-10	$A = \pi D^2/4 = \pi(9)^2/4 = 63.62\ \text{m}^2$
Step B	*Compute Flow*
(3-4)	$Q = AV = 63.62 \times 4 = 254.5\ \text{m}^3/\text{s}$

EXAMPLE OF A TWO-DIMENSIONAL FLOW REDUCED TO ONE-DIMENSIONAL FLOW An artificial canal is 100 ft wide and is of rectangular cross-section. Water flows in this canal to a depth of 25 ft. Measurements made of the velocity profile at a typical section are as follows:

Depth, ft	Velocity, ft/sec
0 (surface)	1.18
5	1.26
10	1.16
15	0.95
20	0.55
25 (bottom)	0

Estimate the average velocity of the water as it flows past this section.*

---

* In order to solve this problem, it is necessary to evaluate $\int U\,dA$ from test data without knowing the functional relationship between the local velocity and area. One method for doing this is to plot the data and draw a smooth curve of the velocity profile and measure the enclosed area mechanically. Another method is to use curve-fitting techniques to derive an equation that best represents the data and integrate this equation. A third method is to use either the trapezoidal rule or Simpson's rule to approximate the integral. Because the intent here is to illustrate streamline and streamtube concepts rather than to obtain maximum numerical accuracy, the trapezoidal rule is used to solve this problem.

Step A *Data Reduction*

Fig. 3-10 (1) *apply the trapezoidal rule.* Divide the velocity profile into five evenly spaced depths Δy based on the six measured velocities. The measured velocities may then be considered *streamlines* and the spaces between them as *streamtubes.* Assume that the velocity of a streamtube is the average of its bounding streamline velocities.

(3-3) (2) *volume flow rate.* For each streamtube, the volume flow rate is $Q = \int U\,dA = U \int dA = U\,\Delta y W$.
The total volumetric flow rate is

$$\Sigma Q = (U_A + U_B + U_C + U_D + U_E)\,\Delta y W \quad \textbf{(a)}$$

$$\Sigma Q = [\tfrac{1}{2}(U_0 + U_5) + \tfrac{1}{2}(U_5 + U_{10}) + \tfrac{1}{2}(U_{10} + U_{15}) + \tfrac{1}{2}(U_{15} + U_{20}) + \tfrac{1}{2}(U_{20} + U_{25})]\,\Delta y W \quad \textbf{(b)}$$

$$\Sigma Q = [\tfrac{1}{2}(U_0 + U_{25}) + U_5 + U_{10} + U_{15} + U_{20}]\,\Delta y W \quad \textbf{(c)}$$

Step B *Compute Average Velocity*

(3-4)
$$V = \frac{\Sigma Q}{A} = \frac{\Sigma Q}{y_o W} = \frac{\Sigma Q}{5\Delta y W}$$

Substituting for ΣQ for (c) in Eq. (3-4),

$$V = [\tfrac{1}{2}(U_0 + U_{25}) + U_5 + U_{10} + U_{15} + U_{20}]\,\Delta y W/5\Delta y W \quad \textbf{(d)}$$

$$V = [\tfrac{1}{2}(U_0 + U_{25}) + U_5 + U_{10} + U_{15} + U_{20}]/5$$

$$V = [\tfrac{1}{2}(1.18 + 0) + 1.26 + 1.16 + 0.95 + 0.55]/5$$

$$V = 0.902 \text{ ft/sec}$$

EXAMPLE OF THREE-DIMENSIONAL LAMINAR FLOW REDUCED TO ONE-DIMENSIONAL FLOW Experiments with the flow of viscous fluids in circular conduits indicate that when viscous forces predominate and laminar flow takes place, the velocity profile is a paraboloid of

FIGURE 3-10 Two-dimensional flow example

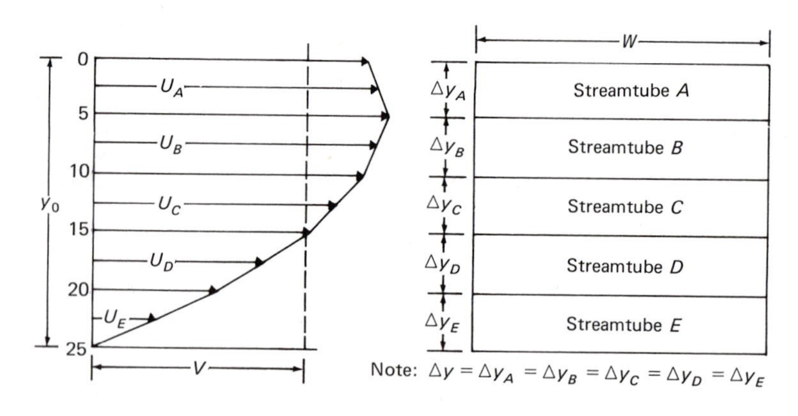

revolution (this will be demonstrated on a theoretical basis when we study the flow of fluids in pipes) with the maximum velocity at the center of the conduit. Derive a relationship between the average velocity and the center line or maximum velocity.

Step A	*Data Reduction*	
Table A-10	*Circle,* $A = \pi r^2 = \pi r_o^2$, $dA = 2\pi r\,dr$	
Table A-10	*Parabola* (horizontal), $y^2 = \left(\frac{a^2}{b}\right)x$	**(a)**
Fig. 3-11	$y = r$, $b = U_m$, $a = r_o$, $x = U_m - U$	**(b)**

Substituting in (a),

$$y^2 = \left(\frac{a^2}{b}\right)x = r^2 = \left(\frac{r_o^2}{U_m}\right)(U_m - U) \qquad \textbf{(b)}$$

Which reduces to

$$U = U_m\left[1 - \left(\frac{r}{r_o}\right)^2\right] \qquad \textbf{(c)}$$

Step B	*Compute Average Velocity*
(3-4)	$V = \frac{1}{A}\int U\,dA$

Substituting data from Step A,

$$V = \frac{1}{A}\int U\,dA = \frac{1}{\pi r_o^2}\int_0^{r_o}\left\{U_m\left[1 - \left(\frac{r}{r_o}\right)^2\right]\right\}(2\pi r\,dr)$$

$$V = \frac{2\pi U_m}{\pi r_o^2}\int_0^{r_o}\left[1 - \left(\frac{r}{r_o}\right)^2\right]r\,dr = \frac{2U_m}{r_o^2}\left[\frac{r^2}{2} - \frac{r^4}{4r_o^2}\right]_0^{r_o}$$

$$V = \frac{U_m}{2} \quad \text{or} \quad \frac{V}{U_m} = 1/2$$

FIGURE 3-11 Three-dimensional flow example

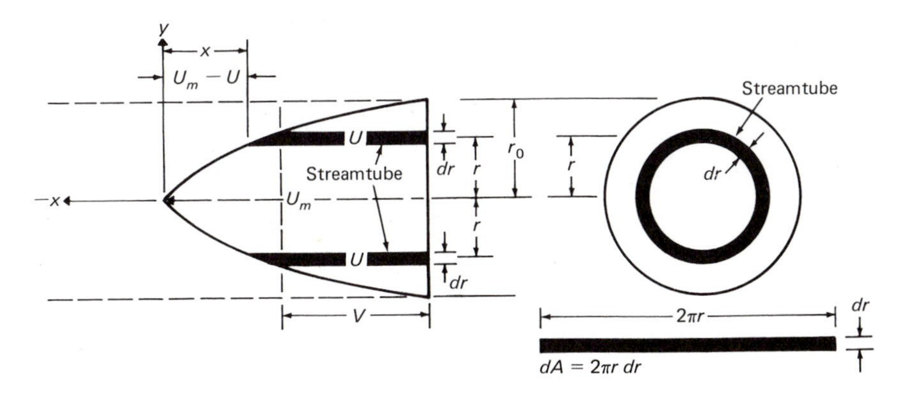

Step C *Check by Geometry*

Table A-10 $$Paraboloid,\ \text{vol} = \frac{\pi b r^2}{2},\quad Q = \frac{\pi U_m r_o^2}{2}$$

(3-4) $$V = Q/A = \frac{\pi U_m r_o^2/2}{\pi r_o^2} = \frac{U_m}{2}$$

or

$$\frac{V}{U_m} = 1/2$$

3-5 CONTINUITY EQUATION

Mass Flow Rate

Consider the volume of fluid $ds\,dA$ moving in a streamtube with a velocity of U as shown in Fig. 3-12. By definition, $m = \rho$ vol or $dm = \rho\ ds\ dA$. Dividing by the time dt for this volume to move the distance ds,

$$d\dot{m} = d\left(\frac{dm}{dt}\right) = \frac{\rho ds\,dA}{dt} = \rho U\,dA \tag{3-5}$$

where $\dot{m}$ is the mass flow rate. Integrating Eq. (3-5) and substituting $AV = \int U\,dA$ from Eq. (3-4),

$$\int d\dot{m} = \dot{m} = \int \rho U\,dA = \rho A V \tag{3-6}$$

FIGURE 3-12 Mass flow rate

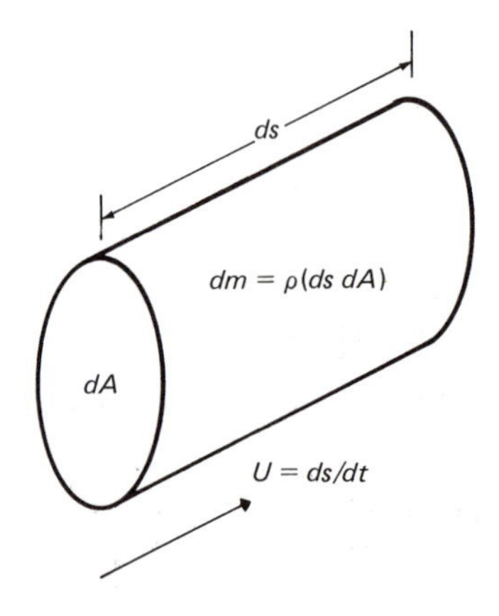

Continuity Equation

This equation is a special case of the general physical law of the conservation of mass. It may be stated simply that the mass flow rate entering a system is equal to the mass rate of storage in the system plus the mass flow rate leaving the system. Consider the flow system shown in Fig. 3-13. Fluid is being supplied to the tank by means of the pipe at the rate $\dot{m}_1 = \rho_1 A_1 V_1$ and leaves the system at the rate of $\dot{m}_2 = \rho_2 A_2 V_2$. If the amount supplied is greater than that leaving, then the tank level s will rise and fluid will be stored in the tank at the rate of $\dot{m}_s = \rho A\, ds/dt$. We can now state:

$$\textit{mass entering} = \textit{mass rate of storage} + \textit{mass rate leaving}$$

$$\dot{m}_1 \quad = \quad \dot{m}_s \quad + \quad \dot{m}_2$$

$$\rho_1 A_1 V_1 \quad = \quad \rho A\left(\frac{ds}{dt}\right) \quad + \quad \rho_2 A_2 V_2 \tag{3-7}$$

Steady State

If the amount supplied is equal to the amount removed, then ds is zero or there is no storage. Equation (3-7) becomes

FIGURE 3-13 Continuity equation

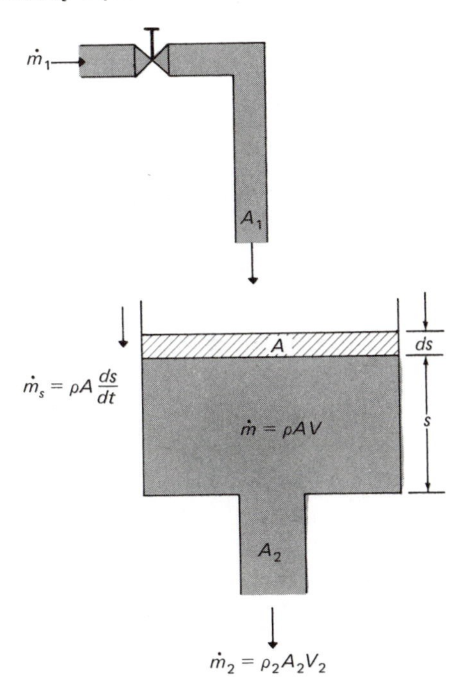

$$\dot{m} = \rho_1 A_1 V_1 = \rho_2 A_2 V_2 = \cdots = \rho_n A_n V_n = \rho AV \quad \textbf{(3-8)}$$

The mass flow rate is constant any place in a steady state system. For compressible fluids, it is sometimes convenient to use a differential form of Eq. (3-8), which may be obtained by writing it in logarithmic form and differentiating, noting that $\dot{m}$ is a constant.

$$\log_e \rho + \log_e A + \log_e V = \log_e \dot{m} \quad \textbf{(3-9)}$$

$$\frac{d\rho}{\rho} + \frac{dA}{A} + \frac{dV}{V} = 0 \quad \textbf{(3-10)}$$

EXAMPLE OF INCOMPRESSIBLE FLOW A 12-in. schedule 40 steel pipe reduces to 6-in. schedule 40 pipe and then expands to 8 in., as shown in Fig. 3-14. If the average velocity in the 12-in. pipe is 4 m/s, compute the average velocity in the 6- and 8-in. pipes, for *any incompressible fluid.*

Step A *Data Reduction*

Pipe Properties

Table A-11

Size, in.	Internal diameter, mm
6	154.1
8	202.7
12	303.3

Step B *Develop a Relationship*

(3-8) $\dot{m} = \rho_{12} A_{12} V_{12} = \rho_6 A_6 V_6 = \rho_8 A_8 V_8$

Definition Incompressible fluid $\rho_{12} = \rho_6 = \rho_8$ **(a)**
Dividing Eq. (3-8) by ρ,

$$\frac{\dot{m}}{\rho} = A_{12} V_{12} = A_6 V_6 = A_8 V_8 \quad \textbf{(b)}$$

Table A-10 $A = \frac{\pi D^2}{4}$, substituting in (b),

FIGURE 3-14 Incompressible flow example

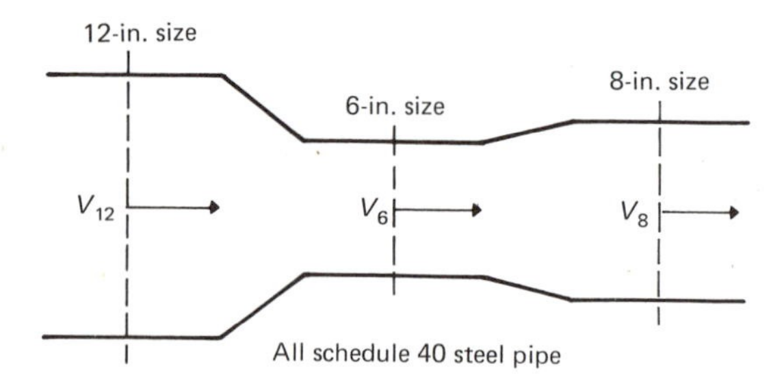

$$\left(\frac{\pi D_{12}^2}{4}\right)V_{12} = \left(\frac{\pi D_6^2}{4}\right)V_6 = \left(\frac{\pi D_8^2}{4}\right)V_8$$

Which reduces to

$$D_{12}^2 V_{12} = D_6^2 V_6 = D_8^2 V_8 \tag{c}$$

Solving (c) for V_6 and V_8,

$$V_6 = V_{12}\left(\frac{D_{12}}{D_6}\right)^2 \tag{d}$$

$$V_8 = V_{12}\left(\frac{D_{12}}{D_8}\right)^2 \tag{e}$$

Step C *Compute Velocities*

$$V_6 = V_{12}\left(\frac{D_{12}}{D_6}\right)^2 = 4\left(\frac{303.3}{154.1}\right)^2 = 15.50 \text{ m/s} \tag{d}$$

$$V_8 = V_{12}\left(\frac{D_{12}}{D_8}\right)^2 = 4\left(\frac{303.3}{202.7}\right)^2 = 8.956 \text{ m/s} \tag{e}$$

Note that for incompressible flow in circular pipes, the velocity varies inversely with the square of the diameter.

EXAMPLE OF COMPRESSIBLE FLOW Air discharges from a 12-in. standard steel pipe through a 4-in. inside diameter nozzle into the atmosphere, as shown in Fig. 3-15. The pressure in the duct is 20 psia, and atmospheric pressure is 14.7 psia. The temperature of the air in the duct just upstream of the nozzle is 150°F, and the duct velocity is 18 ft/sec. For *isentropic flow,* compute (a) the mass flow rate and (b) the velocity in the nozzle jet.

Step A *Data Reduction*

(1) *unit conversions*

(B-89) $p_p = 20 \times 144 = 2880 \text{ lbf/ft}^2$, $p_J = 14.7 \times 144 = 2117 \text{ lbf/ft}^2$

(B-28) $T_p = t_F + 459.67 = 150 + 459.67 = 609.67°\text{R}$

FIGURE 3-15 Nozzle flow example

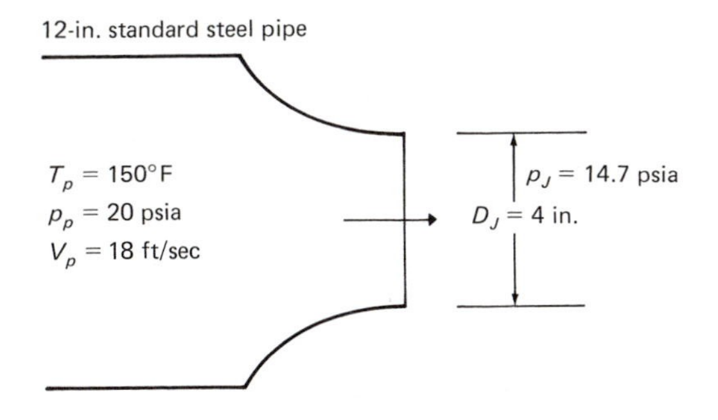

(1-11) (2) *proportionality constant*

$$g_c = 32.17 \text{ lbm-ft/lbf-sec}^2$$

(3) *fluid properties*, air

Table A-7

$$k = 1.400 \qquad R = 53.36 \text{ ft-lbf/lbm°R}$$

(1-44)

$$\rho_p = \frac{p_p}{g_c R T_p} = \frac{2880}{32.17 \times 53.36 \times 609.67}$$

$$\rho_p = 2.752 \times 10^{-3} \text{ slugs/ft}^3$$

(1-47)

$$T_J = T_p \left(\frac{p_J}{p_p}\right)^{(n-1)/n} = T_p \left(\frac{p_J}{p_p}\right)^{(k-1)/k}$$

$$= 609.67 \left(\frac{2117}{2880}\right)^{(1.400-1)/1.400} = 558.35\text{°R}$$

(1-44)

$$\rho_J = \frac{p_J}{g_c R T_J} = \frac{2117}{32.17 \times 53.36 \times 558.35}$$

$$\rho_J = 2.208 \times 10^{-3} \text{ slugs/ft}^3$$

Table A-11 (4) *geometry*, 12 in. standard $A_p = 0.7854 \text{ ft}^2$

Table A-10

$$A_J = \frac{\pi D_J^2}{4} = \frac{\pi (4/12)^2}{4} = 0.08727 \text{ ft}^2$$

Step B *Compute Mass Flow Rate*

(3-9)

$$\dot{m} = \rho_p A_p V_p = (2.752 \times 10^{-3})(0.7854)(18)$$

$$\dot{m} = 0.03891 \text{ slugs/ft}^3$$

(1-19)

$$\dot{M} = \dot{m} g_c = 0.0389 \times 32.17 = 1.252 \text{ lbm/sec}$$

Step C *Compute Nozzle Jet Velocity*

(3-8)

$$V_J = \frac{\dot{m}}{\rho_J A_J}$$

$$V_J = \frac{0.03891}{2.208 \times 10^{-3} \times 0.08727}$$

$$V_J = 201.9 \text{ ft/sec}$$

3-6 CORRECTION FOR KINETIC ENERGY

Kinetic Energy

The kinetic energy of a body was shown in Chapter 1 to be

(1-29)

$$\text{KE} = \frac{mV^2}{2}$$

For a streamtube, the kinetic energy KE flowing past a section is the sum of the kinetic energies of all the streamtubes, or

$$\dot{KE} = \int \frac{U^2}{2}\, d\dot{m} = \frac{1}{2}\int U^2\, (\rho U\, dA) = \frac{\rho}{2}\int U^3\, dA \tag{3-11}$$

For one-dimensional flow $U = V =$ a constant, so that Eq. (3-11) becomes

$$\dot{KE}_{\text{one-dimensional}} = \frac{\rho}{2}\int V^3\, dA = \frac{\rho A V^3}{2} \tag{3-12}$$

Correction

Let α (alpha) be a correction factor to reduce the kinetic energy of two- and three-dimensional flows to one-dimensional flow, or

$$\alpha = \frac{\dot{KE}}{\dot{KE}_{\text{one-dimensional}}} = \frac{\rho/2 \int U^3\, dA}{\rho A V^3/2}$$

$$\alpha = \frac{1}{A}\int \left(\frac{U}{V}\right)^3 dA \tag{3-13}$$

Values of α for three-dimensional flow range from 2 for laminar flow (see example to follow) to nearly unity for turbulent flow. As the fluid velocity increases, the value of alpha approaches unity. The total kinetic energy is small compared with other terms except at high-velocity gas flows where $\alpha \approx 1$. Because of this and the fact that the true value of α is not always known, this correction is often neglected.

EXAMPLE OF CORRECTION OF KINETIC ENERGY FOR THREE-DIMENSIONAL LAMINAR FLOW Determine the value of α for three-dimensional laminar flow using the information developed in the example problem in Sec. 3-4.

Step A *Data Reduction*

Circle

Table A-10 $A = \pi r_o^2,\ dA = 2\pi r\, dr$

Sec. 3-4 $U = U_m\left[1 - \left(\frac{r}{r_o}\right)^2\right],\ V = \frac{U_m}{2}$

or

$$\frac{U}{V} = \frac{U_m}{U_m/2}\left[1 - \left(\frac{r}{r_o}\right)^2\right] = 2\left[1 - \left(\frac{r}{r_o}\right)^2\right] \tag{a}$$

Step B *Compute Correction Factor α*

(3-13) $\alpha = \frac{1}{A}\int \left(\frac{U}{V}\right)^3 dA$

Substituting for (U/V), A and dA from Step A

$$\alpha = \frac{1}{\pi r_o^2} \int_0^{r_o} \left\{ 2\left[1 - \left(\frac{r}{r_o}\right)^2\right]\right\}^3 2\pi r\, dr \qquad \textbf{(b)}$$

Integrate by substitution*

$$\alpha = -8 \int \left[1 - \left(\frac{r}{r_o}\right)^2\right]^3 \left(\frac{-2r}{r_o^2}\right) dr$$

$$\alpha = -8\left[1/4\left(1 - \frac{r^2}{r_o^2}\right)^4\right]_o^{r_o}$$

$$\alpha = -2\left[\left(1 - \frac{r_o^2}{r_o^2}\right)^4 - \left(1 - \frac{0}{r_o^2}\right)^4\right]$$

$$\alpha = 2$$

PROBLEMS

1. Water at 68°F flows in a 2-in. Type L seamless copper water tube with a mass flow rate of 5 lbm/sec. Compute (a) average velocity and (b) the volumetric flow rate in gallons per minute.
2. Two-dimensional flow takes place between two parallel walls separated by a space 1 m normal to the flow. The velocity profile is parabolic with the maximum velocity U_m at the center of 12 m/s and zero velocity at the walls. Compute the volume flow rate for a width of 10 m.
3. Do the same as for Problem 2 except that the velocity profile is semicircular.
4. Do the same as for Problem 2 except that the velocity profile is half of an ellipse.
5. In an open channel flow the velocity profile is given by a parabola with the maximum velocity U_m located at one-fifth the channel depth and zero velocity at the bottom. If the surface velocity $U_s = 1.18$ ft/sec, calculate the maximum and average velocities.
6. Water flows in a 10-in. 250-psi cast iron pipe. Measurements of the velocity profile indicate the following distribution:

Station	Location r/r_o	Velocity, ft/sec
1	0 (center line)	2.5
2	0.2	2.4
3	0.4	2.1
4	0.6	1.6
5	0.8	0.9
6	1 (wall)	0

Compute: (a) average velocity and (b) volume rate of flow.

---

* Let $\mu = [1 - (r/r_o)^2]$; then $du = -2r/r_o^2 dr$ and $\int u^3 du = u^4/4 = 1/4[1 - (r/r_o)^2]^4$.

7. Velocity profile in smooth circular pipes may be empirically expressed by

$$\frac{U}{U_m} = \left(1 - \frac{r}{r_o}\right)^a$$

where U is the streamtube velocity, U_m the maximum velocity at the pipe center line, r is the radius where the velocity is U, and r_o is the radius of the pipe. The value of the exponent a varies from $1/5$ to $1/10$ depending upon the flow conditions, $1/7$ being used for wide ranges of turbulent flow. Derive a general expression for V/U_m and determine the numerical value when $a = 1/7$.

8. Air flows in a 10-in. schedule 40 wrought iron pipe. Measurements of velocity profile indicate the following distribution:

Station	Location, r/r_o	Velocity, m/s
1	0 (center line)	10.0
2	0.2	9.7
3	0.4	9.5
4	0.6	8.8
5	0.8	7.9
6	0.9	7.2
7	0.95	6.5
8	1 (wall)	0

Compute (a) average velocity and (b) volume rate of flow.

9. Ethyl alcohol at 68°F flows in a 6-in. schedule 80 wrought iron pipe which enlarges to an 8-in. schedule 80 pipe. The rate of flow is 90 lbm/sec. Determine the average velocity of the alcohol in the 6-in. and 8-in. sections.

10. An 8-in. schedule 40 steel pipe expands to a 12-in. schedule 40 pipe. Carbon dioxide at 120°F flows isothermally through this system at a mass rate of 5 slugs/min. At a section of the 8-in. pipe the pressure is 60 psia. At another section in the 12-in. pipe the pressure is 40 psia. What is the average velocity in each section?

11. An 8-in. schedule 40 steel pipe expands to a 12-in. schedule 40 pipe. Carbon dioxide flows isentropically through this system at a mass rate of 5 slugs/min. At section 1 in the 8-in. pipe the pressure is 60 psia and the temperature 120°F. At section 2 in the 12-in. pipe the pressure is 40 psia. What is the average velocity in each section?

12. An ideal gas flows in a constant area duct. At one section of this duct the temperature is 15°C, pressure 310 kPa, and velocity 15 m/s. At a second section the temperature is 25°C and the pressure 240 kPa. What is the velocity in the second section?

13. An ideal gas flows isothermally in a constant area duct. At one section the pressure is 45 psia and the velocity 15 ft/sec. At a second section the velocity is 24 ft/sec. What is the pressure in the second section?
14. In an open channel flow the velocity profile is linear. The velocity at the surface is U_s and at the bottom the velocity is zero. Compute the kinetic energy correction factor α (alpha).
15. In an open channel flow the velocity distribution is given by

$$\frac{U}{U_s} = \left(\frac{y}{y_0}\right)^{1/7}$$

where U is the local velocity at a distance y above the bottom of the channel, U_s is the surface velocity and y_0 the distance from the surface. Compute the kinetic energy correction factor α.

REFERENCES

Dull, Raymond W., *Mathematics for Engineers*, 2d ed. McGraw-Hill Book Company, New York, 1941.

Rouse, Hunter and Simon Ince, *History of Hydraulics*. Iowa Institute of Hydraulic Research, State University of Iowa, 1957.

Vennard, John K., *Elementary Fluid Mechanics*, 3d ed. John Wiley and Sons, New York, 1954.

"Fluid Velocity Measurement," Chap. 3, part 5, *Measurement of Quantities of Materials, Instrument and Apparatus Supplements*, Power Test Code, American Society of Mechanical Engineers, New York, PTC 19.5;3-1965.

4 Fluid Dynamics

4-1 INTRODUCTION

Fluid dynamics is that branch of fluid mechanics that deals with energy and force. This chapter considers the equation of motion, the energy equation, and the impulse-momentum equation. The continuity equation was developed in Chapter 3 as a special case of the principle of the conservation of mass. The *equation of motion* is an application of Newton's second law to fluid flow in a streamtube. The *energy equation* is a special case of the principle of the conservation of energy. The *impulse-momentum equation* was developed in Chapter 1 as a special case of the equation of motion.

The *equation of motion* was first developed in 1750 by Leonhard Euler and is sometimes called the Euler equation, although Euler's equations were written for a frictionless fluid. Euler's equation laid the groundwork for an analytical approach to the study of fluid dynamics. The introduction of viscous effects allows for a more general interpretation of the equation and makes it more applicable to the solution of practical problems.

The *energy equation* for steady flow is simply an accounting of all of the energy entering or leaving a control volume. Although an energy equation may be developed to consider all forms of energy, in fluid mechanics, chemical, electrical and atomic energies are not normally considered.

The *impulse-momentum equation* along with the *continuity equation* and *energy equation* provides a third basic tool for the solution of fluid flow problems. Sometimes its application leads to the solution of problems that cannot be solved by the energy principle alone;

more often it is used in conjunction with the energy principle to obtain more comprehensive solutions of engineering problems.

4-2 EQUATION OF MOTION

Derivation

Consider the fluid element flowing steadily in a streamtube shown in Fig. 4-1. This element has a length of dL, an area normal to the motion of dA, and a perimeter of dP. The elemental mass is $\rho\, dA\, dL$. The increase in elevation of this mass is dz, and the motion of the element is upward.

Forces tending to change the velocity U of this fluid mass are

(1) *Pressure* forces on the ends of the element.

$$dF_p = p\,dA - (p + dp)\,dA = -dp\,dA \tag{4-1}$$

(2) *Gravity* force due to the component of weight in the direction of motion.

$$dF_g = -\rho g\,dA\,dL\left(\frac{dz}{dL}\right) = -\rho g\,dA\,dz \tag{4-2}$$

(3) *Friction* force on the outer surface of the element.

$$dF_f = -\tau\,dP\,dL \tag{4-3}$$

FIGURE 4-1 Element of a streamtube

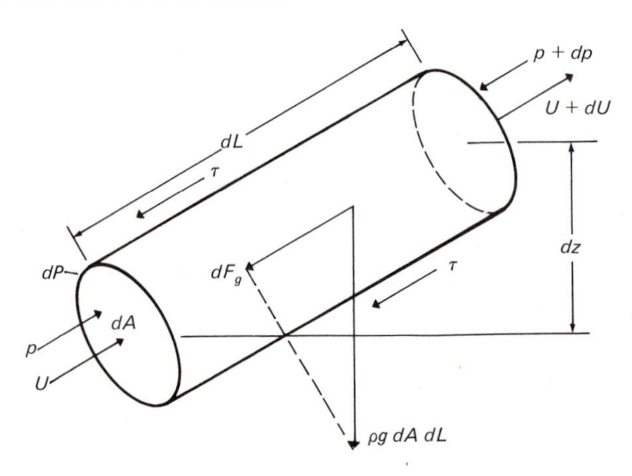

(4) The combined force becomes

$$\Sigma\, dF = dF_p + dF_g + dF_f = -dp\, dA - \rho g\, dA\, dz - \tau\, dP\, dL$$
$$= -dA\,(dP + \rho g\, dz + \tau\, dP\, dL/dA) \qquad \text{(4-4)}$$

Application of Newton's second law:

$$\text{Force} = \text{mass} \times \text{acceleration}$$
$$\Sigma\, dF = (\rho\, dA\, dL)\,(dU/dt) \qquad \text{(4-5)}$$

Substituting from Eq. (4-4) for $\Sigma\, dF$ and noting that $dL/dt = U$, Eq. (4-5) may be written as

$$\Sigma\, dF = -dA(dp + \rho g\, dz + \tau\, dP\, dL/dA) = \rho\, dA\, U\, dU$$

which reduces to

$$-dp - \rho g\, dz - \tau\, dL\left(\frac{dP}{dA}\right) = \rho U\, dU \qquad \text{(4-6)}$$

Equating (4-6) to zero and multiplying by $-v = -1/\rho g_c$ from Eq. (1-36),

$$v\, dp + \frac{g}{g_c}\, dz + v\tau\, dL\left(\frac{dP}{dA}\right) + \frac{U\, dU}{g_c} = 0 \qquad \text{(4-7)}$$

Rearranging the order of terms in Eq. (4-7) for later comparison with the energy equation,

$$\frac{g}{g_c}\, dz + \frac{U\, dU}{g_c} + v\, dp + v\tau\, dL\left(\frac{dP}{dA}\right) = 0 \qquad \text{(4-8)}$$

4-3 HYDRAULIC RADIUS

Definition / Fluid flow area/shear perimeter
Symbol / R_h
Dimensions / L
Units / U.S.: ft SI: m

In the derivation of the equation of motion, the shape of the streamtube cross-section was not specified. The ratio of the fluid flow area (dA) to shear area $dP\,dL$ is a function of streamtube flow shape and is a constant for a given geometry. The hydraulic radius R_h may be used to define this ratio per unit of length as follows:

$$\text{Hydraulic radius} = R_h = \frac{dA}{dP} = \frac{\text{fluid flow area}}{\text{shear perimeter}} = \frac{A}{P} \tag{4-9}$$

The hydraulic radius is used to compute flow losses in noncircular flow passages and circular conduits flowing partly full of liquids. For this reason, it is important to relate the hydraulic radius to the diameter D of a circular pipe:

$$R_h = \frac{\pi D^2/4}{\pi D} = \frac{D}{4} \tag{4-10}$$

The equivalent diameter D_e is

$$D_e = 4R_h \tag{4-11}$$

EXAMPLE A liquid flows in the rectangular duct shown in Fig. 4-2 to a depth of 2 ft. If the duct is 6 ft wide and 3 ft deep, compute the hydraulic radius R_h and the equivalent diameter D_e.

Step A *Derive an Equation for R_h*

Fig. 4-2 Fluid flow area, $A = bh$

Fig. 4-2 Shear perimeter, $P = h + b + h = 2h + b$

(4-9) $$R_h = \frac{A}{P} = \frac{bh}{2h + b} \tag{a}$$

FIGURE 4-2 Notation for hydraulic radius example

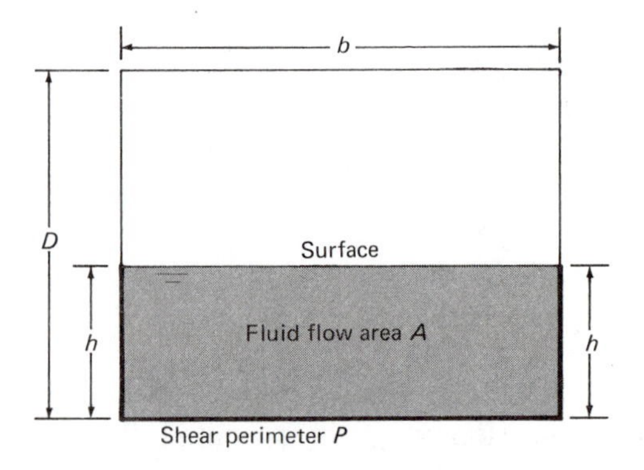

Step B *Compute R_h and D_e*

$$R_h = \frac{6 \times 2}{2 \times 2 + 6} = 1.2 \text{ ft} \tag{a}$$

$$D_e = 4R_h = 4 \times 1.2 = 4.8 \text{ ft} \tag{4-11}$$

Values of the fluid flow area A and the hydraulic radius R_h and equivalent diameter D_e for various cross-sections are given in Table A-14.

4-4 ONE-DIMENSIONAL STEADY-FLOW EQUATION OF MOTION

When the flow is one-dimensional, $V = U$. Substituting this value of U and the definition of hydraulic radius of Eq. (4-9) in Eq. (4-7),

$$\frac{g}{g_c}\,dz + \frac{V\,dV}{g_c} + v\,dp + \frac{v\tau\,dL}{R_h} = 0 \tag{4-12}$$

Integrating Eq. (4-12) between sections 1 and 2,

$$\frac{g}{g_c}(z_2 - z_1) + \frac{V_2^2 - V_1^2}{2g_c} + \int_1^2 v\,dp + \frac{1}{R_h}\int_1^2 v\tau\,dL = 0 \tag{4-13}$$

Let

$$\frac{1}{R_h}\int_1^2 v\tau\,dL = H_f$$

where H_f is the energy "lost" due to friction. Substituting this in Eq. (4-13) results in

$$\frac{g}{g_c}(z_2 - z_1) + \frac{V_2^2 - V_1^2}{2g_c} + \int_1^2 v\,dp + H_f = 0 \tag{4-14}$$

For an *incompressible* fluid ($v_1 = v_2$), Eq. (4-14) becomes

$$\frac{g}{g_c}(z_2 - z_1) + \frac{V_2^2 - V_1^2}{2g_c} + v(p_2 - p_1) + H_f = 0 \tag{4-15}$$

For frictionless flow of an incompressible fluid ($H_f = 0$) Eq. (4-15) reduces to

$$\frac{g}{g_c}(z_2 - z_1) + \frac{V_2^2 - V_1^2}{2g_c} + v(p_2 - p_1) = 0 \tag{4-16}$$

Multiplying Eq. (4-16) by g_c/g and noting from Eq. (1-36) that $v = 1/\rho g_c$ results in

$$(z_2 - z_1) + \frac{V_2^2 - V_1^2}{2g} + \frac{(p_2 - p_1)}{\rho g} = 0 \tag{4-17}$$

This is the equation proposed by Daniel Bernoulli in his "Hydrodynamica" published in 1738.

EXAMPLE A 12-in. Type L seamless copper water tube reduces to a 6-in. tube and then expands to an 8-in. tube. Water at 68°F flows steadily and without friction through this system. At section 1 in the 12-in. tubing, the pipe center line is 10 ft above the datum, the pressure 20 psia, and the velocity 4 ft/sec. At section 2 in the 6-in. tubing, the center line is 15 ft above the datum. At section 3 in the 8-in. tubing, the center line is 20 ft above the datum. Find the velocity and pressure at sections 2 and 3 of Fig. 4-3.

Step A *Data Reduction*

(1) *unit conversion*

(B-89) $p_1 = 144 \times 20 = 2880$ lbf/ft²

(2) *gravity*

(B-50) Assume $g = g_o = 32.17$ ft/sec²

(3) *fluid properties,* water at 68°F

Table A-1 $\rho = 1.937$ slugs/ft³

FIGURE 4-3 Notation for Bernoulli equation example

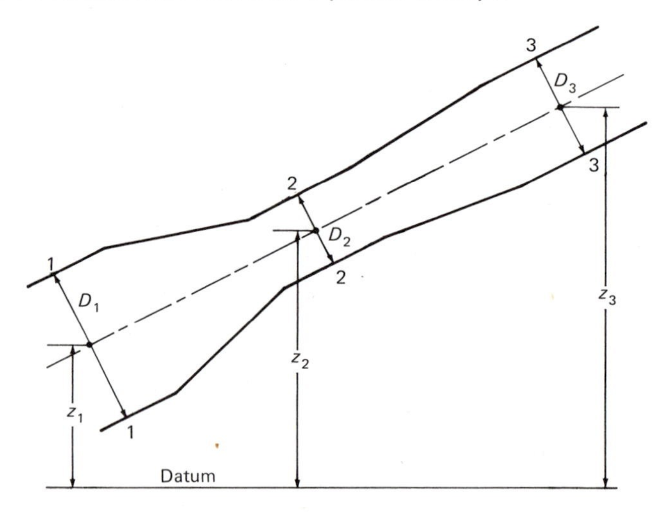

(4) *tube properties,* seamless copper type L

Table A-13

Section	Size, in.	Area, ft²
1	12	0.7295
2	6	0.1863
3	8	0.3255

(5) *volume flow rate*

(3-4) $$Q = A_1V_1 = 0.7295 \times 4 = 2.918 \text{ ft}^3/\text{sec}$$

Step B *Compute Conditions in the 6-In. Tube*

(3-4) $$V_2 = Q/A_2 = 2.918/0.1863 = 15.66 \text{ ft/sec}$$

(4-17) $$(z_2 - z_1) + \frac{V_2^2 - V_1^2}{2g} + \frac{(p_2 - p_1)}{\rho g} = 0$$

Solving Eq. (4-17) for p_2,

$$p_2 = \rho g\left[\left(z_1 - z_2\right) + \frac{V_1^2 - V_2^2}{2g}\right] + p_1 \quad \text{(a)}$$

$$p_2 = (1.937 \times 32.17)\left[\left(10 - 15\right) + \frac{4^2 - 15.66^2}{2 \times 32.17}\right] + 2880$$

$$p_2 = 2346 \text{ lbf/ft}^2$$

(B-50) $$p_2 = \frac{2346}{144} = 16.29 \text{ psia}$$

Step C *Compute Conditions in the 8-In. Tube*

(3-4) $$V_3 = \frac{Q}{A_3} = \frac{2.918}{0.3255} = 8.965 \text{ ft/sec}$$

$$p_3 = \rho g\left[(z_1 - z_3) + \frac{V_1^2 - V_3^2}{2g}\right] + p_1 \quad \text{(a)}$$

$$p_3 = (1.937 \times 32.17)\left[\left(10 - 20\right) + \frac{4^2 - 8.965^2}{2 \times 32.17}\right] + 2880$$

$$p_3 = 2195 \text{ lbf/ft}^2$$

(B-50) $$p_3 = \frac{2195}{144} = 15.24 \text{ psia}$$

4-5 SPECIFIC ENERGY

Definition / Energy per unit mass
Dimensions / $FLM^{-1} = (MLT^{-2})LM^{-1} = L^2T^2$
Units / U.S.: ft-lbf/lbm SI: m · N/kg

Before developing the energy equation, a general discussion of energy is in order. Two sets of energy will be considered. The first is the energy of the fluid at a section, and the second is the energy added to or taken from the fluid between sections. The total energy possessed by a fluid at a section is dependent upon the net energy added to or taken from the fluid between it and a prior section, but the individual energies are independent of their counterparts at the prior section. For this reason, fluid energies are called *point* functions. The energies added to or taken from the fluid between sections depend upon the manner or process, and these transitional energies are called *path* functions because of their dependence upon the process undergone. The total amount of energy in a system cannot be measured but must be referenced to some arbitrary datum. In fluid mechanics, we are interested in *energy change,* and any convenient datum may be used.

4-6 SPECIFIC POTENTIAL ENERGY

The potential energy of a fluid mass is the energy possessed by it due to its elevation relative to some arbitrary datum, as stated in Sec. 1-8. It is equivalent to the work that would be required to lift it from the datum to its elevation in the absence of friction.

The change in specific potential energy (ΔSPE) may be computed from Eq. (1-23) as follows:

$$\Delta \text{SPE} = \frac{\Delta \text{PE}}{M} = \frac{mg}{M} \int_1^2 dz = \frac{mg}{M} (z_2 - z_1) \tag{4-18}$$

From Eq. (1-19) $M = mg_c$, substituting in Eq. (4-18) results in

$$\Delta \text{SPE} = \frac{g}{g_c} (z_2 - z_1) \tag{4-19}$$

Note that Eq. (4-19) is the same as the first term of Eq. (4-13).

EXAMPLE A piping system has an increase in elevation of 100 m between inlet and outlet. Compute the specific potential energy change if (a) the system is on the surface of the earth where the gravity is standard, (b) the system is on the surface of the moon, and (c) the system is in a space laboratory orbiting the earth at an elevation of 1000 km.

Step A *Data Reduction*

(1) *gravity*

(B-51) Earth, $g = g_o = 9.807 \text{ m/s}^2$

Table 1-3 moon, $g = 1.620 \text{ m/s}^2$

Table 1-3	(2) *Earth's radius* $r = 6371$ km
(1-9)	(3) *proportionality constant* $g_c = 1\ \text{kg}\cdot\text{m/N}\cdot\text{s}^2$

Step B *Compute Change on Earth's Surface*

(4-19) $$\Delta\text{SPE} = \frac{g}{g_c}(z_2 - z_1) = (9.807/1)(100) = 980.7\ \text{m}\cdot\text{N/kg}$$

Step C *Compute Change on Moon's Surface*

(4-19) $$\Delta\text{SPE} = \frac{g}{g_c}(z_2 - z_1) = (1.620/1)(100) = 162.0\ \text{m}\cdot\text{N/kg}$$

Step D *Compute Change in Space Laboratory*

(1-17) $$g = \frac{g_\phi r^2}{(r+z)^2} = \frac{g_o r^2}{(r+z)^2}$$

$$g = \frac{(9.807)(6371)^2}{(6371+1000)^2} = 7.327\ \text{m/s}^2$$

(4-19) $$\Delta\text{SPE} = \frac{g}{g_c}(z_2 - z_1) = (7.327/1)(100) = 732.7\ \text{m}\cdot\text{N/kg}$$

4-7 SPECIFIC KINETIC ENERGY

The kinetic energy of a fluid mass is the energy possessed by it due to its motion, as stated in Sec. 1-8. It is equivalent to the work required to impart the motion from rest in the absence of friction.

The change in specific kinetic energy (ΔSKE) may be computed from Eq. (1-29) as follows:

$$\Delta\text{SKE} = \Delta\text{KE}/M = \frac{m}{M}\int_1^2 V\,dV = \frac{m}{M}\left(\frac{V_2^2 - V_1^2}{2}\right) \tag{4-20}$$

From Eq. (1-19) $M = mg_c$, substituting in Eq. (4-20) results in

$$\Delta\text{SKE} = \frac{V_2^2 - V_1^2}{2g_c} \tag{4-21}$$

Note that Eq. (4-21) is the same as the second term of Eq. (4-13).

Equation (4-21) may be used only for one-dimensional flow. As was shown in Sec. 3-6, the correction factor, α (alpha), should be applied for two- and three-dimensional flows. Application of Eq. (3-13) to Eq. (4-21) results in

$$\Delta\text{SKE} = \frac{\alpha_2 V_2^2 - \alpha_1 V_1^2}{2g_c} \tag{4-22}$$

where α_1 and α_2 are the kinetic energy correction factors for the velocity distributions at sections 1 and 2 respectively.

4-8 SPECIFIC INTERNAL ENERGY

The internal energy of a body is the sum total of the *kinetic* and *potential* energies of its molecules, apart from any kinetic or potential energy of the body as a whole. The total kinetic internal energy is due primarily to the translation, rotation, and vibration of its molecules. The potential internal energy is due to the bonding or attractive forces that hold the molecules in a phase.

The potential internal energy decreases as a substance changes from solid to liquid to gaseous phases as the bonding forces decrease. In the gas phase, the internal energy is mainly kinetic. As the ideal gas state is approached and molecular activity increases with temperature increase, the internal energy becomes wholly kinetic, and thus the internal energy of an ideal gas is a pure temperature function.

The symbol for specific internal energy is u, and the change in specific internal energy is given by

$$\Delta u = \int_1^2 du = u_2 - u_1 \tag{4-23}$$

Units

For the SI system, the joule per kilogram or newton metre per kilogram is used. For the U.S. customary units, conventional practice is to use the British thermal unit per pound mass (Btu/lbm). For fluid mechanics, it will be necessary to convert the Btu to ft-lbf.

4-9 SPECIFIC FLOW WORK

Flow work is the amount of mechanical energy required to "push" or force a flowing fluid across a section boundary. Consider the steady-flow system shown in Fig. 4-4. Fluid enters the system at section 1, where the flow area is A_1 and the pressure is p_1, and leaves at section 2, where the flow area is A_2 and the pressure p_2. The force acting to prevent the fluid from crossing a section boundary is

$$F = pA \tag{4-24}$$

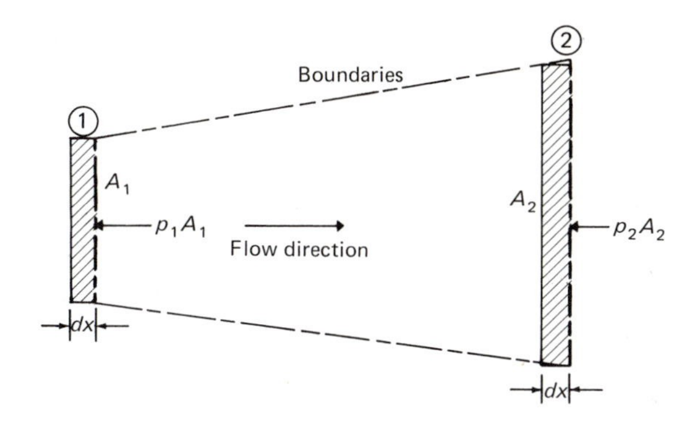

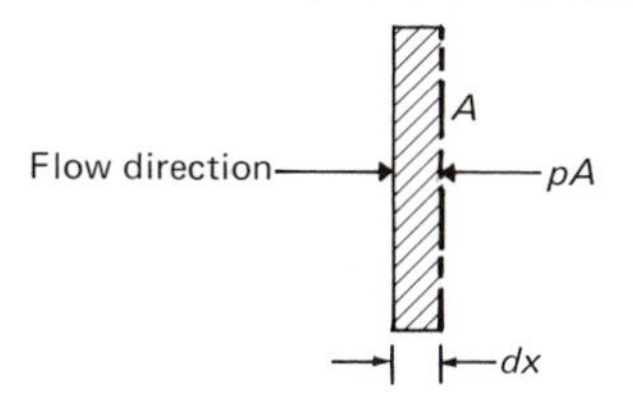

FIGURE 4-4 Flow work

where p is the pressure at the section boundary and A is the flow area. Substituting Eq. (4-24) in Eq. (1-21)

$$\text{ME} = \int F\,dx = \text{FW} = \int pA\,dx \tag{4-25}$$

where FW is the flow work. Noting that $A\,dx$ is the volume (Vol) of fluid being "pushed" across a section boundary and from Eq. (1-36) Vol $= Mv$, Eq. (4-25) may be written as

$$\Delta\text{FW} = \int_1^2 d(p\,\text{Vol}) = \int_1^2 d(pMv) = M\int_1^2 d(pv) \tag{4-26}$$

where ΔFW is the change in flow work. Dividing Eq. (4-26) by M,

$$\Delta\text{SFW} = \frac{\Delta\text{FW}}{M} = \int_1^2 d\,(pv) = p_2v_2 - p_1v_1 \tag{4-27}$$

EXAMPLE Methane flows steadily in a system at a rate of 2 lbm/sec. At the system inlet, section 1, the pressure is 450 psia and the temperature is 50°F. At the system outlet, the pressure is 300 psia. The path of the system is defined by $pv^{1.2} = C$. Compute the change in system specific flow work.

Step A	*Data Reduction*
	(1) *unit conversions*
(B-89)	$p_1 = 450 \times 144 = 64{,}800 \text{ lbf/ft}^2$
(B-89)	$p_2 = 300 \times 144 = 43{,}200 \text{ lbf/ft}^2$
(B-28)	$T_1 = t_F + 459.67 = 50 + 459.67 = 509.67°\text{R}$
	(2) *fluid properties,* methane
Table A-7	$R = 96.34 \text{ ft-lbf/lbm°R}$
(1-42)	$v_1 = \dfrac{RT_1}{p_1} = 96.34 \times 509.67/64{,}800$
	$v_1 = 0.7577 \text{ ft}^3\text{/lbm}$
(1-46)	$v_2 = v_1 \left(\dfrac{p_1}{p_2}\right)^{1/n} = 0.7577\left(\dfrac{64{,}800}{43{,}200}\right)^{1/1.2}$
	$v_2 = 1.062 \text{ ft}^3\text{/lbm}$
Step B	*Compute Change in Specific Flow Work*
(4-27)	$\Delta\text{SFW} = p_2 v_2 - p_1 v_1$
	$= 43{,}200 \times 1.062 - 64{,}800 \times 0.7577$
	$\Delta\text{SFW} = -3220 \text{ ft-lbf/lbm}$
	The negative sign indicates that more energy was required to "push" the gas across the system inlet than across the outlet.

4-10 SPECIFIC ENTHALPY

It is sometimes desirable to combine certain fluid properties to obtain a new one. Enthalpy is a defined property combining internal energy, pressure, and specific volume.

The symbol for specific enthalpy is h, and specific enthalpy is defined by the following equation:

$$h = u + pv \tag{4-28}$$

The change in specific enthalpy becomes

$$\Delta h = \int_1^2 dh = \int_1^2 du + \int_1^2 d(pv)$$

$$= u_2 - u_1 + p_2 v_2 - p_1 v_1 \tag{4-29}$$

Units

For the SI system, the joule per kilogram or newton metre per kilogram is used. For the U.S. system, the British thermal unit per pound mass (Btu/lbm) is used. For fluid mechanics, it will be necessary to convert the Btu to ft-lbf.

4-11 SHAFT WORK

Definition

Shaft work is that form of mechanical energy which crosses the boundaries of a system by being transmitted through the shaft of a machine. The result of this transmission is to increase or decrease the total amount of energy stored in a fluid.

Shaft work is mechanical energy in *transition* and cannot be stored as such in a fluid. For example, consider a pump pumping water from a lower level to a higher one. While the pump is in operation, shaft work is transmitted to the water, and this increase in energy causes the water to rise to a higher elevation. After the pump has stopped, the amount of energy added to the fluid less losses is now stored in the water in the form of increased mechanical potential energy.

Because the first engines built by man were made to extract work from the fluid energy, conventional practice is to call shaft work done *by* a fluid *positive work,* and work done *on* a fluid *negative work.* Shaft work may also be classed as *steady flow* or *nonflow* according to the type of machine and process.

Nonflow Shaft Work

PROCESS Consider the cylinder and piston arrangement shown in Fig. 4-5. As the piston advances from the state point 1 to point 2, the fluid in the cylinder expands and work is done *by the fluid.* If the piston were made to retract, then the fluid would be compressed and work would be done *on the fluid.*

EQUATIONS The force exerted by the fluid on the piston of Fig. 4-5 is given by

$$F = pA \tag{4-24}$$

Substituting Eq. (4-24) in Eq. (1-21),

$$\text{ME} = \int F\,dx = MW_{nf} = \int pA\,dx \tag{4-30}$$

where W_{nf} is the specific shaft work. Since the area of the piston is a constant, then $A\,dx = d(\text{Vol})$ and from Eq. (1-36) $\text{Vol} = Mv$, then $A\,dx = M\,dv$. Substituting in Eq. (4-30),

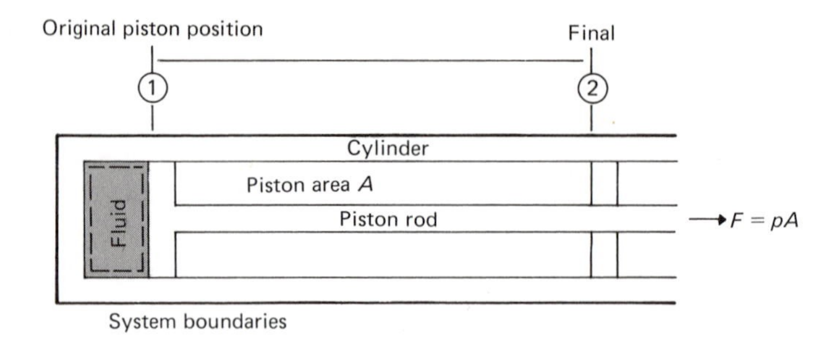

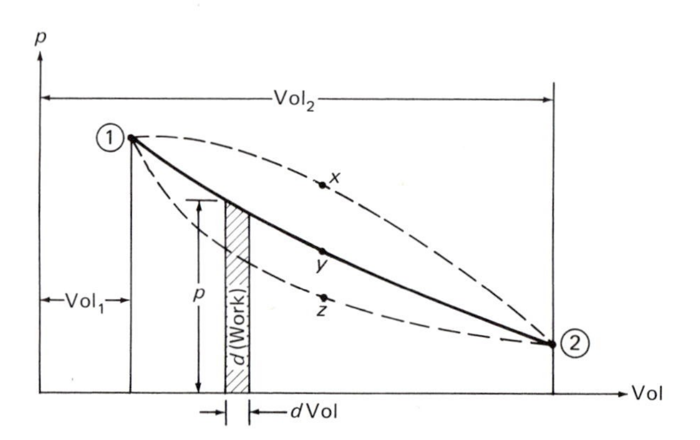

FIGURE 4-5 Nonflow shaft work

$$MW_{nf} \int_1^2 p\, d(\text{Vol}) = M \int_1^2 p\, dv$$

or

$$W_{nf} = \int_1^2 p\, dv \tag{4-31}$$

FUNCTION Equation (4-31) is a mathematical statement that the shaft work is the area "under" the pressure-volume curve of Fig. 4-5. There are an infinite number of ways that the fluid can change from state 1 to state 2. Shown in Fig. 4-5 are three curves which represent the *paths* of three possible *processes*. Path 1-y-2 was chosen to represent the actual path of the process or state change. Had path 1-x-2 been chosen, the amount of work would have been greater; if 1-z-2, the work would have been less. For this reason, shaft work is called a *path function*. Before Eq. (4-31) can be integrated, the pressure–specific volume relationship must be known.

Steady-Flow Shaft Work

EQUATIONS From Eq. (1-21),

$$\mathrm{ME} = \int F\,dx = \mathrm{SW} = M \int dW_{sf} \tag{4-32}$$

where SW is the steady-flow shaft work, W_{sf} is the steady-flow shaft work per unit mass. Because the differential of shaft work is inexact,

$$\int dW_{sf} \neq W_{sf_2} - W_{sf_1}$$

the symbol δ is used in place of d. Equation (4-32) may be written for specific shaft work as follows:

$$W_{sf} = \int \delta W_{sf} \tag{4-33}$$

4-12 HEAT AND ENTROPY

Heat is that form of thermal energy which crosses the boundaries of a system without the transfer of mass as a result of a difference in temperature between the system and its surroundings. The effect of this transfer is to increase or decrease the total amount of energy stored in a fluid. Heat is thermal energy in *transition,* and like shaft work it cannot be stored as such in a fluid. Because the first devices made by man were to produce shaft work by adding heat, heat added to a substance is positive, and heat rejected is negative. *Entropy* is that fluid property required by the second law of thermodynamics to describe the path of a reversible process. Entropy is defined by the following equation:

$$q = \int T\,ds \tag{4-34}$$

where q is the heat transferred and s is the entropy per unit mass.

Process

Heat may also be expressed as

$$q = \int \delta q \tag{4-35}$$

where q is the heat transferred per unit mass. Note that the symbol δ (delta) is used in place of d to remind us that the differential of

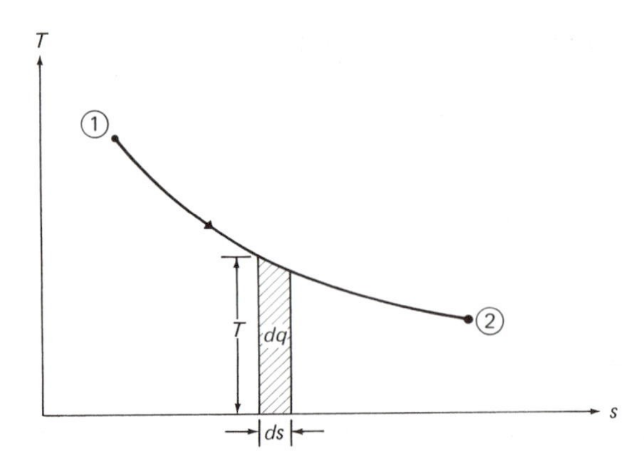

FIGURE 4-6 Temperature-entropy plane

heat transfer is inexact

$$\int dq \neq q_2 - q_1$$

Equations (4-34) and (4-35) may be combined as follows:

$$\int \delta q = q = \int T\,ds \tag{4-36}$$

Equation (4-36) is a mathematical statement; heat is the area "under" the temperature-entropy curve of Fig. 4-6. As with shaft work, there are an infinite number of ways that the heat can be transferred from point 1 to 2, so that heat, like shaft work, is a *path function*. The relation between temperature and entropy must be established before Eq. (4-36) can be integrated.

Units

In the SI system, the joule per kilogram or newton metre per kilogram is used for heat and the joule/kilogram kelvin is used for entropy. In the U.S. system, the British thermal unit per pound mass is used for heat, and the British thermal unit/pound mass degree Rankine is used for entropy. For fluid mechanics, it will be necessary to convert the Btu to ft-lbf and the Btu/lbm-°R to ft-lbf/lbm-°F.

4-13 STEADY-FLOW ENERGY EQUATION

The steady-flow energy equation is readily derived by the application of the principles of conservation of energy to a thermodynamic system. The following forms of energy are considered.

STORED IN FLUID

(4-19) *Potential energy* $\Delta \text{SPE} = \frac{g}{g_c}(z_2 - z_1)$

(4-21) *Kinetic energy* $\Delta \text{SKE} = \frac{V_2^2 - V_1^2}{2g_c}$

(4-23) *Internal energy* $\Delta u = u_2 - u_1$

(4-27) *Flow work* $\Delta \text{SFW} = p_2 v_2 - p_1 v_1$

IN TRANSITION

(4-33) *Shaft work* $\int \delta W_{sf} = W_{sf}$

(4-36) *Heat transfer* $\int \delta q = q$

The basic requirement for the satisfaction of the principle of conservation of energy may be stated:

$$\Sigma \textit{ energy entering system} - \Sigma \textit{ energy stored in system} = \Sigma \textit{ energy leaving system} \quad \textbf{(4-37)}$$

In a steady-flow system, the energy stored in the system does not change with time, so that for any given period of time, Eq. (4-37) reduces to

$$\Sigma \textit{ energy in} = \Sigma \textit{ energy out} \quad \textbf{(4-38)}$$

Equation (4-38) may be modified to show the types of energy as follows:

Energy stored in entering fluid
+ energy in transition added to system
= energy in transition removed from system
+ energy stored in fluid leaving **(4-39)**

Consider the block diagram of Fig. 4-7. The fluid enters the system through section 1 transporting with it its stored energy

$$\frac{g}{g_c} z_1 + \frac{V_1^2}{2g_c} + u_1 + p_1 v_1$$

and leaves the system at section 2, removing its stored energy

$$\frac{g}{g_c} z_2 + \frac{V_2^2}{2g_c} + u_2 + p_2 v_2$$

Since heat (q) added to a system is considered positive, the arrow shows heat being added between sections 1 and 2. In a like manner,

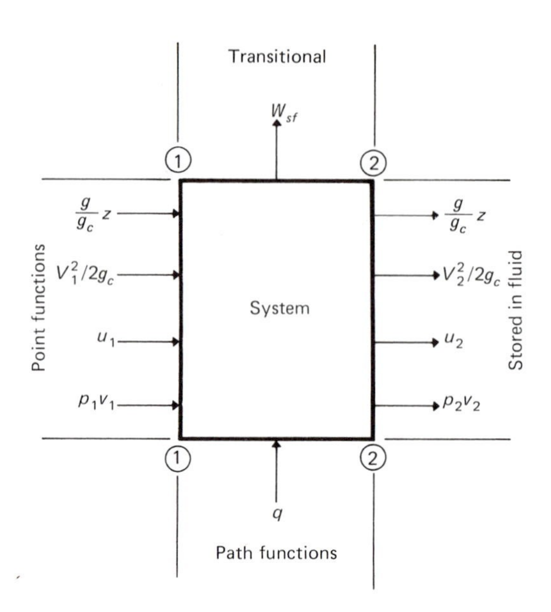

FIGURE 4-7 Steady-flow energy diagram

the steady-flow shaft work (W_{sf}) is shown to be leaving between sections 1 and 2 because work done by the fluid is considered positive.

Application of Fig. 4-7 to Eq. (4-39) results in

$$\left(\frac{g}{g_c} z_1 + \frac{V_1^2}{2g_c} + u_1 + p_1v_1\right) + q = W_{sf} + \left(\frac{g}{g_c} z_2 + \frac{V_2^2}{2g_c} + u_2 + p_2v_2\right) \tag{4-40}$$

Equation (4-40) may be written as

$$q = W_{sf} + \frac{g}{g_c}(z_2 - z_1) + \frac{V_2^2 - V_1^2}{2g_c} + u_2 - u_1 + p_2v_2 - p_1v_1 \tag{4-41}$$

EXAMPLE Test data from a steady-flow air compressor are as follows:

$p_1 = 14.79$ psia, $p_2 = 99.76$ psia, $t_1 = 69.27$°F,
$t_2 = 362.0$°F, $v_1 = 13.24$ ft^3/lbm, $v_2 = 3.049$ ft^3/lbm,
$V_1 = 185.2$ ft/sec, $V_2 = 42.63$ ft/sec, $u_1 = 90.68$ Btu/lbm, $u_2 = 140.9$ Btu/lbm.

If the heat transferred out of the system is 16.73 Btu/lbm, and the outlet is 10 ft above the inlet, find the steady-flow work for each pound mass of air.

Step A *Data Reduction*

(1) *unit conversions*

(B-89) $p_1 = 144 \times 14.79 = 2130$ lbf/ft^2

(B-89) $p_2 = 144 \times 99.76 = 14{,}365$ lbf/ft^2

(B-13) $u_1 = 778.2 \times 90.68 = 70{,}567$ ft-lbf/lbm

(B-13) $u_2 = 778.2 \times 140.9 = 109{,}648$ ft-lbf/lbm

(B-13) $q = 778.2(-16.73) = -13{,}019$ ft-lbf/lbm

(2) *proportionality constant*

(1-11) $g_c = 32.17$ lbm-ft/lbf-sec^2

Step B *Compute the Change in Each Form of Stored Energy Per Unit Mass*

(1) *potential*

(4-19)
$$\frac{g(z_2 - z_1)}{g_c} = \frac{(32.17)(10)}{32.17}$$
$$= 10 \text{ ft-lbf/lbm (assume } g = g_o)$$

(2) *kinetic*

(4-21)
$$\frac{V_2^2 - V_1^2}{2g_c} = \frac{42.63^2 - 185.2^2}{2 \times 32.17}$$
$$= -505 \text{ ft-lbf/lbm}$$

(3) *internal*

(4-23)
$$u_2 - u_1 = 109{,}648 - 70{,}567$$
$$= 39{,}081 \text{ ft-lbf/lbm}$$

(4) *flow work*

(4-27)
$$p_2v_2 - p_1v_1 = 14{,}365 \times 3.049 - 2130 \times 13.24$$
$$= 15{,}598 \text{ ft-lbf/lbm}$$

Step C *Compute Steady-Flow Shaft Work*

(4-41)
$$q = W_{sf} + \frac{g}{g_c}(z_2 - z_1) + \frac{V_2^2 - V_1^2}{2g_c} + u_2 - u_1 + p_1v_1 - p_2v_2$$

$$W_{sf} = q - \frac{g}{g_c}(z_2 - z_1) - \frac{V_2^2 - V_1^2}{2g_c} - (u_2 - u_1)$$
$$- (p_2v_2 - p_1v_1) \qquad \text{(a)}$$

$$W_{sf} = (-13{,}019) - (10) - (-505) - (39{,}081)$$
$$- (15{,}598) = -67{,}203 \text{ ft-lbf/lbm} \qquad \text{(a)}$$

(work done *on* air)

4-14 RELATION OF MOTION AND ENERGY EQUATIONS

The equation of motion was derived in Sec. 4-2 without consideration of steady-state shaft work. Had shaft work been considered, the resulting one-dimensional equation of motion (4-14) would have been

$$W_{sf} + \frac{g}{g_c}(z_2 - z_1) + \frac{V_2^2 - V_1^2}{2g_c} + \int_1^2 v\,dp + H_f = 0 \tag{4-42}$$

The energy equation (4-41) may be written as follows by noting that

$$p_2v_2 - p_1v_1 = \int_1^2 d(pv) = \int_1^2 v\,dp + \int_1^2 p\,dv:$$

$$q = W_{sf} + \frac{g}{g_c}(z_2 - z_1) + \frac{V_2^2 - V_1^2}{2g_c} + u_2 - u_1 + \int_1^2 v\,dp + \int_1^2 p\,dv \tag{4-43}$$

As the equations are now written, the equation of motion (4-42) contains no thermal energy terms and the energy equation (4-43) contains no term for friction. If Eq. (4-42) is subtracted from Eq. (4-43) the following is obtained.

$$q = u_2 - u_1 + \int_1^2 p\,dv - H_f$$

FIGURE 4-8 Nonflow system energy equation

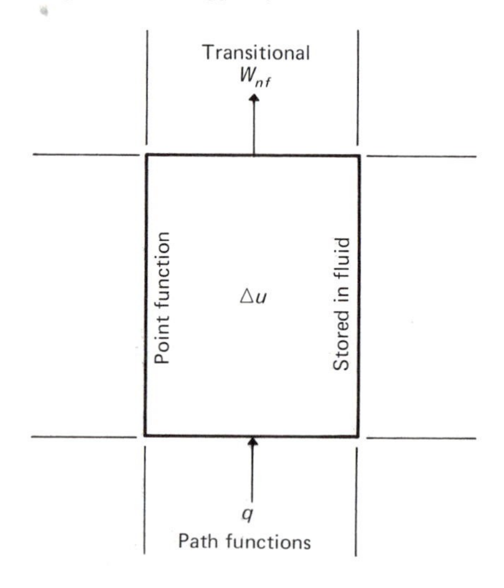

or

$$H_f = u_2 - u_1 + \int_1^2 p\,dv - q \tag{4-44}$$

For an incompressible fluid, $v_1 = v_2$ or $dv = 0$, so that Eq. (4-44) reduces to

$$H_f = u_2 - u_1 - q \tag{4-45}$$

Equation (4-45) indicates that no energy is "lost" due to friction but is simply converted into some other form that is either removed from the system as heat transfer and/or increases the internal energy of the fluid.

4-15 TEMPERATURE RELATIONS FOR INTERNAL ENERGY AND ENTHALPY

Nonflow Energy Equation

Consider the horizontal piston and cylinder arrangement shown in Fig. 4-5. Fluid does not cross the system boundaries so that no flow work is performed, nor can there be any change in kinetic energy. Because the cylinder is horizontal, there is no change in potential energy. Of the six forms of energy considered in Sec. 4-13 for steady-flow equation, only three: internal energy, shaft work, and heat transfer, need be considered for a nonflow system.

From Fig. 4-8, application of the principle of conservation of energy leads to

$$q = \Delta u + W_{nf} \tag{4-46}$$

Noting from Eq. (4-23) that

$$\Delta u = u_2 - u_1 \tag{4-23}$$

and from Eq. (4-31) that

$$W_{nf} = \int_1^2 p\,dv \tag{4-31}$$

the nonflow equation may be written as

$$q = u_2 - u_1 + \int_1^2 p\,dv \tag{4-47}$$

Specific Heat

Definition / Heat transferred per unit mass per unit temperature difference
Symbol / c_x
Dimensions / $FLM^{-1}\theta^{-1}$ or $L^2T^{-2}\theta^{-1}$
Units / U.S.: Btu/lbm-°R SI: J/kg · K

For fluid mechanics, it will be necessary to convert Btu/lbm-°R to ft-lbf/lbm-°R. For ideal gases with constant specific heats, the following relationship applies:

$$q = c_x(T_2 - T_1) \tag{4-48}$$

where c_x is the specific heat, the subscript x denoting the type of process. Equating Eqs. (4-48) and (4-47),

$$q = c_x(T_2 - T_1) = u_2 - u_1 + \int_1^2 p\,dv$$

or

$$c_x(T_2 - T_1) = u_2 - u_1 + \int_1^2 p\,dv \tag{4-49}$$

Internal Energy

Consider an isometric or constant volume process. For this process, $c_x = c_v$ (specific heat at constant volume) and $v_2 = v_1$ or $v_2 - v_1 = 0$, or $dv = 0$. Equation (4-49) becomes

$$c_v(T_2 - T_1) = \Delta u = u_2 - u_1 \tag{4-50}$$

Values of c_v are given in Table A-7. Because the internal energy of an ideal gas is a pure function of temperature, Eq. (4-50) may be applied to *any ideal gas regardless of the process.*

Enthalpy

Consider an isobaric or constant pressure process. For this process $c_x = c_p$ (specific heat at constant pressure) and $p_2 = p_1 = p$. For such a process

$$\int_1^2 p\,dv = p(v_2 - v_1) = p_2v_2 - p_1v_1$$

so that Eq. (4-49) may be written as

$$c_p(T_2 - T_1) = u_2 - u_1 + p_2v_2 - p_1v_1 \tag{4-51}$$

Values of c_p are given in Table A-7. The right-hand terms of Eq. (4-51) are identical to those of Eq. (4-29):

$$\text{(4-29)} \qquad \Delta h = u_2 - u_1 + p_2v_2 - p_1v_1$$

Since the right-hand terms of both equations are identical, it follows that their left-hand sides are equal, or

$$c_p(T_2 - T_1) = \Delta h = h_2 - h_1 \tag{4-52}$$

The relation of Eq. (4-52) may be applied to *any ideal gas regardless of the process.*

Ratio of Specific Heats

The ratio of the specific heat at constant pressure to that at constant volume is called the ratio of specific heats. This ratio is the exponent that describes the path of an isentropic (constant entropy) process. For this reason, the ratio of specific heats is also called the isentropic exponent. In mathematical form,

$$k = \frac{c_p}{c_v} \tag{4-53}$$

Values of k are given in Table A-7.

Specific Heat Relations

A useful relation may be developed by substituting from Eq. (4-50)

$$c_v(T_2 - T_1) = u_2 - u_1$$

and from the equation of state (1-42)

$$p_1v_1 = RT_1; \; p_2v_2 = RT_2$$

in Eq. (4-51):

$$c_p(T_2 - T_1) = c_v(T_2 - T_1) + RT_2 - RT_1$$

which reduces to

$$c_p = c_v + R$$

or

$$c_p - c_v = R \tag{4-54}$$

Dividing Eq. (4-54) by c_v and noting from Eq. (4-53) that

(4-53)
$$k = \frac{c_p}{c_v}$$

$$\frac{c_p}{c_v} - \frac{c_v}{c_v} = \frac{R}{c_v} = k - 1 = \frac{R}{c_v}$$

or

$$c_v = \frac{R}{(k-1)} \tag{4-55}$$

Again using the relationship $k = c_p/c_v$ from Eq. (4-54),

$$c_p = kc_v = \frac{kR}{k-1} \tag{4-56}$$

EXAMPLE Check the values of R, c_v, and c_p in SI units for nitrogen in Table A-7 using only the molecular weight and the value of k.

Step A *Data Reduction*

Sec. 1-14 (1) *universal gas constant*

$\bar{R} = 8\,314$ J/kg · mol · K

(2) *properties of nitrogen*

Table A-7 $M = 28.01$, $k = 1.404$

Step B *Compute Gas Constant*

(1-43) $R = \dfrac{\bar{R}}{M} = \dfrac{8\,314}{28.01} = 296.8$ (vs. 296.2) J/kg · K

Step C *Compute Constant Volume Specific Heat*

(4-55) $c_v = \dfrac{R}{k-1} = \dfrac{296.8}{1.404 - 1}$

$= 734.7$ (vs. 734.8) J/kg · K

Step D *Compute Constant Pressure Specific Heat*

(4-53) $c_p = kc_v = 1.404 \times 734.7 = 1032$ (vs. 1032) J/kg · K

4-16 FRICTIONLESS COMPRESSIBLE FLOW

For compressible fluids, only the special case of the frictionless adiabatic flow of ideal gases will be considered. This is isentropic flow, and from Sec. 1-13 the path of this process is $pv^k =$ a constant. For isentropic flow in passages, the equations and methods of Chapter 5 should be used. For frictionless flow,

$$H_f = 0,$$

Eq. (4-14) reduces to

$$\frac{g}{g_c}(z_2 - z_1) + \frac{V_2^2 - V_1^2}{2g_c} + \int_1^2 v\,dp = 0 \tag{4-57}$$

The third term of Eq. (4-57) may be integrated by noting from Eq. (1-46) that $v = v_1(p_1/p)^{1/k}$, so that

$$\int_1^2 v\,dp = v_1 p_1^{1/k}\int_1^2 p^{-1/k}\,dp = v_1 p_1^{1/k}\left[\left(\frac{k}{k-1}\right)p^{(k-1)/k}\right]_1^2$$

$$= p_1 v_1\left(\frac{k}{k-1}\right)\left[\left(\frac{p_2}{p_1}\right)^{(k-1)/k} - 1\right] \tag{4-58}$$

Substituting Eq. (4-58) in Eq. (4-57),

$$\frac{g}{g_c}(z_2 - z_1) + \frac{V_2^2 - V_1^2}{2g_c} + p_1 v_1\left(\frac{k}{k-1}\right)\left[\left(\frac{p_2}{p_1}\right)^{(k-1)/k} - 1\right] \tag{4-59}$$

From the equation of state $p_1 v_1 = RT_1$ (1-42), and from Eq. (1-47)

$$\frac{T_2}{T_1} = \left(\frac{p_2}{p_1}\right)^{(k-1)/k} \tag{1-47}$$

and from Eq. (4-56)

$$c_p = \frac{kR}{k-1} \tag{4-56}$$

Substitution in Eq. (4-59) results in

$$\frac{g}{g_c}(z_2 - z_1) + \frac{V_2^2 - V_1^2}{2g_c} + RT_1\left(\frac{k}{k-1}\right)\left(\frac{T_2}{T_1} - 1\right)$$

$$= \frac{g}{g_c}(z_2 - z_1) + \frac{V_2^2 - V_1^2}{2g_c} + c_p(T_2 - T_1) = 0 \tag{4-60}$$

The same result may be arrived at from the energy equation:

$$q = W_{sf} + \frac{g}{g_c}(z_2 - z_1) + \frac{V_2^2 - V_1^2}{2g_c} + u_2 - u_1 + p_2v_2 - p_1v_1 \tag{4-41}$$

For an isentropic process, no heat is transferred, so that $q = 0$, and for no shaft work

$$W_{sf} = 0$$

By definition

$$u_2 - u_1 + p_2v_2 - p_1v_1 = h_2 - h_1 \tag{4-29}$$

With these substitutions, Eq. (4-41) becomes

$$\frac{g}{g_c}(z_2 - z_1) + \frac{V_2^2 - V_1^2}{2g_c} + h_2 - h_1 = 0 \tag{4-61}$$

From Eq. (4-52)

$$h_2 - h_1 = c_p(T_2 - T_1) \tag{4-52}$$

Substituting in Eq. (4-61), we arrive at Eq. (4-60):

$$\frac{g}{g_c}(z_2 - z_1) + \frac{V_2^2 - V_1^2}{2g_c} + c_p(T_2 - T_1) = 0 \tag{4-60}$$

In general, the use of the complete energy equation for compressible fluids provides simple and direct problem solutions. The complete energy equation may be modified by noting from Eqs. (4-29) and (4-52) that

$$u_2 - u_1 + p_2v_2 - p_1v_1 = h_2 - h_1 = c_p(T_2 - T_1)$$

resulting in

$$q = W_{sf} + (z_2 - z_1) + \frac{V_2^2 - V_1^2}{2g_c} + c_p(T_2 - T_1) \tag{4-62}$$

EXAMPLE 1 Stagnation Concept The rocket shown in Fig. 4-9 travels through the U.S. atmosphere at an altitude of 10,000 ft, with a velocity of 900 ft/sec. Find the temperature and pressure at the forward stagnation point. Assume isentropic compression.

Step A *Data Reduction*

(1) *atmospheric properties* at 10,000 ft

Table A-9 $p_o = 1456 \text{ lbf/ft}^2$, $T_o = 483.03°\text{R}$

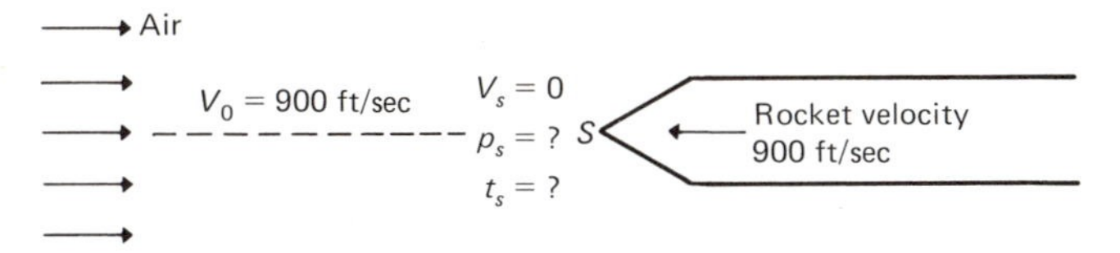

FIGURE 4-9 Notation for stagnation example

(1-11) (2) *proportionality constant*

$$g_c = 32.17 \text{ lbm-ft/lbf-sec}^2$$

(3) *fluid properties,* air

Table A-7 $k = 1.400,\ c_p = 0.2400$ Btu/lbm-°F

(B-15) $c_p = 0.2400 \times 778.2 = 186.8$ ft-lbf/lbm-°F

Step B *Evaluate Each Term of the General Energy Equation in Terms of Fig. 4-9*

(4-62)
$$q = W_{sf} + \frac{g}{g_c}(z_s - z_o) + \frac{V_s^2 - V_o^2}{2g_c} + c_p(T_s - T_o)$$

(1) *heat transfer*

q; compression is assumed to be isentropic or frictionless adiabatic $\therefore q = 0$

(2) *steady-flow work*

W_{sf}; none in this system $\therefore W_{sf} = 0$

(3) *potential energy change*

$$\frac{g}{g_c}(z_s - z_o)$$

On a streamline to the stagnation point, path is horizontal

$$\therefore \frac{g}{g_c}(z_s - z_o) = 0$$

(4) *kinetic energy change*

$$\frac{V_s^2 - V_o^2}{2g_c}$$

By definition, the velocity at the stagnation point is zero, or $V_s = 0$, because the fluid cannot cross a solid boundary.

$$\therefore \frac{V_s^2 - V_o^2}{2g_c} = -\frac{V_o^2}{2g_c} = \frac{-(900)^2}{2 \times 32.17} = -12{,}589 \text{ ft-lbf/lbm}$$

(5) *enthalpy change*

$$c_p(T_s - T_o) = 186.8(T_s - 483.03)$$

Step C *Solve General Energy Equation for T_s*

(4-62)
$$q = W_{sf} + \frac{g}{g_c}(z_s - z_o) + \frac{V_s^2 - V_o^2}{2g_c} + c_p(T_s - T_o)$$

$$0 = 0 + 0 - 12{,}589 + 186.8(T_s - 483.03)$$

$$T_s = \frac{12{,}589}{186.8} + 483.03 = 550.42°\text{R}$$

(B-28)
$$t_s = 550.42 - 459.67 = 90.75°\text{F}$$

Step D *Solve for p_s*

(1-45)
$$p_s = p_o\left(\frac{T_s}{T_o}\right)^{k/(k-1)} = 1456\left(\frac{550.42}{483.03}\right)^{1.400/(1.400-1)}$$

$$p_s = 2300 \text{ lbf/ft}^2$$

(B-87)
$$p_s = \frac{2300}{144} = 15.97 \text{ psia}$$

EXAMPLE 2 Maximum Work Carbon dioxide is to be used in a horizontal gas turbine. The CO_2 expands from 400 kPa and 30°C to 175 kPa. The flow of CO_2 is 1 kg/s. What is the maximum power that can be obtained from this turbine?

Step A *Data Reduction*

(1) *unit conversion*

(B-27)
$$T_1 = 30 + 273.15 = 303.15 \text{ K}$$

(2) *fluid properties*, carbon dioxide

Table A-7
$$k = 1.304, \quad c_p = 814.7 \text{ J/(kg} \cdot \text{K)}$$

Step B *Evaluate Each Term of the General Energy Equation*

(4-62)
$$q = W_{sf} + \frac{g}{g_c}(z_2 - z_1) + \frac{V_2^2 - V_1^2}{2g_c} + c_p(T_2 - T_1)$$

(1) *heat transfer*

q; for maximum work, the process should be isentropic, $\therefore q = 0$

(2) *steady-flow work*, W_{sf}, to be computed

(3) *potential energy change*

$$\frac{g}{g_c}(z_2 - z_1),$$

Turbine is horizontal,

$$z_2 = z_1 \therefore \frac{g}{g_c}(z_2 - z_1) = 0$$

(4) *kinetic energy change*

$$\frac{V_2^2 - V_1^2}{2g_c}$$

In a well-designed system, the gas should enter and leave at the same velocity

$$\therefore \frac{V_2^2 - V_1^2}{2g_c} = 0$$

(5) *enthalpy change*

$c_p(T_2 - T_1)$ for an isentropic process:

(1-47) $$T_2 = T_1\left(\frac{p_2}{p_1}\right)^{\frac{k-1}{k}} = 303.15\left(\frac{175}{400}\right)^{\frac{1.304-1}{1.304}} = 250.01 \text{ K}$$

$$c_p(T_2 - T_1) = 814.7(303.15 - 250.01) = 43\,292 \text{ J/kg}$$

Step C *Solve General Energy Equation for W_{sf}*

(4-61) $$q = W_{sf} + \frac{g}{g_c}(z_2 - z_1) + \frac{V_s^2 - V_o^2}{2g_c} + c_p(T_2 - T_1)$$

$$0 = W_{sf} + 0 + 0 + 43\,293$$

$$W_{sf} = -43\,293 \text{ J/kg (work on fluid)}$$

Step D *Compute Power*

$$P_{sf} = \dot{m}\, W_{sf} = 1(43\,293) = 43.29 \text{ kW}$$

4-17 FLOW WITH FRICTION

Incompressible Fluids

Integration of Eq. (4-42) for an incompressible ($v_1 = v_2$) fluid yields

$$W_{sf} + \frac{g}{g_c}(z_2 - z_1) + \frac{V_2^2 - V_1^2}{2g_c} + v(p_2 - p_1) + H_f = 0 \tag{4-63}$$

From the relation

$$H_f = \frac{1}{R_h}\int_1^2 v\tau\, dL$$

established in Sec. 4-4 for an incompressible fluid ($v_1 = v_2$), and constant wall shear stress τ, $H_f = (v\tau/R_h)(L_2 - L_1)$, Eq. (4-63) may be written as

$$W_{sf} + \frac{g}{g_c}(z_2 - z_1) + \frac{V_2^2 - V_1^2}{2g_c} + v(p_2 - p_1) + \frac{v\tau}{R_h}(L_2 - L_1) = 0 \tag{4-64}$$

Compressible Fluids

For ideal gases, Eq. (4-62) may be used. For nonideal gases whose properties are known, Eq. (4-62) may be modified by noting from Eq. (4-52) that

$$h_2 - h_1 = c_p (T_2 - T_1) \qquad \text{(4-52)}$$

which results in

$$q = W_{sf} + \frac{g}{g_c}(z_2 - z_1) + \frac{V_2^2 - V_1^2}{2g_c} + h_2 - h_1 \qquad \text{(4-65)}$$

Note: It is very important to realize that all the energy and motion equations describe the same thing, an accounting of energy. The variations are for convenience only. In the examples that follow, there are many other ways to arrive at a correct solution in addition to those given.

EXAMPLE 1 Incompressible Flow with Friction Water at 68°F flows steadily through the 6-in., 250-psi cast iron piping system shown in Fig. 4-10. At section 1, the pressure is 60 psia and the pipe center line is 10 ft above a datum. At section 2, the pressure is 150 psia and the pipe center line is 110 ft above a datum. Between sections 1 and 2 are 1000 ft of 6-in. pipe whose unit shear stress at the wall, τ_o, is 0.55 lbf/ft². Also between sections 1 and 2 is a pump that imparts 90 hp to the fluid. Calculate the flow of water in gallons per minute.

FIGURE 4-10 Notation for incompressible-flow example

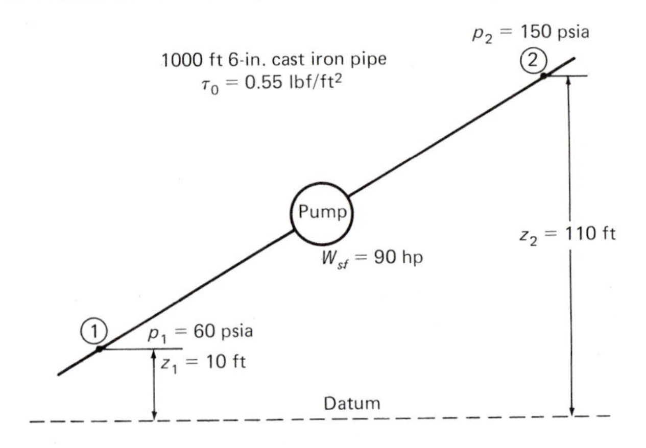

Step A *Data Reduction*

(1) *unit conversions*

(B-52) $W_{sf} = 90 \times 550 = 49{,}500$ ft-lbf/sec

(B-89) $p_1 = 144 \times 60 = 8{,}640$ lbf/ft^2

$p_2 = 150 \times 144 = 21{,}600$ lbf/ft^2

(2) *gravity*

(B-50) Assume $g = g_o = 32.17$ ft/sec^2

(3) *proportionality constant*

(1-11) $g_c = 32.17$ lbm-ft/lbf-sec^2

(4) *fluid properties,* water at 68°F

Table A-1 $\rho = 1.937$ slugs/ft^3

(1-37) $$v = \frac{1}{\rho g_c} = \frac{1}{1.937 \times 32.17} = 0.01605 \text{ ft}^3\text{/lbm}$$

(5) *pipe properties,* 6-in. 250 psi cast iron pipe

Table A-12 $D = 0.5117$ ft

(6) *hydraulic radius,* circular channel

Table A-14 $$R_h = \frac{D}{4} = \frac{0.5117}{4} = 0.1279 \text{ ft}$$

Step B *Evaluate Each Term of Eq. (4-64)*

(4-64) $$W_{sf} + \frac{g}{g_c}(z_2 - z_1) + \frac{V_2^2 - V_1^2}{2g_c} + v(p_2 - p_1) + \frac{v\tau}{R_h}(L_2 - L_1) = 0$$

(1) *steady-flow work,* W_{sf}

$$\frac{-49{,}500}{\dot{m} g_c} \text{ (done on fluid)}$$

(2) *potential energy change*

$$\frac{g}{g_c}(z_2 - z_1) = \frac{32.17}{32.17}(110 - 10) = 100 \text{ ft-lbf/lbm}$$

(3) *kinetic energy change*

$$\frac{V_2^2 - V_1^2}{2g_c}$$

(3-8) $\dot{m} = \rho_1 A_1 V_1 = \rho_2 A_2 V_2$

For an incompressible fluid $\rho_1 = \rho_2$, and for a constant diameter duct, $A_1 = A_2$. Thus from Eq. (3-8), $V_1 = V_2$ and the change in kinetic energy is zero.

(4) *flow work*

$$v(p_2 - p_1)$$

$$0.01605(21{,}600 - 8{,}640) = 208.01 \text{ ft-lbf/lbm}$$

(5) *friction*

$$\frac{v\tau}{R_h}(L_2 - L_1)$$

$$[(0.01605)(0.55)/(0.1279)](1000) = 69.02 \text{ ft-lbf/lbm}$$

Step C — *Solve Eq. (4-64) for Mass Flow Rate*

(4-64)

$$W_{sf} + \frac{g}{g_c}(z_2 - z_1) + \frac{V_2^2 - V_1^2}{2g_c} + v(p_2 - p_1) + \frac{v\tau}{R_h}(L_2 - L_1) = 0$$

$$\frac{-49{,}500}{\dot{m}g_c} + 100 + 0 + 208.01 + 69.02 = 0$$

$$49{,}500 = 377.03\ \dot{m}g_c, \quad \dot{m}g_c = 131.29 \text{ lbm/sec}$$

$$\dot{m} = \frac{131.29}{32.17} = 4.081 \text{ slugs/sec}$$

Step D — *Compute Volumetric Flow Rate*

(3-8)

$$\frac{\dot{m}}{\rho} = AV = Q = \frac{4.081}{1.937} = 2.107 \text{ ft}^3/sec$$

(B-43)

$$\text{GPM} = \frac{2.107(60)}{231/1728} = 945.7 \text{ GPM}$$

EXAMPLE 2 Compressible Flow with Friction Air flows steadily through a horizontal tube at a constant temperature of 32°C. The mass flow of air is 1 kg/s. At section 1, the diameter is 50 mm and the pressure 345 kPa. At section 2, the diameter is 75 mm and the pressure is 359 kPa. What is the value of the "lost head" H_f due to friction?

Step A — *Data Reduction*

(1) *unit conversions*

(B-27)

$$T_1 = T_2 = 32 + 273.15 = 305.15 \text{ K}$$

(2) *proportionality constant*

(1-9)

$$g_c = 1 \text{ kg} \cdot \text{m/N} \cdot \text{s}$$

(3) *fluid properties*, air

Table A-7

$$c_p = 1\,005 \text{ J/kg} \cdot \text{K}, \ k = 1.400, \ R = 287.1 \text{ J/kg} \cdot \text{K}$$

(1-42)

$$v_1 = \frac{RT_1}{p_1} = \frac{287.1 \times 303.15}{345\,000} = 0.252\,3 \text{ m}^3/\text{kg}$$

(1-36) $$\rho_1 = \frac{1}{v_1 g_c} = \frac{1}{(0.252\,3)\,(1)} = 3.964 \text{ kg/m}^3$$

(1-42) $$v_2 = \frac{RT_2}{p_2} = \frac{287.1 \times 303.15}{359\,000}$$

$$= 0.242\,4 \text{ m}^3/\text{kg}$$

(1-36) $$\rho_2 = \frac{1}{v_2 g_c} = \frac{1}{(0.242\,4)\,(1)} = 4.125 \text{ kg/m}^3$$

(4) *channel properties*

Table A-10 $$A_1 = \frac{\pi D_1^2}{4} = \frac{\pi (50)^2}{4} = 1\,963 \text{ mm}^2$$

Table A-10 $$A_2 = \frac{\pi D_2^2}{4} = \frac{\pi (75)^2}{4} = 4\,418 \text{ mm}^2$$

(5) *fluid velocities*

(3-8) $$V_1 = \frac{\dot{m}}{\rho_1 A_1} = \frac{1}{(3.964)\,(1\,963 \times 10^{-6})} = 128.5 \text{ m/s}$$

(3-8) $$V_2 = \frac{\dot{m}}{\rho_2 A_2} = \frac{1}{(4.125)\,(4\,418 \times 10^{-6})} = 54.87 \text{ m/s}$$

Step B *Evaluate Each Term of the General Energy Equation*

(4-62) $$q = W_{sf} + \frac{g}{g_c}\,(z_2 - z_1) + \frac{V_2^2 - V_1^2}{2g_c} + c_p\,(T_2 - T_1)$$

(1) *heat transfer, q,* to be computed

(2) *steady-flow work, W_{sf},* none in system

$\therefore W_{sf} = 0$

(3) *potential energy change*

$$\frac{g}{g_c}\,(z_2 - z_1),$$

horizontal system

$$\therefore \frac{g}{g_c}\,(z_2 - z_1) = 0$$

(4) *kinetic energy change*

$$\frac{V_2^2 - V_1^2}{2g_c} = \frac{54.87^2 - 128.5^2}{2 \times 1} = -6\,763 \text{ J/kg}$$

(5) *enthalpy change*

$c_p\,(T_2 - T_1)$

For a constant temperature process

$T_1 = T_2 \therefore c_p\,(T_2 - T_1) = 0$

Step C *Compute Heat Transfer*

(4-62)

$$q = W_{sf} + \frac{g}{g_c}(z_2 - z_1) + \frac{V_2^2 - V_1^2}{2g_c} + c_p(T_2 - T_1)$$

$$q = 0 + 0 - 6\,763 + 0$$

$$q = -6\,763 \text{ J/kg (out of tube)}$$

Step D *Evaluate Each Term of Eq. (4-44)*

(4-44)

$$H_f = u_2 - u_1 + \int_1^2 p\,dv - q$$

(1) *friction*

H_f to be computed

(2) *internal energy change*

$u_2 - u_1$

(4-50)

$$u_2 - u_1 = c_v(T_2 - T_1)$$

since the process is constant temperature

$T_1 = T_2 \therefore u_2 - u_1 = 0$

(3) *volume change*

$$\int_1^2 p\,dv$$

From the equation of state (1-42), $p = \frac{RT}{v}$.

Substituting for a constant temperature process,

$$\int_1^2 p\,dv = RT\int_1^2 \frac{dv}{v} = RT\log_e \frac{v_2}{v_1}$$

$$= 287.1 \times 305.15 \log_e \frac{0.242\,4}{0.252\,3}$$

$$= -3\,507 \text{ J/kg}$$

(4) *heat transfer*

q from Step C

$q = -6\,763$ J/kg

Step E *Compute Friction*

(4-44)

$$H_f = u_2 - u_1 + \int_1^2 p\,dv - q$$

$$H_f = 0 + (-3\,507) - (-6\,763)$$

$$H_f = 3\,256 \text{ J/kg}$$

EXAMPLE 3 Non-Ideal Gas Flow Steam flows through a horizontal nozzle. At inlet, the velocity is 1000 ft/sec and the enthalpy is 1320 Btu/lbm. At outlet, the enthalpy is 1200 Btu/lbm. Between the inlet and outlet there is a transfer of 5 Btu/lbm of heat. What is the outlet velocity?

Step A *Data Reduction*

(1) *unit conversions*

(B-11) $h_1 = 1320 \times 778.2 = 1{,}027{,}224$ ft-lbf/lbm

(B-11) $h_2 = 1200 \times 778.2 = 933{,}840$ ft-lbf/lbm

(B-11) $q = -5 \times 778.2 = -3{,}891$ ft-lbf/lbm

(2) *proportionality constant*

(1-11) $g_c = 32.17$ lbm-ft/lbf-sec^2

Step B *Evaluate Each Term of Eq. (4-65)*

(4-65)
$$q = W_{sf} + \frac{g}{g_c}(z_2 - z_1) + \frac{V_2^2 - V_1^2}{2g_c} + h_2 - h_1$$

(1) *heat transfer*

$q = -3{,}891$ ft-lbf/lbm

(2) *steady-flow work*

W_{sf}, none in system $\therefore W_{sf} = 0$

(3) *potential energy change*

$$\frac{g}{g_c}(z_2 - z_1)$$

horizontal system $z_1 = z_2 \therefore \frac{g}{g_c}(z_2 - z_1) = 0$

(4) *kinetic energy change*

$$\frac{V_2^2 - V_1^2}{2g_c} = \frac{V_2^2 - (1000)^2}{2 \times 32.17} = \frac{V_2^2}{64.34} - 15{,}542$$

(5) *enthalpy change*

$h_2 - h_1 = 933{,}840 - 1{,}027{,}224 = -93{,}384$ ft-lbf/lbm

Step C *Compute Outlet Velocity*

(4-65)
$$q = W_{sf} + \frac{g}{g_c}(z_2 - z_1) + \frac{V_2^2 - V_1^2}{2g_c} + h_2 - h_1$$

$$-3{,}891 = 0 + 0 + \frac{V_2^2}{64.34} - 15{,}542 + (-93{,}384)$$

$$V_2 = \sqrt{64.34 \times 105{,}035} = 2600 \text{ ft/sec}$$

4-18 IMPULSE MOMENTUM EQUATION

The impulse momentum equation is used to calculate the forces exerted on a solid boundary by a moving stream. It was derived in Sec. 1-8 as an application of Newton's second law. This resulted in

$$F = \dot{m}(V_2 - V_1) \qquad \text{(1-34)}$$

Substituting ΣF (the summation of all forces acting) for F and from Eq. (3-8) $\dot{m} = \rho AV$, yields

$$\Sigma F = \dot{m}(V_2 - V_1) = \rho AV(V_2 - V_1) \qquad \text{(4-66)}$$

In the application of Eq. (4-66), it must be remembered that velocity is a vector and as such has both magnitude and direction. The impulse-momentum equation is often used in conjunction with the continuity and energy equations to solve engineering problems. Because of the wide variety of applications possible, some examples are given to illustrate methods of attack.

In general, the "free body" method is used to compute the forces involved on the boundaries on a control volume. The symbol F is used for the force exerted by the boundaries on the fluid. There is an equal but opposite force exerted by the fluid on the boundaries.

EXAMPLE 1 Compressible Fluid in a Constant Area Duct Carbon dioxide flows steadily through a horizontal 6-in. schedule 40 wrought iron pipe at a mass rate of flow of 11 kg/s. At section 1, the pressure is 827 kPa and the temperature 38°C. At section 2, the pressure is 552 kPa and the temperature is 43°C. Find the friction force opposing the motion. (See Fig. 4-11.)

Step A *Data Reduction*

(1) *unit conversions*

(B-27) $T_1 = 38 + 273.15 = 311.15$ K

(B-27) $T_2 = 43 + 273.15 = 316.15$ K

FIGURE 4-11 Notation for Example 1

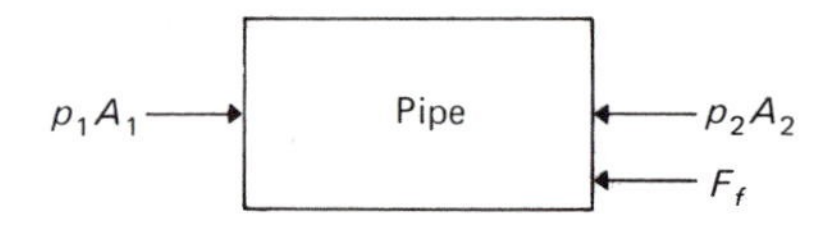

(2) *proportionality constant*

(1-9) $g_c = 1\ \text{kg} \cdot \text{m/N} \cdot \text{s}^2$

(3) *fluid properties,* carbon dioxide

Table A-7 $R = 188.9\ \text{J/(kg} \cdot \text{K)}$

(1-44) $$\rho_1 = \frac{p_1}{g_c R T_1} = \frac{827\,000}{(1)(188.9)(311.15)} = 14.07\ \text{kg/m}^3$$

(1-44) $$\rho_2 = \frac{p_2}{g_c R T_2} = \frac{552\,000}{(1)(188.9)(316.15)} = 9.243\ \text{kg/m}^3$$

(4) *pipe properties,* 6-in. schedule 40

Table A-11 $A = A_1 = A_2 = 18\,650\ \text{mm}^2$

(5) *fluid velocities*

(3-8) $$V_1 = \frac{\dot{m}}{\rho_1 A_1} = \frac{11}{(14.07)(18\,650 \times 10^{-6})} = 41.92\ \text{m/s}$$

(3-8) $$V_2 = \frac{\dot{m}}{\rho_2 A_2} = \frac{11}{(9.243)(18\,650 \times 10^{-6})} = 63.81\ \text{m/s}$$

Step B *Compute Friction Force*

Fig. 4-11 Balance of forces from free body diagram

(4-66) $\Sigma F = p_1 A_1 - p_2 A_2 - F_f = \dot{m}(V_2 - V_1)$

Solving for F_f noting $A_1 = A_2 = A$,

$$F_f = A(p_1 - p_2) - \dot{m}(V_2 - V_1) \quad \text{(a)}$$

$$F_f = (18\,650 \times 10^{-6})(827\,000 - 552\,000) - 11(63.81 - 41.92) \quad \text{(a)}$$

$$F_f = 4\,888\ \text{N}$$

EXAMPLE 2 Vertical Liquid Jet The vertical nozzle shown in Fig. 4-12 discharges a circular jet of 86°F water at a mass flow rate

FIGURE 4-12 Notation for Example 2

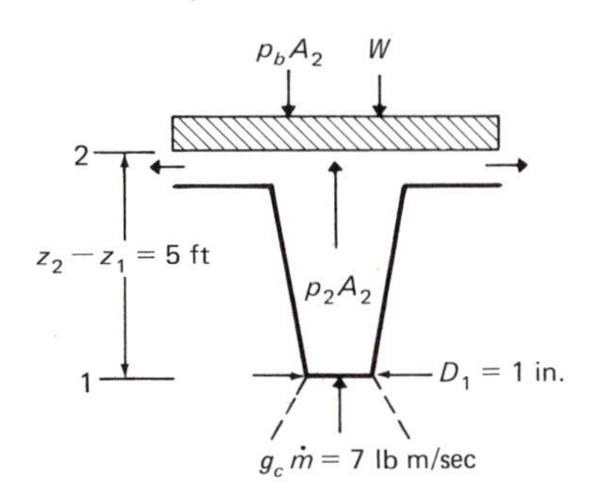

of 7 lbm/sec. The diameter of the jet is 1 in. A large circular disk is held by the impact of the jet in a horizontal position 5 ft above the nozzle. For frictionless flow, what is the weight of the disk?

Step A — *Data Reduction*

(1) *proportionality constant*

(1-11) $g_c = 32.17$ lbm-ft/lbf-sec^2

(2) *unit conversion*

(1-19) $$\dot{m} = \frac{M}{g_c} = \frac{7}{32.17} = 0.2176 \text{ slugs/sec}$$

(B-55) $$D = \frac{1}{12} = \frac{1}{12} \text{ ft}$$

(3) *gravity*

(B-50) Assume $g = g_o = 32.17$ ft/sec^2

(4) *fluid properties,* water at 86°F

Table A-1 $\rho = 1.932$ slugs/ft^3

(5) *geometric,* circular nozzle

Table A-10 $$A_1 = \frac{\pi D_1^2}{4} = \frac{\pi (1/12)^2}{4} = 5.454 \times 10^{-3} \text{ ft}^2$$

(6) *velocity*

(3-8) $$V_1 = \frac{\dot{m}}{\rho_1 A_1} = \frac{0.2176}{(1.932)\,(5.454 \times 10^{-3})} = 20.65 \text{ ft/sec}$$

Step B — *Compute V_2 Using the Bernoulli Equation*

(4-17) $$(z_2 - z_1) + \frac{V_2^2 - V_1^2}{2g} + \frac{p_2 - p_1}{\rho g} = 0$$

For an open jet

$p_b = p_1 = p_2$

so that Eq. (4-17) reduces to

$$(z_2 - z_1) + \frac{V_2^2 - V_1^2}{2g} = 0 \qquad \textbf{(a)}$$

Solving Eq. (a) for V_2,

$$V_2 = \sqrt{V_1^2 - 2g\,(z_2 - z_1)} \qquad \textbf{(b)}$$

$$V_2 = \sqrt{(20.65)^2 - 2 \times 32.17 \times 5} = 10.23 \text{ ft/sec}$$

Step C — *Apply Impulse Momentum Equation across Boundary 2*

(4-66) $$\Sigma F = p_2 A_2 - p_b A_2 - W = \dot{m}\,(0 - V_2)$$

Solving for W, again noting that $p_2 = p_b$,

$$W = \dot{m}V_2 = 0.2176 \times 10.23 = 2.226 \text{ lbf}$$

EXAMPLE 3 Incompressible Flow Through a Reducing Bend Carbon tetrachloride flows steadily at 20°C through a 90° reducing bend. The mass flow rate is 56.7 kg/s, the inlet diameter is 152.4 mm, and the outlet is 76.2 mm. The inlet pressure is 344.75 kPa, and the barometric pressure is 101.33 kPa. For frictionless flow, find the magnitude and direction of the force required to "anchor" this bend in a horizontal position. (See Fig. 4-13.)

Step A *Data Reduction*

(1) *proportionality constant*

(1-9) $g_c = 1 \text{ kg} \cdot \text{m/N} \cdot \text{s}^2$

(2) *fluid properties,* carbon tetrachloride at 20°C

Table A-1 $\rho = 1\,594 \text{ kg/m}^3$

(3) *geometric*

Table A-10 $$A_1 = \frac{\pi D_1^2}{4} = \frac{\pi (152.4 \times 10^{-3})^2}{4} = 0.0182\,4 \text{ m}^2$$

Table A-10 $$A_2 = \frac{\pi D_2^2}{4} = \frac{\pi (76.2 \times 10^{-3})^2}{4} = 0.004\,560 \text{ m}^2$$

(4) *fluid velocities*

(3-8) $$V_1 = \frac{\dot{m}}{\rho_1 A_1} = \frac{56.7}{(1\,594)\,(0.0182\,4)} = 1.950 \text{ m/s}$$

FIGURE 4-13 Notation for Example 3

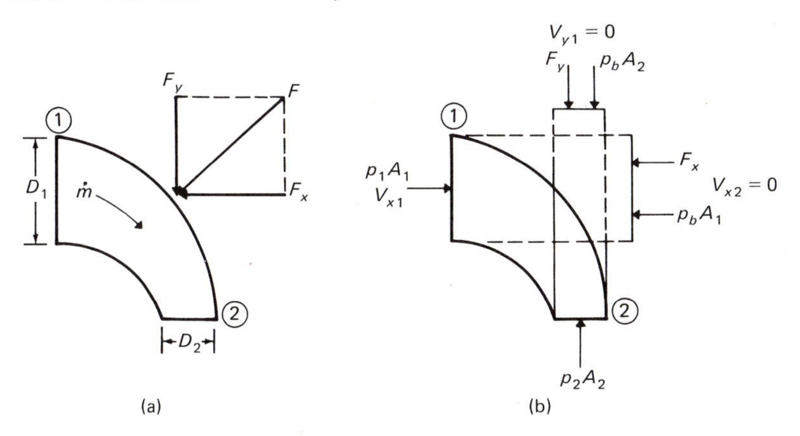

(3-8)

$$V_2 = \frac{\dot{m}}{\rho_2 A_2} = \frac{56.7}{(1\,594)\,(0.004\,560)} = 7.801 \text{ m/s}$$

Step B *Compute p_2 Using the Bernoulli Equation*

(4-17)

$$(z_2 - z_1) + \frac{V_2^2 - V_1^2}{2g} + \frac{p_2 - p_1}{\rho g} = 0$$

For a horizontal bend $z_2 = z_1$ and Eq. (4-17) reduces to

$$\frac{V_2^2 - V_1^2}{2g} + \frac{p_2 - p_1}{\rho g} = 0 = \frac{V_2^2 - V_1^2}{2} + \frac{p_2 - p_1}{\rho} \qquad \text{(a)}$$

Solving (a) for p_2,

$$p_2 = p_1 + \frac{\rho\,(V_1^2 - V_2^2)}{2} \qquad \text{(b)}$$

$$p_2 = 344\,750 + \frac{1\,594[(1.950)^2 - (7.801)^2]}{2}$$

$$p_2 = 299\,279 = 299.279 \text{ kPa}$$

Step C *Apply Impulse Momentum Equation*

Fig. 4-13

Since velocity is a vector, it is necessary to reduce the forces into their X and Y components, as shown in Fig. 4-13. In the X direction, the velocity changes from 1.950 m/s to 0 m/s, and in the Y direction from 0 to 7.801 m/s.

(4-66)

$$\Sigma F_x = p_1 A_1 - p_b A_1 - F_x = \dot{m}\,(V_{x_2} - V_{x_1})$$

Solving for F_x,

$$F_x = (p_1 - p_b)\,A_1 - \dot{m}\,(V_{x_2} - V_{x_1}) \qquad \text{(a)}$$

$$F_x = (344\,750 - 101\,330)\,0.0182\,4 - 56.7\,(0 - 1.950)$$

$$F_x = 4\,551 \text{ N}$$

(4-66)

$$\Sigma F_y = p_b A_2 + F_y - p_2 A_2 = \dot{m}\,(V_{y_1} - V_{y_2})$$

Solving for F_y,

$$F_y = (p_2 - p_b)\,A_2 + \dot{m}\,(V_{y_1} - V_{y_2}) \qquad \text{(b)}$$

$$F_y = (299\,279 - 101\,330)(0.004\,560) + 56.7(7.801 - 0) = 1\,345 \text{ N}$$

Step D *Resolve Forces*

(2-38)

$$F = \sqrt{F_x^2 + F_y^2} = \sqrt{(4\,551)^2 + (1\,345)^2}$$

$$F = 4\,746 \text{ N}$$

$$\theta = \tan^{-1}\left(\frac{F_y}{F_x}\right) = \tan^{-1}\left(\frac{1\,345}{4\,551}\right)$$

$$\theta = 16^\circ\ 27'\ 52''$$

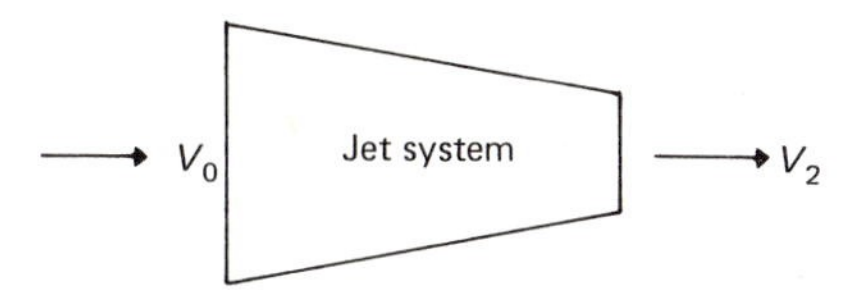

FIGURE 4-14 Notation for jet propulsion

4-19 JET PROPULSION

The propulsive force exerted by a jet may be calculated by use of the impulse momentum equation, as was shown in Sec. 1-18, where a fluid enters a jet system with a velocity of V_o (relative to the body) and leaves with a velocity of V_2 (relative to the body). In the notation of Fig. 4-14, if the jet system is moving through a still fluid, then V_o is the velocity of the body and V_2 is its jet velocity. The thrust T produced for a given mass flow rate of $\dot{m}$ from Eq. (4-66) is

$$T = \dot{m}(V_2 - V_o) \tag{4-67}$$

The useful power developed, P_o, is

$$P_o = TV_o = \dot{m}(V_2 - V_o)V_o \tag{4-68}$$

The *minimum* energy needed to change the kinetic energy produces a minimum power input of P_i:

$$P_i = \frac{\dot{m}}{2}(V_2^2 - V_o^2) \tag{4-69}$$

The *maximum* propulsive efficiency, E, is

$$E = \frac{P_o}{P_i} = \frac{(\dot{m})(V_2 - V_o)V_o}{(\dot{m}/2)(V_2^2 - V_o^2)} = \frac{2V_o}{V_2 + V_o}$$

$$E = \frac{2}{1 + V_2/V_o} \tag{4-70}$$

For 100% efficiency $V_2 = V_o$ and no thrust is produced!

EXAMPLE The jet airplane shown in Fig. 4-15 flies at a constant altitude of 50,000 ft and a speed of 360 mph. Heat in the amount of 325 Btu/lbm is added to the air in the propulsion system. The system mass flow rate is 12.5 lbm/sec of air. Hot gases leave the

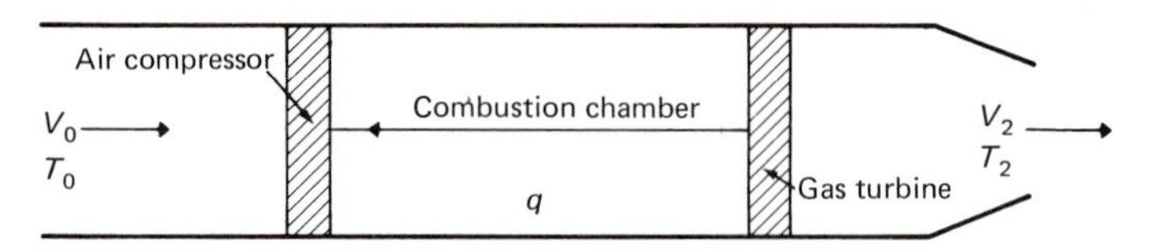

FIGURE 4-15 Notation for jet aircraft example

gas turbine at 1,200°F. Assuming that the hot gas has the same properties as air and neglecting the weight of fuel, determine (a) thrust produced, (b) maximum propulsive efficiency, (c) system efficiency.

Step A *Data Reduction*

(1-11) (1) *proportionality constant*

$$g_c = 32.17 \text{ lbm-ft/lbf-sec}^2$$

(2) *fluid properties,* assumed to be air

Table A-7 $c_p = 0.2400$ Btu/lbm °R

(3) *atmospheric properties,* 50,000 ft

Table A-9 $T_o = 389.97$ °R

(4) *unit conversions*

(1-19) $$\dot{m} = \dot{M}/g_c = \frac{12.5}{32.17} = 0.3886 \text{ slugs/ft}^3$$

(B-11) $$q = 325 \times 778.2 = 252{,}915 \text{ ft-lbf/lbm}$$

(B-15) $$c_p = 0.2400 \times 778.2 = 186.8 \text{ ft-lbf/lbm °R}$$

(B-28) $$T_2 = 1{,}200 + 459.67 = 1{,}659.67 \text{ °R}$$

(B-65/54) $$V_o = 360\left(\frac{5{,}280}{3{,}600}\right) = 528 \text{ ft/sec}$$

Step B *Evaluate Each Term of Eq. (4-62)*

(4-62) $$q = W_{sf} + \frac{g}{g_c}(z_2 - z_1) + \frac{V_2^2 - V_1^2}{2g_c} + c_p(T_2 - T_1)$$

Writing Eq. (4-62) in the notation of Fig. 4-15,

$$q = W_{sf} + \frac{g}{g_c}(z_2 - z_o) + \frac{V_2^2 - V_o^2}{2g_c} + c_p(T_2 - T_o) \qquad \textbf{(a)}$$

(1) *heat transfer*

$$q = 252{,}915 \text{ ft-lbf/lbm}$$

(2) *shaft work* W_{sf}. The work of the turbine is equal to that of the compressor, or the work done on the gas is equal to that done by the gas $\therefore W_{sf} = 0$.

(3) *potential energy change*

$$\frac{g}{g_c}(z_2 - z_o)$$

for a constant altitude (horizontal flight)

$$z_2 = z_o \therefore \frac{g}{g_c}(z_2 - z_o) = 0$$

(4) *kinetic energy change*

$$\frac{V_2^2 - V_o^2}{2g_c} = \frac{V_2^2 - 528^2}{(2 \times 32.17)} = \frac{V_2^2}{64.34} - 4{,}333$$

(5) *enthalpy change*

$$c_p(T_2 - T_o) = 186.8(1659.67 - 389.97)$$
$$= 237{,}180 \text{ ft-lbf/lbm}$$

Step C — *Solve Eq. (a) for V_2*

$$q = W_{sf} + \frac{g}{g_c}(z_2 - z_o) + \frac{V_2^2 - V_o^2}{2g_c} + c_p(T_2 - T_o) \qquad \textbf{(a)}$$

$$252{,}915 = 0 + 0 + \frac{V_2^2}{64.34} - 4{,}333 + 237{,}180$$

$$V_2 = \sqrt{64.34 \times 20{,}068} = 1136 \text{ ft/sec}$$

Step D — *Compute Thrust*

(4-67)

$$T = \dot{m}(V_2 - V_o) = 0.3886(1136 - 528) = 236.3 \text{ lbf}$$

Step E — *Compute Maximum Propulsive Efficiency*

(4-70)

$$E = \frac{2}{1 + V_2/V_o} = \frac{2}{1 + 1136/528}$$
$$= 0.6346 \quad \text{or} \quad 63.46\%$$

Step F — *Compute System Efficiency*

(4-68)

(1) *useful power* $= P_o = TV_o = 236.3 \times 528$
$= 124{,}766$ ft-lbf/sec

(2) *power supplied* $= P_s = \dot{M}q = 12.5 \times 252{,}915$ **(b)**
$P_s = 3{,}161{,}438$ ft-lbf/sec

(3) *system efficiency*

$$E_s = \frac{\text{useful power}}{\text{power supplied}} = \frac{P_o}{P_s} = \frac{124{,}766}{3{,}161{,}438} \qquad \textbf{(c)}$$
$$= 0.03946 \quad \text{or} \quad 3.95\%$$

4-20 PROPELLERS

Although propellers for ships and aircraft cannot be designed with the energy and impulse-momentum relations alone, application of these relations to problems leads to some of the laws that characterize their operation.

Slipstream Analysis

The stream of fluid passing through the propeller of Fig. 4-16 is called the slipstream. Fluid approaches the slipstream with a velocity of V_o and leaves with a velocity V_2. Within the propeller boundary, the velocity is V_1 and work in the amount of W is added between sections A and B by the propeller. The inlet and outlet pressure are p_o, and the pressure at section A is p_A and at B is p_B. For this analysis, we will use the Bernoulli equation for an incompressible fluid. Writing the Bernoulli equation (4-17) for a horizontal slipstream $(z_2 - z_1)$ results in

$$0 + \frac{V_2^2 - V_1^2}{2g} + \frac{p_2 - p_1}{\rho g} = 0$$

which reduces to

$$p_1 - p_2 = \frac{\rho}{2}(V_2^2 - V_1^2) \tag{4-71}$$

FIGURE 4-16 Notation for slipstream analysis

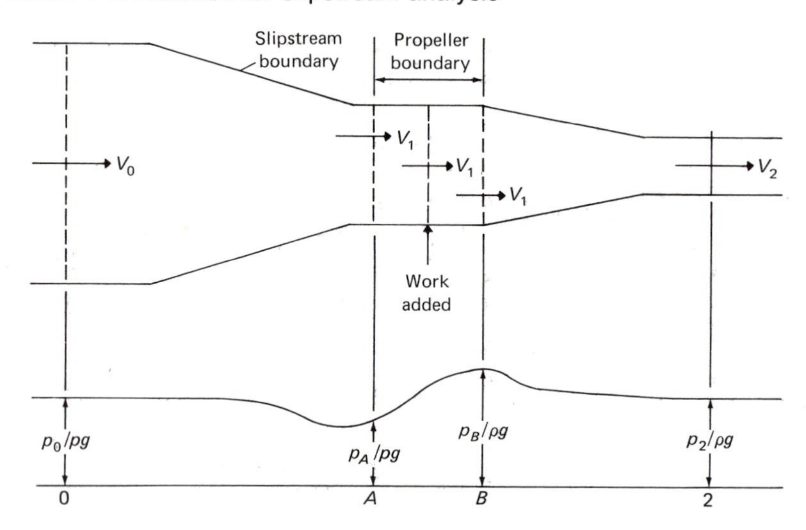

Between sections 0 and A. Application of Eq. (4-71) yields

$$p_o - p_A = \frac{\rho}{2}(V_1^2 - V_o^2) \qquad \textbf{(4-72)}$$

Between sections B and 2. In a like manner,

$$p_B - p_o = \frac{\rho}{2}(V_2^2 - V_1^2) \qquad \textbf{(4-73)}$$

Adding Eqs. (4-71) and (4-72),

$$p_B - p_A = \frac{\rho}{2}(V_2^2 - V_o^2) \qquad \textbf{(4-74)}$$

When the fluid in the slipstream is isolated, it is noted that between sections 0 and 2 the only force acting is that exerted by the propeller on the fluid. This force results in the creation of the pressure difference $(p_B - p_A)$ over the propeller area A. This force must also be equal to the force created by the change in momentum per second of the fluid between sections 0 and 2. Therefore

$$\textbf{(4-66)} \qquad \Sigma F = (p_B - p_A)A = \rho A V_1(V_2 - V_o)$$

which reduces to

$$p_B - p_A = \rho V_1(V_2 - V_o) \qquad \textbf{(4-75)}$$

Letting Eq. (4-74) equal Eq. (4-75),

$$p_B - p_A = \frac{\rho}{2}(V_2^2 - V_o^2) = \rho V_1(V_2 - V_o)$$

which reduces to

$$V_1 = \frac{V_2 + V_o}{2} \qquad \textbf{(4-76)}$$

The maximum propeller efficiency from Eq. (4-70) becomes

$$E_p = \frac{2}{1 + V_2/V_o} = \frac{2V_o}{V_o + V_2} = \frac{V_o}{V_1} \qquad \textbf{(4-77)}$$

EXAMPLE OF A PROPELLER CALCULATION A propeller 355.6 mm in diameter drives a torpedo through 20°C sea water at 11.32 m/s. The ideal propeller efficiency is 75%. Determine (a) useful power and (b) power added to the water.

Step A	*Data Reduction*
	(1) *fluid properties,* sea water at 20°C
Table A-1	$\rho = 1\ 025\ \text{kg/m}^3$
	(2) *geometric*
Table A-10	$A = \frac{\pi D^2}{4} = \frac{\pi(355.6 \times 10^{-3})^2}{4} = 0.099\ 31\ \text{m}^2$
Step B	*Calculate Velocities and Mass Flow Rates*
(4-77)	$E_p = \frac{V_o}{V_1} = 0.75,\ V_1 = \frac{V_o}{0.75} = \frac{11.32}{0.75} = 15.09\ \text{m/s}$
(4-76)	$V_1 = \frac{V_2 + V_o}{2},\ V_2 = 2V_1 - V_o = 2 \times 15.09 - 11.32$
	$V_2 = 18.86\ \text{m/s}$
(3-8)	$\dot{m} = \rho A V_1 = (1\ 025)(0.099\ 31)15.09 = 1\ 536\ \text{kg/s}$
Step C	*Compute powers*
(4-68)	$P_0 = \dot{m}(V_2 - V_o)V_o = (1\ 536)(18.86 - 11.32)(11.32) = 131\ 101\ \text{W} = 131.1\ \text{kW}$
(4-69)	$P_i = \frac{\dot{m}}{2}(V_2^2 - V_o^2) = \left(\frac{1\ 536}{2}\right)(18.86)^2 - (11.32)^2 = 174\ 764\ \text{W} = 174.8\ \text{kW}$
Check	$E_p = \frac{P_u}{P_i} = \frac{131.1}{174.8} = 0.75$ or 75%

4-21 FLOW IN A CURVED PATH

In Sec. 2-11, the effects of rotating a fluid mass were explored. This type of rotating produces a "forced vortex," so called because the fluid is forced to rotate because energy is supplied from some external source. When a fluid flows through a bend, it is also rotated around some axis, but the energy required to produce this rotation is supplied from the energy already in the fluid mass. This is called a "free vortex" because it is "free" of outside energy.

Consider the fluid mass $\rho(r_o - r_i)\,dA$ of Fig. 4-17 being rotated as it flows through a bend of outer radius r_o, inner radius r_i with a velocity of V. Application of Newton's second law to this mass results in

$$\Sigma\, dF = (p_o dA - p_i dA) = [\rho(r_o - r_i)\, dA]\left[\frac{V^2}{(r_o + r_i)/2}\right] \tag{4-78}$$

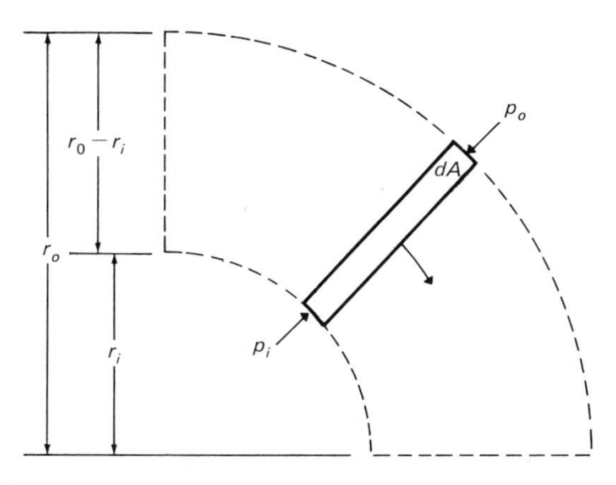

FIGURE 4-17 Notation for flow in a curved path

which reduces to

$$p_o - p_i = 2\left(\frac{r_o - r_i}{r_o + r_i}\right)\rho V^2 \tag{4-79}$$

where p_o is the pressure at the outer wall and p_i is the inner wall pressure.

Because of the difference in fluid pressure between the inner and outer walls of the bend, secondary flows are set up, and this is the primary cause of friction loss of bends. These secondary flows set up turbulence that requires 50 or more straight pipe diameters downstream to dissipate. Thus, this loss does not take place in the bend, but in the downstream system. These losses may be reduced by the use of splitter plates, which help minimize the secondary flows by reducing $r_o - r_i$ and hence $p_o - p_i$.

For one-dimensional flow, Eq. (3-4) $V = Q/A$ (note that this is the flow area, not the area shown in Fig. 4-17). Substituting in Eq. (4-79) for V,

$$p_o - p_i = 2\left(\frac{r_o - r_i}{r_o + r_i}\right)\frac{\rho Q^2}{A^2} \tag{4-80}$$

For a bend of *square cross-section* $A = (r_o - r_i)^2$ and Eq. (4-80) becomes

$$p_o - p_i = 2\left(\frac{r_o - r_i}{r_o + r_i}\right)\frac{\rho Q^2}{(r_o - r_i)^2} \tag{4-81}$$

which reduces to

$$p_o - p_i = \frac{2\rho Q^2}{(r_o^2 - r_i^2)} \tag{4-82}$$

For a bend of circular cross-section,

$$A = \frac{\pi D^2}{4} = \frac{\pi}{4}(r_o - r_i)^2$$

and Eq. (4-80) becomes

$$p_o - p_i = 2\left(\frac{r_o - r_i}{r_o + r_i}\right)\frac{\rho Q^2}{\pi/4\,(r_o - r_i)^2} \quad \textbf{(4-83)}$$

which reduces to

$$p_o - p_i = \frac{8\rho Q^2}{\pi\,(r_o^2 - r_i^2)} \quad \textbf{(4-84)}$$

EXAMPLE Benzene at 68°F flows at a rate of 8 ft³/sec in a square horizontal duct. This duct makes a turn of 90° with an inner radius of 1 ft and an outer radius of 2 ft. Assume frictionless flow and calculate the difference in pressure between the inner and outer wall.

Step A *Data Reduction*

Fluid properties, benzene at 68°F

Table A-1 $\rho = 1.705$ slugs/ft³

Step B *Compute* $p_o - p_i$

(4-82) $$p_o - p_i = \frac{2\rho Q^2}{r_o^2 - r_i^2} = \frac{2 \times 1.705 \times 8^2}{2^2 - 1^2} = 72.75 \text{ lbf/ft}^2$$

(B-89) $$p_o - p_i \frac{72.75}{144} = 0.5052 \text{ psi}$$

4-22 FORCES ON MOVING BLADES

Consider the fluid jet whose area is A issuing from a nozzle with a velocity of V, as shown in Fig. 4-18. The fluid jet impinges on the blade which is moving in the direction of the jet and turns the jet through an angle of θ degrees.

Assuming that the flow is without friction, the jet enters and leaves the blade with a velocity of $(V - v)$ with respect to it. In the x direction, the velocity V_{x_1} is $(V - v)$ and $V_{x_2} = (V - v)\cos\theta$. In the y direction, V_{y_1} is zero, and $V_{y_2} = (V - v)\sin\theta$.

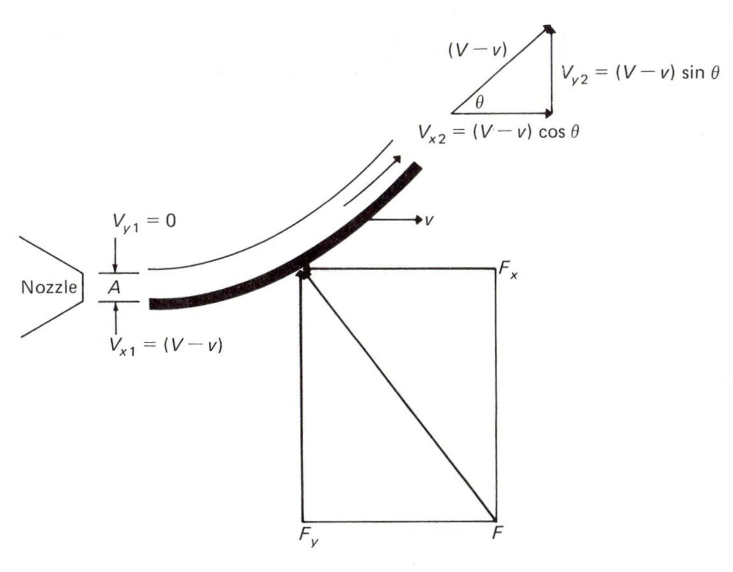

FIGURE 4-18 Notation for blade study

Application of Eq. (4-66) in x direction,

$$F_x = (\rho A V_{x_1})(V_{x_2} - V_{x_1}) = \rho A(V - v)[(V - v)\cos\theta - (V - v)]$$

$$F_x = \rho A(V - v)^2(\cos\theta - 1) \tag{4-85}$$

And in the y direction,

$$F_y = (\rho A V_{x_1})(V_{y_2} - V_{y_1}) = \rho A(V - v)[(V - v)\sin\theta - 0]$$

$$F_y = \rho A(V - v)^2 \sin\theta \tag{4-86}$$

The combined force from Eqs. (4-85) and (4-86) substituted in Eq. (2-38) is

$$F = \sqrt{F_x^2 + F_y^2} = \sqrt{[\rho A(V - v)^2(\cos\theta - 1)]^2 + [\rho A(V - v)^2 \sin\theta]^2}$$

$$F = \rho A(V - v)^2 \sqrt{(\cos\theta - 1)^2 + \sin^2\theta}$$

$$F = \rho A(V - v)^2 \sqrt{1 - 2\cos\theta + \cos^2\theta + \sin^2\theta}$$

$$F = \rho A(V - v)^2 \sqrt{2 - 2\cos\theta} = 2\rho A(V - v)^2 \sin(\theta/2) \tag{4-87}$$

EXAMPLE A jet of glycerine at 68°F and diameter of 1½ in. is deflected through an angle of 80° by a single vane. The jet velocity is 35 ft/sec and the blade moves away from the nozzle at

10 ft/sec in the direction of the entering jet. Assume frictionless flow and calculate the total force acting on the vane.

Step A *Data Reduction*

(1) *unit conversion*

(B-55) $D = (1.5/12) = 0.125$ ft

(2) *fluid properties*, glycerine at 68°F

Table A-1 $\rho = 2.447$ slugs/ft^3

(3) *geometric*

Table A-10 $A = \dfrac{\pi D^2}{4} = \dfrac{\pi\,(0.125)^2}{4} = 0.012272$ ft^2

Step B *Compute Force on Blade*

(4-87)
$$F = 2\rho A(V - v)^2 \sin(\theta/2)$$
$$F = 2(2.447)(0.012272)(35 - 10)^2 \sin(80°/2)$$
$$F = 24.13 \text{ lbf}$$

PROBLEMS

1. A fluid flows between two concentric cylinders, filling the annular space between them. The outside diameter of the smaller cylinder is 2 ft and the inside diameter of the larger is 3 ft. Compute the hydraulic radius and equivalent diameter for this flow condition.
2. A liquid flows to a depth of 8 m in a horizontal triangular duct. This duct is 10 m high and 20 m wide at the top. Compute the hydraulic radius and equivalent diameter for this flow condition.
3. Water at 68°F flows without friction through a 12-in. schedule 40 steel pipe. At section 1, the pressure is 100 psia and the elevation of the datum is 500 ft. What is the pressure at section 2, which is 400 ft above the datum?
4. Hydraulic turbines installed at the Boulder power plant have a rated capacity of 86 000 kw under a head of 145 m when supplied with 66.5 m^3/s of water at 20°C.
   (a) What is the overall efficiency of these turbines, and
   (b) What would be the ratings of these turbines on the planet Jupiter?
5. What is the maximum power that can be expected from a water turbine located 1000 ft below a dam if the available water supply is 165 ft^3/sec?

6. A liquid flows through a pump at a rate of 0.125 m^3/s. The pump inlet is 300 mm in diameter, the outlet 225 mm. What is the change in the kinetic energy of the liquid as it flows through the pump?
7. Prove that the change in flow work for an isothermal process of any ideal gas is zero.
8. Sea water at 68°F enters a pump with a vacuum of 6 in. of mercury and leaves at 20 lb/in^2 gage. What is the increase in flow work?
9. Compute the nonflow shaft work required to compress 1 lbm of air from 700 kPa to 1400 kPa at a constant temperature of 30°C.
10. Compute the nonflow shaft work required to compress 1 lbm of air from 100 psia and 70°F to 200 psia by an isentropic process.
11. Air enters a system at a temperature of 60°C and leaves at 40°C. What is the change in internal energy?
12. Hydrogen enters a system at 600°F and leaves with an increased internal energy of 10 Btu/lbm. What is the exit temperature?
13. Compute the change in enthalpy for Problem 12.
14. Ten kg of carbon dioxide is heated until the temperature is increased 100°C. What is the increase in enthalpy?
15. Five pounds of carbon monoxide are heated in a nonflow process from 100 psia and 500°F to 100 psia and 750°F. How much heat was added?
16. Five kilograms of carbon monoxide at 700 kPa and 500°C is heated in a rigid sealed container until the pressure is doubled. How much heat was added?
17. A submarine moving in sea water at 68°F and at a depth of 1000 ft has a pressure of 462 psig on its forward stagnation point. Estimate the speed of the submarine in knots.
18. An airplane flying at an altitude of 10,000 m has a pressure of 28 kPa on its forward stagnation point. Estimate the air speed of this airplane.
19. The temperature at the forward stagnation point of an airplane is 100°F. At a point A on the wing of this aircraft, the temperature is 84°F. Estimate the velocity of the air at point A assuming that no heat transfer takes place.
20. A large tank has a 51-mm circular hole in its bottom. When this tank is filled with a liquid to a depth of 15 m, the velocity of the jet issuing from the hole is found to be 15.86 m/s. Compute the loss due to friction.
21. Air flows steadily through a 14-in. schedule 30 horizontal steel pipe with a mass rate of flow of 2.878 lbm/sec. At section 1, the pressure is 14.8 psia and the temperature 60°F. At section 2, the pressure is 14.5 psia and the temperature is 269°F. Compute (a) heat transfer and (b) friction loss.

22. Air flows steadily through a 6-in. schedule 40 steel pipe with a flow rate of 10 kg/s. At section 1, the pressure is 825 kPa, and the temperature 40°C. At section 2, the pressure is 525 kPa and the temperature is 45°C. What is the friction force opposing the motion?
23. Water at 68°F flows through a 180° horizontal bend in a 10-in. standard steel pipe at a steady rate of 2400 gal/min. What is the magnitude and direction of the force required to "anchor" this bend if the inlet pressure is 115 psia and the outlet pressure 114 psia?
24. A turbojet engine similar to the one shown in Fig. 4-15 has an inlet diameter of 3 ft and an exhaust diameter of 1.5 ft. Air flow to this engine is 600,000 lbm/hr, and fuel is supplied at a rate of 12,000 lbm/hr. Static inlet temperature is 60°F, and the pressure is 14.7 psia. The exhaust pressure is 20 psia, and exhaust temperature is 800°F. *Do not* neglect the weight of the fuel and assume that the exhaust gases have the same properties as air. Find (a) the thrust produced, (b) the useful horsepower if the engine were in an aircraft at the same air speed as the inlet velocity, and (c) the maximum propulsive efficiency for this flight condition.
25. A solid-fueled rocket produced a jet 152.4 mm in diameter with a velocity of 450 m/s and a density of 0.772 kg/m^3. The rocket velocity in horizontal steady flight is 335 m/s. Compute (a) the thrust produced and (b) useful power.
26. An airplane propeller must produce a thrust of 2000 lbf to drive an airplane at 175 mph through still air whose density is 0.002 slugs/ft^3. What size ideal propeller is necessary if it operates at 80% efficiency?
27. A helicopter weighing 8 900 N is hovering at a certain level in air whose density is 1.24 kg/m^3. The propeller diameter is 6 m. Compute (a) useful power and (b) power delivered by the propeller.
28. A 90° bend occurs in a horizontal plane of a passage 3 ft wide. The inner wall radius is 1 ft, and the radius of the outer wall is 4 ft. The outer wall pressure is 25 psia, and the velocity is 10 ft/sec. Compute the pressure on the inner wall if 68°F water is flowing.
29. A jet of 20°C sea water 50 mm in diameter has an absolute velocity of 50 m/s. This jet strikes a large flat plate which is moving away from the nozzle with an absolute velocity of 15 m/s. The plate is at an angle of 45° with the jet. Compute the total force acting on the plate surface.

REFERENCES

Binder, R. C., *Fluid Mechanics,* 4th ed. Prentice-Hall Inc., Englewood Cliffs, N.J., 1962.

Daugherty, R. L. and A. C. Ingersoll, *Fluid Mechanics – with Engineering Applications,* 5th ed. McGraw-Hill Book Company, New York, 1954.

Hughes, W. F. and J. A. Brighton, *Theory and Problems of Fluid Dynamics.* Schaum Publishing Company, New York, 1967.

Shames, I. H., *Mechanics of Fluids.* McGraw-Hill Book Company, New York, 1962.

Streeter, V. L., *Fluid Mechanics,* 5th ed. McGraw-Hill Book Company, New York, 1971.

5 Compressible Flow

5-1 INTRODUCTION

This chapter is a continuation of the material on ideal gases presented in Chapter 4. It is concerned with some of the effects of the elasticity of ideal gases. The scope of this chapter is limited to the development of concepts needed for the understanding of compressible flow through pipes and flow meters covered in later chapters.

Because the potential energy changes in ideal gas systems are usually small compared with other energy changes, all systems in this chapter will be assumed to be *horizontal,* and thus $z_2 - z_1 = 0$. It is further assumed that the flow is *one-dimensional.*

5-2 AREA–VELOCITY RELATIONS

Incompressible Fluids

The continuity equation in differential form, Eq. (3-10) may be written as

$$\frac{dA}{A} = -\frac{dV}{V} - \frac{d\rho}{\rho} \qquad \text{(3-10)}$$

For an incompressible fluid, $d\rho = 0$, so that Eq. (3-10) reduces to

$$\frac{dA}{A} = -\frac{dV}{V} \qquad \text{(5-1)}$$

Inspection of Eq. (5-1) indicates the following:

(1) If area increases, velocity decreases.
(2) If area decreases, velocity increases.
(3) If area is constant, velocity is constant.
(4) There are no critical values.

Compressible Fluids

The equation of motion (4-12) for a horizontal system ($dz = 0$) and for frictionless flow ($\tau = 0$), becomes

$$\frac{V\,dV}{g_c} + v\,dp = 0 \tag{5-2}$$

Substituting the defining equation (1-36) $v = 1/\rho g_c$ in Eq. (5-2) and simplifying,

$$\frac{V\,dV}{g_c} + \frac{dp}{\rho g_c} = 0 \quad \text{or} \quad V\,dV + dp/\rho = 0 \tag{5-3}$$

Substituting Eq. (1-58) of Sec. 1-16, $\rho = E/c^2$ and Eq. (1-49) $dp = -E\,dv/v$ in Eq. (5-3) and dividing by V^2,

$$\frac{V\,dV}{V^2} + \frac{dp/\rho}{V^2} = \frac{dV}{V} + \frac{(-E\,dv/v)/(E/c^2)}{V^2}$$

or

$$-\frac{dv}{v} = \left(\frac{V}{c}\right)^2 \frac{dV}{V} \tag{5-4}$$

The defining equation (1-36) may be written in logarithmic form:

$$\log_e v = -\log_e \rho - \log_e g_c \tag{5-5}$$

and differentiated, noting that g_c is a constant,

$$\frac{dv}{v} = -\frac{d\rho}{\rho} \tag{5-6}$$

Substituting in the continuity equation (3-10),

$$\frac{dA}{A} = -\frac{dV}{V} + \frac{d\rho}{\rho} = -\frac{dV}{V} - \frac{dv}{v} \tag{5-7}$$

Substituting the value of $-dv/v$ from Eq. (5-4) in Eq. (5-7),

$$\frac{dA}{A} = -\frac{dV}{V} + \left(\frac{V}{c}\right)^2 \frac{dV}{V} = -\frac{dV}{V}\left[1 - \left(\frac{V}{c}\right)^2\right] \tag{5-8}$$

The ratio of actual velocity V to the speed of sound c is known as the Mach number, **M**, named in honor of Ernst Mach, an Austrian physicist.

$$\mathbf{M} = \frac{V}{c} \tag{5-9}$$

Substituting from Eq. (5-9) in Eq. (5-8),

$$\frac{dA}{A} = -\frac{dV}{V}(1 - \mathbf{M}^2) \tag{5-10}$$

Analysis of Eq. (5-10) leads to the following conclusions:

(1) $V < c$ $\quad \mathbf{M} < 1 \quad \frac{dA}{A} \propto -\frac{dV}{V}$ *Velocity Subsonic*
If area increases, velocity decreases—same for incompressible flow.

(2) $V = c$ $\quad \mathbf{M} = 1 \quad \frac{dA}{A} \propto 0\frac{dV}{V}$ *Velocity sonic**
Sonic velocity can exist only where the change in area is zero, i.e., at the *end* of a convergent passage.

(3) $V > c$ $\quad \mathbf{M} > 1 \quad \frac{dA}{A} \propto +\frac{dV}{V}$ *Velocity supersonic*
If area increases, velocity increases—*reverse* of incompressible flow. Also, supersonic velocity can exist only in the *expanding* portion of a passage *after* a *constriction* where *sonic* (acoustic) velocity existed.

5-3 FRICTIONLESS COMPRESSIBLE FLOW IN HORIZONTAL PASSAGES

General Considerations

Frictionless adiabatic (isentropic) compressible flow of an ideal gas must satisfy the following requirements:

* For viscous fluids sonic velocity will exist only at the end of constant area duct.

(1) *conservation of mass* The continuity equation may be expressed as

$$\dot{m} = VA\rho = V_1 A_1 \rho_1 = V_2 A_2 \rho_2 \tag{3-8}$$

(2) *conservation of energy* The sum of all the energy at a section is the same for all sections. Equation (4-61) for a horizontal passage is

$$\frac{V_1^2}{2g_c} + h_1 = \frac{V_2^2}{2g_c} + h_2 \tag{5-11}$$

(3) *the process relationship* For an ideal gas undergoing an isentropic process:

$$pv^k = p_1 v_1^k = p_2 v_2^k \tag{1-41}$$

(4) *the ideal gas law* The equation of state for an ideal gas is

$$pv = RT \tag{1-42}$$

For convenience in solving engineering problems in U.S. customary units, the continuity equation (3-8) may be modified by substituting from Eq. (1-36) $\rho = 1/vg_c$ and from Eq. (1-19) $m = M/g_c$:

$$\dot{m} = \frac{M}{g_c} = \frac{AV}{vg_c} = \frac{A_1 V_1}{v_1 g_c} = \frac{A_2 V_2}{v_2 g_c}$$

which reduces to

$$\dot{M} = \frac{AV}{v} = \frac{A_1 V_1}{v_1} = \frac{A_2 V_2}{v_2} \tag{5-12}$$

5-4 CONVERGENT NOZZLES

Consider the flow of an ideal gas from a large tank through a convergent nozzle that discharges into the atmosphere or to another large tank as shown in Fig. 5-1. The tanks are large enough so that the gas in them may be assumed to be at stagnation conditions.

In Sec. 5-2, it was demonstrated that in order to meet the requirements of conservation of energy and the conservation of mass, the *maximum* velocity that can exist at the end of a convergent passage is acoustic or sonic velocity.

To determine the temperature ($T_2 = T_c$) that can exist in the nozzle throat, write Eq. (5-11) for $V_1 = 0$ and for $V_2 = V_c$, and substi-

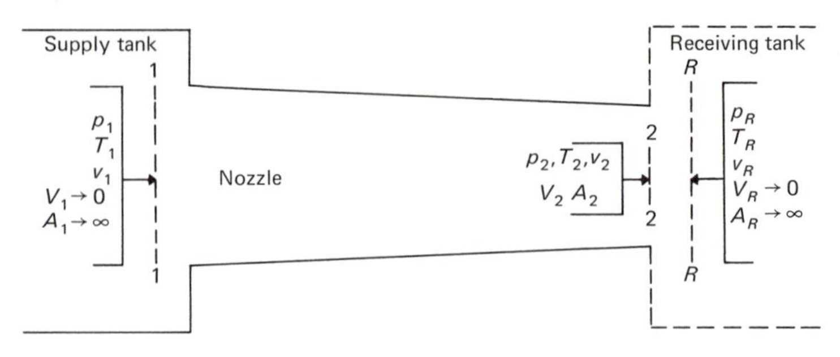

FIGURE 5-1 Notation for convergent nozzle study

tute $c_p(T_1 - T_c) = (h_1 - h_2)$ of Eq. (4-52) and $c_p = Rk/(k-1)$ of Eq. (4-56) and for $V_c = c_2 = \sqrt{kg_cRT_c}$ of Eq. (1-60):

$$\frac{V_2^2}{2g_c} = h_1 - h_2 = \frac{V_c^2}{2g_c} = h_1 - h_c = c_p(T_1 - T_c)$$

$$= \left(\frac{Rk}{k-1}\right)(T_1 - T_c) = \frac{c_2^2}{2g_c} = \frac{kg_cRT_c}{2g_c}$$

which reduces to

$$\frac{T_c}{T_1} = \frac{2}{k+1} \tag{5-13}$$

The critical pressure ratio from Eq. (1-45) is

$$\frac{p_c}{p_1} = \left(\frac{T_c}{T_1}\right)^{k/(k-1)} = \left(\frac{2}{k+1}\right)^{k/(k-1)} \tag{5-14}$$

Equation (5-14) represents the minimum pressure ratio that can exist in the nozzle throat. If p_R/p_1 is less than p_c/p_1 then p_c is the pressure in the throat and the velocity is sonic. If p_R/p_1 is greater than p_c/p_1 then p_R is the pressure in the throat and the velocity is subsonic.

EXAMPLE Air discharges from the large tank shown in Fig. 5-1 in which the temperature and pressure are 100°F and 115 psia respectively through a convergent nozzle whose throat diameter is 1 in. and into a large receiving tank. Compute the temperature, pressure, velocity, Mach number, and mass flow rate of the nozzle jet when the pressure in the receiving tank is (a) 45 psia and (b) 95 psia.

Step A *Data Reduction*

(1) *fluid properties,* air

Table A-7 $k = 1.400,\ c_p = 0.2400$ Btu/lbm-°R,

$R = 53.36$ ft-lbf/lbm-°R

(2) *unit conversions*

(B-28) $T_1 = 100 + 459.67 = 559.67$°R

(B-89) $p_1 = 115 \times 144 = 16{,}560$ lbf/ft^2

(B-89) $p_{R_a} = 45 \times 144 = 6{,}480$ lbf/ft^2

(B-89) $p_{R_b} = 95 \times 144 = 13{,}680$ lbf/ft^2

(B-55) $D = 1/12$ ft

(B-15) $c_p = 0.2400 \times 778.2 = 186.77$ ft-lbf/lbm-°R

(3) *proportionality constant*

(1-11) $g_c = 32.17$ lbm-ft/lbf-sec^2

(4) *geometric,* circle

Table A-10 $$A = \frac{\pi D^2}{4} = \frac{\pi(1/12)^2}{4} = 5.454 \times 10^{-3}\ \text{ft}^2$$

Step B *Determine Throat Conditions*

The minimum pressure that can exist in the throat is

(5-14) $$p_c = p_1\left(\frac{2}{k+1}\right)^{k/(k-1)} = 16{,}560\left(\frac{2}{1.400+1}\right)^{1.400/(1.400-1)}$$

$p_c = 8{,}748$ lbf/ft^2

For part (a), $p_{R_a} = 6{,}480$, which is less than the minimum pressure. ∴ flow is sonic and $p_2 = p_c = 8{,}748$ lbf/ft^2.

For part (b), $p_{R_b} = 13{,}680$, which is greater than the minimum pressure. ∴ flow is subsonic and $p_2 = p_{R_b} = 13{,}680$ lbf/ft^2.

Step C *Solve Part (a)*

(5-13) $$T_2 = T_c = T_1\left(\frac{2}{k+1}\right) = 559.67\left(\frac{2}{1.400+1}\right) = 466.39\text{°R}$$

(1-60) $$V_2 = c_2 = \sqrt{kg_cRT_c} = \sqrt{1.400 \times 32.17 \times 53.36 \times 466.39}$$

$V_2 = 1{,}059$ ft/sec

(5-9) $$\mathbf{M}_2 = \frac{V_2}{c_2} = 1$$

(1-42) $$v_2 = v_c = \frac{RT_c}{p_c} = 53.36 \times \frac{466.39}{8{,}748}$$

$$v_2 = 2.845 \text{ ft}^3/\text{lbm}$$

(5-12) $$\dot{M} = \frac{V_c A_2}{v_c} = \frac{1059 \times 5.454 \times 10^{-3}}{2.845}$$

$$\dot{M} = 2.030 \text{ lbm/sec}$$

Step D *Solve Part (b)*

(1-47) $$T_2 = T_1 \left(\frac{p_2}{p_1}\right)^{(k-1)/k} = 559.67 \left(\frac{13{,}680}{16{,}560}\right)^{(1.400-1)/1.400}$$

$$T_2 = 529.94°\text{R}$$

(4-52) $$h_1 - h_2 = c_p(T_1 - T_2) = 186.67(559.67 - 529.94)$$

$$h_1 - h_2 = 5{,}550 \text{ ft-lbf/lbm}$$

(5-11) $$V_2 = \sqrt{2g_c(h_1 - h_2) + V_1^2} = \sqrt{2 \times 32.17(5550) + 0}$$

$$V_2 = 597.6 \text{ ft/sec}$$

(1-60) $$c_2 = \sqrt{kg_c RT_2}$$

$$= \sqrt{1.400 \times 32.17 \times 53.36 \times 529.97} = 1{,}129 \text{ ft/sec}$$

(5-9) $$\mathbf{M}_2 = \frac{V_2}{c_2} = \frac{597.6}{1{,}129} = 0.5293$$

(1-42) $$v_2 = \frac{RT_2}{p_2} = \frac{53.36 \times 529.94}{13{,}680} = 2.067 \text{ ft}^3/\text{lbm}$$

(5-12) $$\dot{M} = \frac{V_2 A_2}{v_2} = \frac{597.6 \times 5.454 \times 10^{-3}}{2.067} = 1.577 \text{ lbm/sec}$$

5-5 REDUCERS

When a compressible fluid flows through a reducer in a pipeline, the inlet velocity is not zero and the inlet kinetic energy must be taken into account. Consider the flow through a reducer as shown in Fig. 5-2. An equation for the velocity at section 2 may be obtained by writing Eq. (5-11) and substituting from Eq. (5-12),

$$V_1 = V_2 (A_2/A_1)(v_1/v_2)$$

and

(4-52) $$c_p(T_1 - T_2) = h_1 - h_2$$

from Eq. (4-52) and

(4-56) $$c_p = \frac{Rk}{k-1}$$

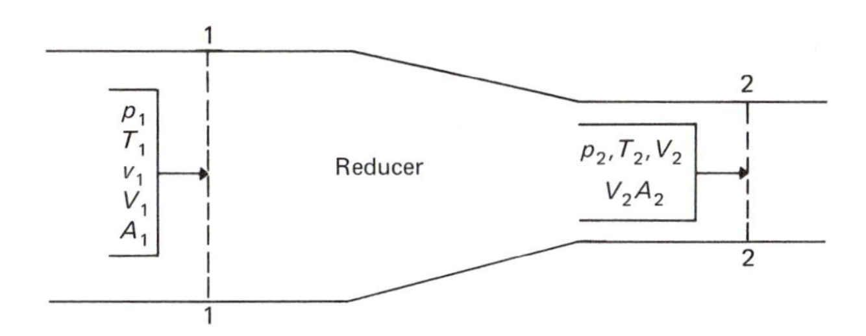

FIGURE 5-2 Notation for reducer study

from Eq. (4-56) to obtain

$$\frac{V_2^2}{2g_c} - \frac{V_1^2}{2g_c} = \frac{V_2^2}{2g_c} - \frac{V_2^2}{2g_c}\left(\frac{A_2}{A_1}\right)^2\left(\frac{v_1}{v_2}\right)^2 = (h_1 - h_2)$$
$$= c_p(T_1 - T_2) = \left(\frac{Rk}{k-1}\right)(T_1 - T_2)$$

or

$$\frac{V_2^2}{2g_c}\left[1 - \left(\frac{A_2}{A_1}\right)^2\left(\frac{v_1}{v_2}\right)^2\right] = \left(\frac{RkT_1}{k-1}\right)\left(1 - \frac{T_2}{T_1}\right) \tag{5-15}$$

Solving Eq. (5-15) for V_2 and substituting

$$\left(\frac{p_2}{p_1}\right)^{1/k} = \left(\frac{v_1}{v_2}\right) \tag{1-45}$$

from Eq. (1-45) and

$$\left(\frac{p_2}{p_1}\right)^{(k-1)/k} = \frac{T_2}{T_1} \tag{1-47}$$

from Eq. (1-47) and

$$RT_1 = p_1 v_1 \tag{1-42}$$

from Eq. (1-42),

$$V_2 = \sqrt{\frac{2g_c k p_1 v_1}{k-1}\left[\frac{1 - (p_2/p_1)^{(k-1)/k}}{1 - (A_2/A_1)^2 (p_2/p_1)^{2/k}}\right]} \tag{5-16}$$

To develop an equation for the mass flow rate through the reducer note that from Eq. (5-12):

$$\dot{M} = \frac{V_2 A_2}{v_2} \tag{5-12}$$

and substitute V_2 from Eq. (5-16).

$$\dot{M} = \frac{V_2 A_2}{v_2} = A_2 \sqrt{\frac{2g_c k p_1 v_1 [1 - (p_2/p_1)^{(k-1)/k}]}{(k-1) v_2^2 [1 - (A_2/A_1)^2 (p_2/p_1)^{2/k}]}} \tag{5-17}$$

From Eq. (1-45):

$$\frac{v_1}{v_2} = \left(\frac{p_2}{p_1}\right)^{1/k} \tag{1-45}$$

substituting in Eq. (5-17),

$$\dot{M} = A_2 \sqrt{\left[\frac{2g_c k}{(k-1)}\right] \frac{(p_1/v_1)(p_2/p_1)^{2/k}[1 - (p_2/p_1)^{(k-1)/k}]}{1 - (A_2/A_1)^2 (p_2/p_1)^{2/k}}} \tag{5-18}$$

Differentiating Eq. (5-18) with respect to (p_2/p_1) and setting

$$d\dot{M}/d(p_2/p_1)$$

equal to zero, we have

$$\left(\frac{p_c}{p_1}\right)^{(1-k)/k} + \left(\frac{k-1}{2}\right)\left(\frac{A_2}{A_1}\right)^2 \left(\frac{p_c}{p_1}\right)^{2/k} = \frac{k+1}{2} \tag{5-19}$$

For the special case of $A_2/A_1 = 0$, Eq. (5-19) reduces to

$$\frac{p_c}{p_1} = \left[\frac{2}{k+1}\right]^{k/(k-1)} \tag{5-14}$$

This of course is the same result obtained from the convergent nozzle discharging from a large tank.

When an incompressible fluid flows without friction through a reducer, the mass flow rate $\dot{M}_i$ may be obtained by writing the Bernoulli equation (4-16) for a horizontal reducer:

$$\frac{V_2^2 - V_1^2}{2g_c} + v_1(p_2 - p_1) = 0 \tag{5-20}$$

Substituting

$$\dot{M}_i = \frac{V_1 A_1}{v_1} = \frac{V_2 A_2}{v_2}$$

from Eq. (5-12), noting that for an incompressible fluid $v_2 = v_1$,

$$\frac{(\dot{M}_i v_1/A_2)^2 - (\dot{M}_i v_1/A_1)^2}{2g_c} + v_1(p_2 - p_1) = 0$$

which reduces to

$$\dot{M}_i = A_2 \sqrt{\frac{2g_c(p_1 - p_2)/v_1}{1 - (A_2/A_1)^2}} \tag{5-21}$$

The *expansion factor* Y is defined as

$$Y = \frac{\dot{M}}{\dot{M}_i} \tag{5-22}$$

Substituting from Eq. (5-18) for $\dot{M}$ and from Eq. (5-21) for $\dot{M}_i$ in Eq. (5-22) and simplifying results in

$$Y = \frac{\dot{M}}{\dot{M}_i} = \sqrt{\frac{k(p_2/p_1)^{2/k}[1-(p_2/p_1)^{k-1}][1-(A_1/A_2)^2]}{(k-1)[1-(p_2/p_1)][1-(A_2/A_1)^2(p_2/p_1)^{2/k}]}} \tag{5-23}$$

Values of the expansion factor Y are given in Table A-15. In this table

$$r = \frac{p_2}{p_1} \tag{5-24}$$

$$\beta = \frac{D_2}{D_1}, \qquad \beta^4 = \left(\frac{D_2}{D_1}\right)^4 = \left(\frac{A_2}{A_1}\right)^2 \tag{5-25}$$

and r_c is the critical pressure ratio as defined by Eq. (5-19).

The use of the expansion factor from Table A-15 facilitates computation. An expression for the compressible flow rate may be obtained by substituting Eq. (5-21) for $\dot{M}_i$ in Eq. (5-22), and solving for $\dot{M}$,

$$\dot{M} = Y\dot{M}_i = YA_2\sqrt{\frac{2g_c(p_1-p_2)}{v_1[1-(A_2/A_1)^2]}} \tag{5-26}$$

Noting from Eq. (5-25) that $\beta^4 = (A_2/A_1)^2$, Eq. (5-26) may be written as

$$\dot{M} = YA_2\sqrt{\frac{2g_c(p_1-p_2)}{v_1(1-\beta^4)}} \tag{5-27}$$

EXAMPLE Air flows through a pipe 50 mm in diameter that reduces to a diameter of 25 mm. In the 50-mm diameter portion, the temperature is 40°C and the air pressure is 800 kPa. Determine (a) the minimum pressure that can exist in the 25-mm pipe, and the maximum mass flow rate, and (b) the mass flow rate if the air pressure in the 25-mm pipe is 650 kPa.

Step A *Data Reduction*

(1) *unit conversions*

(B-27) $T_1 = 40 + 273.15 = 313.15$ K

(2) *Proportionality constant*

(1-9) $g_c = 1$ kg · m/N · s^2

(3) *fluid properties,* air

Table A-7 $k = 1.400 \quad R = 287.1 \quad J/(kg \cdot K)$

(1-42) $$v_1 = \frac{RT_1}{p_1} = \frac{287.1 \times 313.15}{800\ 000} = 0.112\ 4\ m^3/kg$$

(4) *geometric,* circle

Table A-10 $$A_2 = \frac{\pi D_2^2}{4} = \frac{\pi(25 \times 10^{-3})^2}{4} = 4.909 \times 10^{-4}\ m^2$$

(5-25) $$\beta = \frac{D_2}{D_1} = \frac{25}{50} = 0.5$$

(5) *expansion factor*

$\beta = 0.5,\ k = 1.400$

Table A-15 $r_a = r_c = 0.5362$

Table A-15 $Y_a = Y_c = 0.6973$

(5-24) $$r_b = \frac{p_{2b}}{p_1} = \frac{650\ 000}{800\ 000} = 0.8125$$

Table A-15 $Y_b = 0.8782$ (by linear interpolation)

Step B *Determine Minimum Pressure in 25-mm Pipe*

(5-24) $$p_c = p_1 r_c = 800\ 000 \times 0.5362 = 428\ 960\ N/m^3$$
$$= 429.0\ kPa$$

Step C *Determine Maximum Mass Flow Rate*

(5-27) $$\dot{M}_a = YA_2 \sqrt{\frac{2g_c(p_1 - p_2)}{v_1(1-\beta^4)}} = \dot{M}_c = Y_c A_2 \sqrt{\frac{2g_c(p_1 - p_c)}{v_1(1-\beta^4)}}$$

$$= 0.6973 \times 4.909 \times 10^{-4} \sqrt{\frac{2 \times 1 \times (800\ 000 - 428\ 960)}{0.112\ 4(1 - 0.5^4)}}$$

$$\dot{M}_a = 0.908\ 4\ kg/s = 908.4\ g/s$$

Step D *Determine Mass Flow Rate for $p_2 = 650$ kPa*

(5-27) $$\dot{M}_b = Y_b A_2 \sqrt{\frac{2g_c(p_1 - p_{2b})}{v_1(1-\beta^4)}}$$

$$= 0.8782 \times 4.909 \times 10^{-4} \sqrt{\frac{2 \times 1(800\ 000 - 650\ 000)}{0.112\ 4(1 - 0.5^4)}}$$

$$\dot{M}_b = 0.727\ 4\ g/s = 72.74\ g/s$$

5-6 CONVERGENT-DIVERGENT NOZZLES

The mass flow rate through any section of the convergent-divergent nozzle shown in Fig. 5-3 may be determined by modifying Eq. (5-18) for stagnation inlet conditions ($A/A_1 \rightarrow 0$):

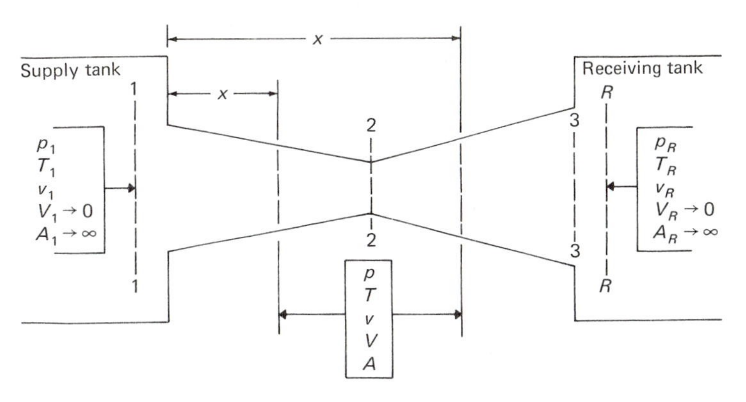

FIGURE 5-3 Notation for convergent-divergent nozzle study

$$\dot{M} = A\sqrt{\frac{2g_c k}{k-1}\left(\frac{p_1}{v_1}\right)\left(\frac{p}{p_1}\right)^{2/k}[1-(p/p_1)^{(k-1)/k}]} \tag{5-28}$$

The area-pressure relations may be established by squaring Eq. (5-28) and equating for sections 2 and 3:

$$\frac{\dot{M}^2}{\dot{M}^2} = \frac{A_2^2[2g_c k/(k-1)]\,(p_2/p_1)^{2/k}[1-(p_2/p_1)^{(k-1)/k}]}{A_3^2[2g_c k/(k-1)]\,(p_3/p_1)^{2/k}[1-(p_3/p_1)^{(k-1)/k}]}$$

This reduces to

$$\left(\frac{p_3}{p_1}\right)^{2/k} - \left(\frac{p_3}{p_1}\right)^{(k+1)/k} = \left(\frac{A_2}{A_3}\right)^2\left[\left(\frac{p_2}{p_1}\right)^{2/k} - \left(\frac{p_2}{p_1}\right)^{(k+1)/k}\right] \tag{5-29}$$

If the velocity in the throat is sonic and critical flow takes place, then from Eq. (5-14):

$$\frac{p_c}{p_1} = \frac{p_2}{p_1} = \left(\frac{2}{k+1}\right)^{k/(k-1)} \tag{5-14}$$

Substituting in Eq. (5-29),

$$\begin{aligned}\left(\frac{p_3}{p_1}\right)^{2/k} - \left(\frac{p_3}{p_1}\right)^{(k+1)/k} &= \left(\frac{A_2}{A_3}\right)^2\left[\left(\frac{2}{k+1}\right)^{2/k[k/(k-1)]} - \left(\frac{2}{k+1}\right)^{[(k+1)/k][k/(k-1)]}\right] \\ &= \left(\frac{A_2}{A_3}\right)^2\left[\left(\frac{2}{k+1}\right)^{2/(k-1)} - \left(\frac{2}{k+1}\right)^{(k+1)/(k-1)}\right]\end{aligned} \tag{5-30}$$

Note that Eq. (5-30) has *two* solutions.

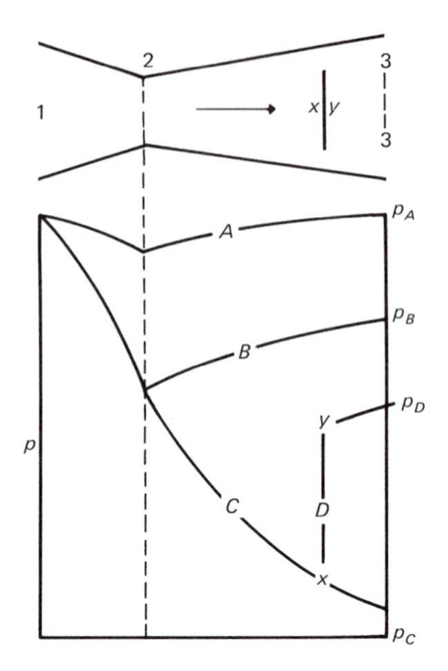

FIGURE 5-4 Pressures in a convergent-divergent nozzle

Figure 5-4 shows the various pressure paths that can exist in a convergent-divergent nozzle. If the flow in the throat is subsonic, then the flow through the entire nozzle must be subsonic from Sec. 5-2. Path *A* is the subsonic path. If the flow in the throat is sonic, then from Eq. (5-30) there are two paths the gas can follow: (1) an isentropic compression (path B), or an isentropic expansion (path C), depending upon the receiver pressure. If $p_R > p_A$, then the flow will be subsonic. If $p_R = p_B$, then the flow will be subsonic in the divergent section. If $p_R = p_C$, then the flow will be supersonic in the divergent section. If $p_B > p_R < p_C$, then a compression shock wave (Sec. 5-7) will be formed. The shock wave will follow path *D*. The flow before the shock wave is supersonic and after the wave subsonic, and the pressure rises from p_x to p_y. If $p_c > p_R$ then path *C* will be followed.

EXAMPLE Air discharges from a large tank in which the pressure is 115 psia, and the temperature is 100°F through a convergent-divergent nozzle whose throat diameter is 1 in. into a large receiving tank whose pressure is maintained at 15 psia. Calculate (a) the nozzle exit area for full isentropic expansion and (b) the exit temperature, velocity, acoustic velocity, Mach number, and mass rate of flow.

Step A	*Data Reduction*
	(1) *fluid properties,* air
Table A-7	$k = 1.400,\ c_p = 0.2400$ Btu/lbm-°R, $R = 53.36$ ft-lbf/lbm-°R
	(2) *unit conversions*
(B-28)	$T_1 = 100 + 459.67 = 559.67°\text{R}$
(B-89)	$p_1 = 115 \times 144 = 16{,}560 \text{ lbf/ft}^2$
(B-89)	$p_3 = 15 \times 144 = 2{,}160 \text{ lbf/ft}^2$
(B-55)	$D_1 = 1/12$ ft
(B-15)	$c_p = 0.2400 \times 778.2 = 186.77$ ft-lbf/lbm-°R
	(3) *proportionality constant*
(1-11)	$g_c = 32.17 \text{ lbm-ft/lbf-sec}^2$
	(4) *geometric,* circle
Table A-10	$A_1 = \dfrac{\pi D_1^2}{4} = \dfrac{\pi(1/12)^2}{4} = 5.454 \times 10^{-3}$ ft

Step B	*Determine Exit Area*
(5-14)	$\dfrac{p_c}{p_1} = \left(\dfrac{2}{k+1}\right)^{k/(k-1)} = \left(\dfrac{2}{1.400+1}\right)^{1.400/(1.400-1)}$
	$\dfrac{p_c}{p_1} = 0.5283$
	$\dfrac{p_3}{p_1} = \dfrac{2{,}160}{16{,}560} = 0.1304$
	$0.1304 < 0.5283$ ∴ throat velocity is sonic and Eq. (5-30) may be applied.
(5-30)	$A_3 = A_2 \sqrt{\dfrac{[2/(k+1)]^{2/(k-1)} - [2/(k+1)]^{(k+1)/(k-1)}}{(p_3/p_1)^{2/k} - (p_3/p_1)^{(k+1)/k}}}$
	$A_3 = 5.454 \times 10^{-3} \sqrt{\dfrac{\left[\dfrac{2}{1.400+1}\right]^{2/(1.400-1)} - \left[\dfrac{2}{1.400+1}\right]^{(1.400+1)/(1.400-1)}}{\left[\dfrac{2{,}160}{16{,}560}\right]^{2/1.400} - \left[\dfrac{2{,}160}{16{,}560}\right]^{(1.400+1)/1.400}}}$
	$A_3 = 9.104 \times 10^{-3} \text{ ft}^2$
Table A-10	$D_3 = \sqrt{\dfrac{4A_3}{\pi}} = \sqrt{\dfrac{4 \times 9.104 \times 10^{-3}}{\pi}}$
(B-55)	$D_3 = 0.1077 \text{ ft} = 0.1077 \times 12 = 1.292$ in.

Step C *Compute Nozzle Exit Conditions*

(1-47) $$T_3 = T_1\left(\frac{p_3}{p_1}\right)^{(k-1)/k} = 559.67(2{,}160/16{,}560)^{(1.400-1)/1.400}$$

$$T_3 = 312.74°\text{R}$$

(4-52) $$h_1 - h_3 = c_p(T_1 - T_3)$$

$$= 186.77(559.67 - 312.74)$$

$$= 46{,}119 \text{ ft-lbf/lbm}$$

(5-11) $$V_3 = \sqrt{2g_c(h_1 - h_3) + V_1^2} = \sqrt{2 \times 32.17 \times 46{,}119 + 0}$$

$$V_3 = 1{,}723 \text{ ft/sec}$$

(1-60) $$c_3 = \sqrt{kg_cRT_3} = \sqrt{1.400 \times 32.17 \times 53.36 \times 312.74}$$

$$c_3 = 866.9 \text{ ft/sec}$$

(5-9) $$\mathbf{M}_3 = V_3/c_3 = 1{,}723/866.9 = 1.987$$

(1-42) $$v_3 = RT_3/p_3 = 53.36 \times 312.74/2{,}160$$

$$v_3 = 7.726 \text{ ft}^3\text{/lbm}$$

(5-12) $$\dot{M} = \frac{V_3A_3}{v_3} = \frac{1{,}723 \times 9.104 \times 10^{-3}}{7.726}$$

$$= 2.030 \text{ lbm/sec}$$

(Note that this checks the value of 2.030 lbm/sec obtained with the convergent nozzle example of Sec. 5-4.)

5-7 COMPRESSION SHOCK WAVE

When sonic flow exists in the throat and supersonic flow in the diverging section of a convergent-divergent nozzle, a compression shock wave will be formed if the requirements for the conservation of mass and energy are not satisfied. This type of wave is associated with large and sudden rises in pressure, density, temperature, and entropy. The shock wave is so thin that for computation purposes it may be considered as a single line, as shown in Figs. 5-4 and 5-5.

Conservation of Energy

The formation of a shock wave does not change the total energy of the system, so that energy relations may be established by Eq. (5-11) and substituting from Eq. (4-52),

(4-52) $$c_p(T_x - T_y) = h_x - h_y$$

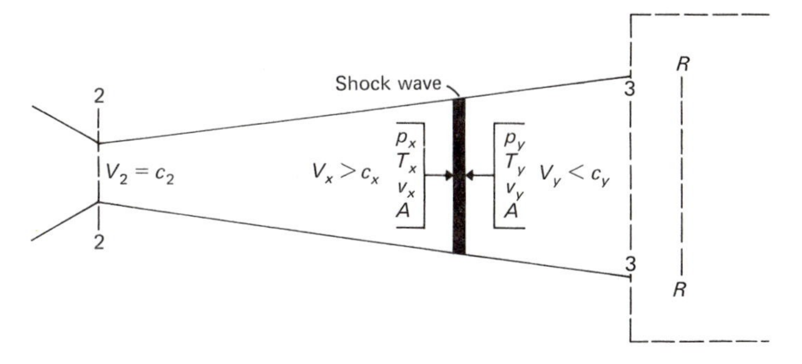

FIGURE 5-5 Notation for compression-shock-wave study

and from Eq. (4-56),

(4-56)
$$c_p = \frac{Rk}{k-1}$$

to obtain

$$\frac{V_y^2}{2g_c} - \frac{V_x^2}{2g_c} = h_x - h_y = c_p(T_x - T_y) = \frac{Rk}{k-1}(T_x - T_y)$$

or

$$V_y^2 - V_x^2 = \frac{2g_c - Rk}{k-1}(T_x - T_y)$$

From Eq. (1-60) $c^2 = kg_cRT$, so that

$$V_y^2 - V_x^2 = \frac{2}{k-1}(c_x^2 - c_y^2)$$

and by definition $\mathbf{M} = V/c$, Eq. (5-9),

$$\frac{V_y^2 + 2c_y^2}{k-1} = \frac{V_x^2 + 2c_x^2}{k-1}$$

or

$$c_y^2[\mathbf{M}_y^2 + 2/(k-1)] = c_x^2\mathbf{M}_x^2 + 2/(k-1)c_y^2/c_x^2 = \frac{kg_cRT_y}{kg_cRT_x}$$

or

$$\frac{T_y}{T_x} = \frac{2 + (k-1)\mathbf{M}_x^2}{2 + (k-1)\mathbf{M}_y^2} \tag{5-31}$$

Conservation of Mass

The continuity equation (5-12) $\dot{M} = AV/v$ may be written in terms of Mach number by noting that the definition of Mach number is $\mathbf{M} = V/c$ and from Eq. (1-60) $c = \sqrt{kg_cRT}$ and from the equation of state (1-42) $v = RT/p$, as follows:

$$\dot{M}/A = \frac{V}{v} = \frac{\mathbf{M}c}{v} = \mathbf{M}\sqrt{kg_cRT/v^2} = \mathbf{M}\sqrt{kg_cRT/(RT/p)^2}$$

$$= \mathbf{M}p\sqrt{kg_c/RT} \tag{5-32}$$

For a constant area across the shock wave, Eq. (5-32) becomes

$$\dot{M}/A = \mathbf{M}_y p_y \sqrt{kg_cT_y} = \mathbf{M}_x p_x \sqrt{kg_cT_x}$$

or

$$\frac{p_y}{p_x} = \frac{\mathbf{M}_x}{\mathbf{M}_y}\sqrt{\frac{T_x}{T_y}} \tag{5-33}$$

Impulse-Momentum Concept

The impulse-momentum equation (4-66) when applied to the shock wave of Fig. 5-5 yields

$$(p_x - p_y)A = [(AV/vg_c)](V_y - V_x) \tag{4-66}$$

or

$$p_y + \frac{V_y^2}{g_cv_y} = p_x + \frac{V_x^2}{g_cv_x}$$

Substituting in the equation above the definition of Mach number $V = \mathbf{M}/c$ (5-9) and from Eq. (1-60):

$$c^2 = kg_cRT \tag{1-60}$$

and from the equation of state (1-42):

$$v = \frac{RT}{p} \tag{1-42}$$

we have

$$p_y + \frac{\mathbf{M}_y^2 kg_cRT_y}{g_cRT_y/p_y} = p_x + \frac{\mathbf{M}_x^2 kg_cRT_x}{g_cRT_x/p_x}$$

or

$$p_y\,(1 + k\,\mathbf{M}_y^2) = p_x\,(1 + k\,\mathbf{M}_x^2)$$

or

$$\frac{p_y}{p_x} = \frac{1 + k\,\mathbf{M}_x^2}{1 + k\,\mathbf{M}_y^2} \tag{5-34}$$

Equations (5-31), (5-33), and (5-34) involve three unknowns: T_y, p_y, and $\mathbf{M}_y$ and may be combined to yield

$$\mathbf{M}_y = \sqrt{\frac{\mathbf{M}_x^2(k-1) + 2}{2k\mathbf{M}_x^2 - k+1}} \tag{5-35}$$

EXAMPLE Air flows from a large tank at 40°C and 800 kPa through a convergent-divergent nozzle into a large receiving tank. The nozzle throat diameter is 25 mm, and the exit diameter is 32.3 mm. A compression shock wave is formed in the divergent cone of the nozzle. Just before this wave, the pressure is 139 kPa. Compute (a) the diameter where the shock wave forms, (b) conditions just before the wave, (c) conditions just after the wave, and (d) exit conditions.

Step A *Data Reduction*

(1) *unit conversions*

(B-27) $T_1 = 40 + 273.15 = 313.15\ \text{K}$

(2) *proportionality constant*

(1-9) $g_c = 1\ \text{kg} \cdot \text{m/N} \cdot \text{s}^2$

(3) *fluid properties,* air

Table A-7 $k = 1.400,\ R = 287.1\ \text{J/kg} \cdot \text{K},\ c_p = 1\,005\ \text{J/kg} \cdot \text{K}$

(4) *geometric,* circle

Table A-10 $A_2 = \dfrac{\pi D_2^2}{4} = \dfrac{\pi(25 \times 10^{-3})^2}{4} = 4.909 \times 10^{-4}\ \text{m}^2$

Table A-10 $A_3 = \dfrac{\pi D_3^2}{4} = \dfrac{\pi(32.3 \times 10^{-3})^2}{4} = 8.194 \times 10^{-4}\ \text{m}^2$

Step B *Determine the Diameter at the Shock Wave*

(5-14) $\dfrac{p_2}{p_1} = \left(\dfrac{2}{k+1}\right)^{k/(k-1)} = \left(\dfrac{2}{1.400+1}\right)^{1.400/(1.400-1)} = 0.5283$

$$\frac{p_x}{p_1} = \frac{139}{800} = 0.1738$$

0.1738 < 0.5283 ∴ throat velocity is sonic and Eq. (5-30) may be applied:

(5-30) $$A_x = A_2 \sqrt{\frac{[2/(k+1)]^{2/(k-1)} - [2/(k+1)]^{(k+1)/(k-1)}}{(p_x/p_1)^{2/k} - (p_x/p_1)^{(k+1)/k}}}$$

$$A_x = 4.909 \times 10^{-4} \sqrt{\frac{[2/(1.400+1)]^{2/(1.400-1)} - [2/(1.400+1)]^{(1.400+1)/(1.400-1)}}{(139/800)^{2/1.400} - (139/800)^{(1.400+1)/1.400}}}$$

$$A_x = 7.070 \times 10^{-4} \text{ m}^2$$

Table A-10 $$D_x = \sqrt{4A_x/\pi} = \sqrt{4 \times 7.070 \times 10^{-4}/\pi}$$

$$D_x = 0.030\,00 \text{ m} = 30.00 \text{ mm}$$

Step C *Compute Conditions Just Before Shock Wave*

(1-47) $$T_x = T_1\left(\frac{p_x}{p_1}\right)^{(k-1)/k} = 313.15\left(\frac{139}{800}\right)^{(1.400-1)/1.400}$$

$$T_x = 189.93 \text{ K}$$

(4-52) $$h_1 - h_x = c_p(T_1 - T_x) = 1\,005(313.15 - 189.93)$$

$$= 123\,836 \text{ J/kg}$$

(5-11) $$V_x = \sqrt{2g_c(h_1 - h_x) + V_1^2} = \sqrt{2 \times 1 \times 123\,836 + 0}$$

$$V_x = 497.7 \text{ m/s}$$

(1-60) $$c_x = \sqrt{kg_cRT_x} = \sqrt{1.400 \times 287.1 \times 189.93}$$

$$c_x = 276.3 \text{ m/s}$$

(5-9) $$\mathbf{M}_x = \frac{V_x}{c_x} = \frac{497.7}{276.3} = 1.801$$

(1-42) $$v_x = \frac{RT_x}{p_x} = \frac{287.1 \times 189.93}{139\,000}$$

$$v_x = 0.392\,3 \text{ m}^3\text{/kg}$$

(5-12) $$\dot{M} = \frac{V_xA_x}{v_x} = \frac{497.7 \times 7.070 \times 10^{-4}}{0.392\,3}$$

$$\dot{M} = 0.897\,0 \text{ kg/s}$$

Step D *Compute Conditions Just After Shock Wave*

(5-35) $$\mathbf{M}_y = \sqrt{\frac{\mathbf{M}_x^2(k-1) + 2}{2k\,\mathbf{M}_x^2 - k+1}}$$

$$= \sqrt{\frac{(1.801)^2(1.400-1) + 2}{2 \times 1.400(1.801)^2 - 1.400+1}} = 0.6163$$

(5-31) $$T_y = \frac{T_x[2 + (k-1)\mathbf{M}_x^2]}{[2 + (k-1)\mathbf{M}_y^2]}$$

$$T_y = \frac{189.93[2 + (1.400-1)(1.801)^2]}{[2 + (1.400-1)(0.6163)^2]}$$

$$T_y = 291.03 \text{ K}$$

(5-33) $$p_y = p_x\left(\frac{\mathbf{M}_x}{\mathbf{M}_y}\right)\sqrt{\frac{T_y}{T_x}}$$

$$p_y = 139\,000\,(1.801/0.6163)\,\sqrt{291.03/189.93}$$

$$p_y = 502\,815 \text{ N/m}^2 = 502.8 \text{ kPa}$$

(1-60) $$c_y = \sqrt{kg_cRT_y} = \sqrt{1.400 \times 1 \times 287.1 \times 291.03}$$

$$c_y = 342.0 \text{ m/s}$$

(5-9) $$V_y = c_y\mathbf{M}_y = 342.0 \times 0.6163 = 210.8 \text{ m/s}$$

(1-42) $$v_y = RT_y/p_y = 287.1 \times 291.03/502\,815$$

$$v_y = 0.166\,2 \text{ m}^3\text{/kg}$$

(5-12) $$\dot{M} = V_yA_y/v_y = 210.8 \times 7.070 \times 10^{-4}/0.166\,2$$

$$\dot{M} = 0.896\,7 \text{ kg/s versus } 0.897\,0 \text{ from Step C}$$

Step E *Compute Exit Conditions*

(5-11) $$\frac{V_y^2 - V_3^2}{2g_c} = h_3 - h_y$$

(4-52) $$h_3 - h_y = c_p\,(T_3 - T_y)$$

(5-12) $$\dot{M} = V_yA_y/v_y = V_3A_3/v_3$$

or

$$V_3 = V_y\,(A_y/A_3)\,(v_3/v_y) \qquad \text{(a)}$$

(1-46) $$v_3/v_y = (T_y/T_3)^{1/(k-1)}$$

Substituting Eqs. (4-52), (5-12), and (1-46) in Eq. (5-11),

$$V_3^2 = V_y^2\left(\frac{A_y}{A_3}\right)^2\left(\frac{T_y}{T_3}\right)^{2/(k-1)} = 2g_cc_p(T_3 - T_y)$$

which may be written as

$$T_3 = T_y + \left(\frac{V_y^2}{2g_cc_p}\right)\left[1 - \left(\frac{A_y}{A_3}\right)^2\left(\frac{T_y}{T_3}\right)^{2/(k-1)}\right]$$

$$T_3 = 291.03 + \left(\frac{210.8^2}{2 \times 1 \times 1\,005}\right)\left[1 - \left(\frac{7.070 \times 10^{-4}}{8.194 \times 10^{-4}}\right)^2\left(\frac{291.03}{T_3}\right)^{2/(1.400-1)}\right]$$

$$T_3 = 291.03 + 22.108\left[1 - 0.74447\left(\frac{291.03}{T_3}\right)^5\right]$$

$T_3 = 298.60$ K (trial and error solution)

(1-45) $$p_3 = p_y\left(\frac{T_3}{T_y}\right)^{k/(k-1)}$$

$$p_3 = 502\,815\left(\frac{298.60}{291.03}\right)^{1.400/(1.400-1)}$$

(B-76) $$p_3 = 550\,098 \text{ N/m}^2 = 550.1 \text{ kPa}$$

(1-42) $$v_3 = RT_3/p_c = 287.1 \times 298.60/550\,098$$

$$v_3 = 0.155\,8 \text{ m}^3\text{/kg}$$

(1-60) $$c_3 = \sqrt{kg_cRT_3} = \sqrt{1.400 \times 1 \times 287.1 \times 298.60}$$

$$c_3 = 346.4 \text{ m/s}$$

$$V_3 = V_y\left(\frac{A_y}{A_3}\right)\left(\frac{v_3}{v_y}\right) \qquad \text{(a)}$$

$$= 210.8(7.070 \times 10^{-4}/8.194 \times 10^{-4})\ (0.155\,8/0.166\,2)$$

$$= 170.5 \text{ m/s}$$

(5-9) $$\mathbf{M}_3 = \frac{V_3}{c_3} = \frac{170.5}{346.4} = 0.4922$$

(5-12) $$\dot{M} = \frac{V_3A_3}{v_3} = \frac{170.5 \times 8.194 \times 10^{-4}}{0.155\,8}$$

$$= 0.896\,7 \text{ kg/s} \quad \text{(check Step D)}$$

PROBLEMS

1. Air discharges from a large tank in which the pressure and temperature are maintained at 800 kPa and 40°C respectively, through a horizontal convergent nozzle whose throat diameter is 25 mm and into a large receiving tank. Compute the temperature, pressure, velocity, Mach number, and mass flow rate of the nozzle jet when the pressure in the receiving tank is (a) 300 kPa and (b) 655 kPa.
2. Make the same computation as in Problem 1 except that the gas is methane.
3. Carbon dioxide discharges from a large tank at a rate of 2 lbm/sec through a horizontal convergent nozzle discharging into the atmosphere. The supply tank temperature and pressure are maintained at 200°F and 100 psia respectively. Compute the diameter of the nozzle.
4. Make the same computation as in Problem 3 except that the gas is carbon monoxide.
5. Air flows through a horizontal tube 2 in. in diameter that reduces to a diameter of 1 in. In the 2-in. portion, the temperature is 100°F and the air pressure is 115 psia. Compute (a) the minimum pressure that can exist in the 1-in. tube and the maximum mass flow rate, and (b) the mass flow rate if the air pressure in the one inch portion is 95 psia.
6. Same as problem 5 except that the gas is methane.
7. Air flows steadily and isentropically through a horizontal piping system. The mass flow rate of the air is 1 kg/s. At section 1, the diameter is 50 mm, the pressure 345 kPa, and the temperature 32°C. At section 2, the diameter is 75 mm. Compute (a) the velocity, (b) the temperature and (c) the pressure at section 2.

8. Same as problem 7 except that the gas is helium.
9. An ideal gas ($k = 1.5$, $c_p = 0.3$ Btu/lbm-°R) is to flow from a large tank through a convergent-divergent nozzle into another large tank. In the supply tank, the gas pressure is 100 psia, and the temperature is 100°F. The gas is to flow at a rate of 2 lbm/sec for a throat Mach number of unity. Design a nozzle and compute exit conditions for (a) isentropic expansion to $\mathbf{M}_3 = 2$, and (b) isentropic compression to $\mathbf{M}_3 = 1/2$.
10. Same as problem 9 except that the gas is helium.
11. Air discharges into the atmosphere from a large supply tank in which the pressure is 800 kPa and the temperature is 40°C through a horizontal convergent-divergent nozzle whose throat diameter is 25 mm. Calculate (a) the nozzle exit area for full isentropic expansion and (b) the exit temperature, velocity, acoustic velocity, Mach number, and mass rate of flow.
12. Make the same computation as in Problem 11 except that the gas is nitrogen.
13. A compression shock wave exists normal to an air flow. Just upstream of the shock wave, the pressure is 200 kPa, the temperature 0°C, and the Mach number 2. Compute the conditions just downstream of the shock wave.
14. Make the same computation as in Problem 13 except that the gas is methane.
15. A compression shock wave exists normal to an air flow. Just downstream of the shock wave, the pressure is 135 psia, the temperature 320°F, and the Mach number 1/2. Compute conditions just upstream of the shock wave.
16. Make the same computation as in Problem 15 except that the gas is helium.

REFERENCES

Benedict, R. P. and W. G. Steltz, *Handbook of Generalized Gas Dynamics.* Plenum Press, Data Division, New York, 1966.

Cambel, A. B. and B. H. Jennings, *Gas Dynamics.* Dover Publications, New York, 1946.

Keenan, J. H. and J. Kaye, *Gas Tables.* John Wiley and Sons, New York, 1946.

Rotty, R. M., *Introduction to Gas Dynamics.* John Wiley and Sons, New York, 1962.

Shapiro, A. *The Dynamics and Thermodynamics of Compressible Fluid Flow,* vols. I and II. Ronald Press Company, New York, 1953.

6 Dimensionless Parameters

6-1 INTRODUCTION

Modern fluid mechanics is based on a combination of theoretical analysis and experimental data. Very often, the engineer is faced with the necessity of obtaining dependable, practical results in situations where for various reasons the flow phenomena cannot be described mathematically and experimental data must be considered. The generation and use of dimensionless parameters provide a powerful and useful tool in (1) reducing the number of variables required for an experimental program, (2) establishing the principles of model design and testing, (3) developing equations, and (4) converting data from one system of units to another. Dimensionless parameters may be generated from (1) *physical equations,* (2) the principles of *similarity,* and (3) *dimensional analysis.*

All physical equations should be dimensionally correct so that a dimensionless parameter may be generated by simply dividing one side of the equation by the other, as will be illustrated. The principles of similarity are used to develop dimensionless parameters for model-prototype relations to insure geometric, kinematic, and dynamic similarity by consideration of dimensions, velocities, and forces involved between the two. *Dimensional analysis* is the mathematics of dimensions and quantities. Two formal methods are used, the Lord Rayleigh's and the Buckingham Π theorem. Lord Rayleigh (1842–1919), who was born John William Strut in Essex, England, popularized the principle of dynamic similarity by introducing in 1899 a generalization of the principle. Edgar Buckingham (1867–1940) was a physicist at the National Bureau of Standards.

In a series of papers published in 1914 and 1915 he brought to American notice the uses of dimensional analysis and presented his Π theorem.

6-2 PHYSICAL EQUATIONS

Good engineering practice demands that all physical equations be dimensionally consistent. All terms in an equation must have the same dimensions. Dissimilar quantities cannot be added or subtracted when forming a true physical equation. For example:

$$\text{coffee} + \text{eggs} + \text{bacon} + \text{toast} = \text{breakfast}$$

may be true, but this is not the type of relationship being considered.

Dimensionless parameters may be derived by simply dividing one side of any physical equation by the other. A minimum of two dimensionless parameters will be formed, one being the inverse of the other.

EXAMPLE Velocity of Sound

$$c = \sqrt{E_s/\rho} \tag{1-59}$$

$$N_1 = \frac{c}{\sqrt{E_{s/\rho}}}, \quad N_2 = \frac{\sqrt{E_s/\rho}}{c} = N_1^{-1}$$

Both N_1 and N_2 are velocity ratios.

6-3 MODELS VS. PROTOTYPES

There are many times when for economic or other reasons it is desirable to determine the performance of a structure or machine by testing another structure or machine. This type of testing is called model testing. The structure or machine being tested is called the *model* and the structure whose performance is to be predicted is called the *prototype*. A *model* may be smaller than, the same size as, or larger than the *prototype*.

Model experiments on airplanes, rockets, missiles, pipes, ships, canals, turbines, pumps, and other structures and machines have resulted in savings that more than justified the expenditure of funds for the design, construction, and testing of the model. Under some situations, the model and prototype may be the same piece of equipment, for example, the calibration of a flow meter with water in a laboratory to predict its performance when metering steam. Many manufacturers of fluid machinery have test equipment that is limited to one or two fluids and are forced to test with what they have available in order to predict performance under other conditions.

6-4 GEOMETRIC SIMILARITY

Geometric similarity exists between model and prototype when the ratios of all corresponding dimensions in the model and prototype are equal (see Fig. 6-1). These ratios may be written

Length
$$\frac{L_{\text{model}}}{L_{\text{prototype}}} = L_{\text{ratio}} = \frac{L_m}{L_p} = L_r \tag{6-1}$$

Area
$$\frac{L^2_{\text{model}}}{L^2_{\text{prototype}}} = L^2_{\text{ratio}} = \frac{L^2_m}{L^2_p} = L^2_r \tag{6-2}$$

EXAMPLE It is desired to use a smaller pipe as a model for a standard 30-in. pipe. Available are sections of 3-in. schedule 40 wrought iron pipe, 10-in. 250 psi cast iron pipe, and 1-in. type K seamless copper water tube. For geometric similarity, which section should be used as the model? What should be the length of the model if the prototype length is 100 ft?

Step A *Data Reduction*

Pipe properties

(1) *wrought iron pipe*

Table A-11 $\epsilon = 150 \times 10^{-6}$ ft (note 5)

Table A-11 3-in. size schedule 40, $D_m = 0.2557$ ft

Table A-11 30-in. size standard, $D_p = 2.438$ ft

(2) *cast iron pipe,* 250 psi

Table A-12 $\epsilon = 850 \times 10^{-6}$ ft (note 3)

Table A-12 10-in. size, $D_m = 0.8517$ ft

(3) *seamless copper water tubing,* Type K

Table A-13 $\epsilon = 5 \times 10^{-6}$ ft (note 4)

Table A-13 1 in. size, $D_m = 0.08292$ ft

Step B *Similarity for Wrought Iron Model*

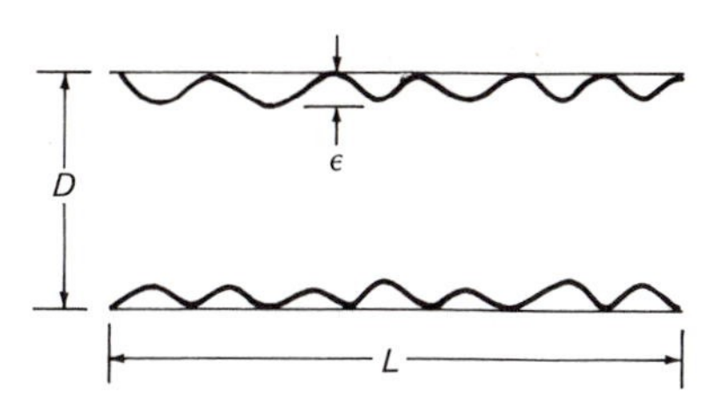

FIGURE 6-1 Notation for geometric similarity

Diameter ratio

(6-1) $$L_{rD} = \frac{D_m}{D_p} = \frac{0.2557}{2.438} = 0.1049$$

Surface roughness ratio

(6-1) $$L_{r\epsilon} = \frac{\epsilon_m}{\epsilon_p} = \frac{150 \times 10^{-6}}{150 \times 10^{-6}} = 1$$

Complete similarity

$L_{rD} = L_{r\epsilon}$

$0.1049 \neq 1$

The only way to attain complete similarity would be to machine the model to a surface roughness of

$0.1049 \times 150 \times 10^{-6} = 15.74 \times 10^{-6}$ ft

Step C *Similarity for Cast Iron Model*

Diameter ratio

(6-1) $$L_{rD} = \frac{D_m}{D_p} = \frac{0.8517}{2.438} = 0.3493$$

Surface roughness ratio

(6-1) $$L_{r\epsilon} = \frac{\epsilon_m}{\epsilon_p} = \frac{850 \times 10^{-6}}{150 \times 10^{-6}} = 5.667$$

Complete similarity

$L_{rD} = L_{r\epsilon} = 5.667$

$0.3493 \neq 5.667$ The only way to attain complete similarity would be to machine the model to a surface roughness of

$0.3493 \times 150 \times 10^{-6} = 52.4 \times 10^{-6}$ ft

Step D *Similarity for Copper Tubing*

Diameter ratio

(6-1) $$L_{rD} = \frac{D_m}{D_p} = \frac{0.08292}{2.438} = 0.03401$$

Surface roughness ratio

$$L_{r\epsilon} = \frac{\epsilon_m}{\epsilon_p} = \frac{5 \times 10^{-6}}{150 \times 10^{-6}} = 0.03333 \quad \text{(6-1)}$$

Complete similarity

$L_{rD} = L_{r\epsilon}$, 0.03401 $\approx$ 0.03333, best of three

Length required

$$L_r = \frac{L_m}{L_p}, \quad \text{(6-1)}$$

$$L_m = L_r L_p = 0.03401 \times 100 = 3.4 \text{ ft}$$

6-5 KINEMATIC SIMILARITY

Kinematic similarity exists between model and prototype when their streamlines are geometrically similar. Comparison of the velocity profiles of Fig. 6-2 (a) with (b) and (c) indicates that (a) and (b) have kinematic similarity, but (a) and (c) and (b) and (c) do not.

Some of the more common kinematic similarity ratios are

Acceleration $$a_r = \frac{a_m}{a_p} = \frac{L_m T_m^{-2}}{L_p T_p^{-2}} = L_r T_r^{-2} \quad \text{(6-3)}$$

Velocity $$V_r = \frac{V_m}{V_p} = \frac{L_m T_m^{-1}}{L_p T_p^{-1}} = L_r T_r^{-1} \quad \text{(6-4)}$$

Volume flow rate $$Q_r = \frac{Q_m}{Q_p} = \frac{L_m^3 T_m^{-1}}{L_p^3 T_p^{-1}} = L_r^3 T_r^{-1} \quad \text{(6-5)}$$

EXAMPLE Ethyl alcohol at 10°C is to flow in a tube with a 300-mm inside diameter and with an average velocity of 15 mm/s. To predict the performance of the 300-mm tube, a geometrically similar 75-mm tube is to be tested using 30°C benzene. If the flow in the 300-mm tube is laminar (viscosity determines the velocity gradient), at what average velocity should the benzene flow in the 75-mm tube for kinematic similarity?

FIGURE 6-2 Notation for kinematic similarity

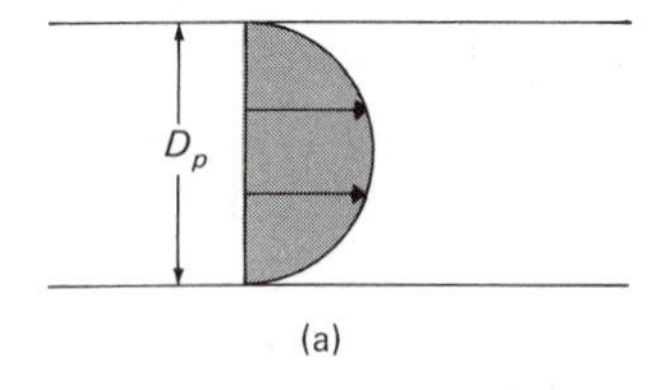

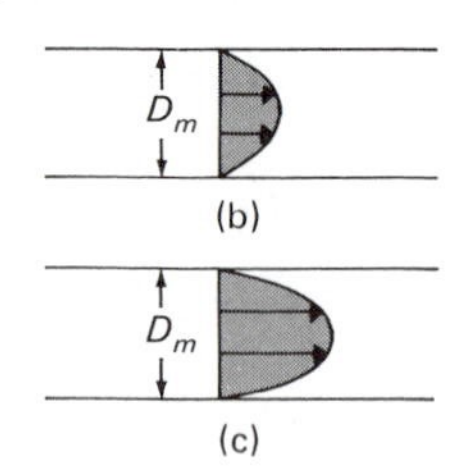

Step A — *Data Reduction*

Fluid properties

(1) *ethyl alcohol* at 10°C

Table A-1: $\rho_p = 797.8\ \text{kg/m}^3$

Table A-4: $\mu_p = 14.47 \times 10^{-4}\ \text{Pa}\cdot\text{s}$

(1-62): $$\nu_p = \frac{\mu_p}{\rho_p} = \frac{14.47 \times 10^{-4}}{797.8} = 1.814 \times 10^{-6}\ \text{m}^2/\text{s}$$

(2) *benzene,* 30°C

Table A-1: $\rho_m = 868.0\ \text{kg/m}^3$

Table A-4: $\mu_m = 5.612 \times 10^{-4}\ \text{Pa}\cdot\text{s}$

(1-62): $$\nu_m = \frac{\mu_m}{\rho_m} = \frac{5.612 \times 10^{-4}}{868.0} = 6.465 \times 10^{-7}\ \text{m}^2/\text{s}$$

Step B — *Develop an Equation for This Application*

When viscosity determines the velocity gradient, then for kinematic similarity:

$$\nu_r = \frac{\nu_m}{\nu_p} = \frac{L_m^2 T_m^{-1}}{L_p^2 T_p^{-1}} = L_r^2 T_r^{-1}$$

or

$$T_r^{-1} = \nu_r L_r^{-2} \qquad \textbf{(a)}$$

Substituting (a) in Eq. (6-4),

$$V_r = L_r T_r^{-1} = L_r(\nu_r L_r^{-2}) = \nu_r L_r^{-1} \qquad \textbf{(b)}$$

Step C — *Compute Velocity in Model*

(b): $$V_m = V_p V_r = V_p(\nu_r L_r^{-1}) = V_p\left(\frac{\nu_m}{\nu_p}\right)\left(\frac{L_p}{L_m}\right)$$

$$= (15 \times 10^{-3})\left(\frac{6.465 \times 10^{-7}}{1.814 \times 10^{-6}}\right)\left(\frac{300}{75}\right)$$

$$= 0.021\,38\ \text{m/s} = 21.38\ \text{mm/s}$$

6-6 DYNAMIC SIMILARITY

To maintain geometric and kinematic similarity, it is necessary to have dynamic or force similarity. Consider the model-prototype relations for the flow around the object shown in Fig. 6-3:

For geometric similarity,

(6-1): $$\frac{D_m}{D_p} = \frac{L_m}{L_p} = L_r$$

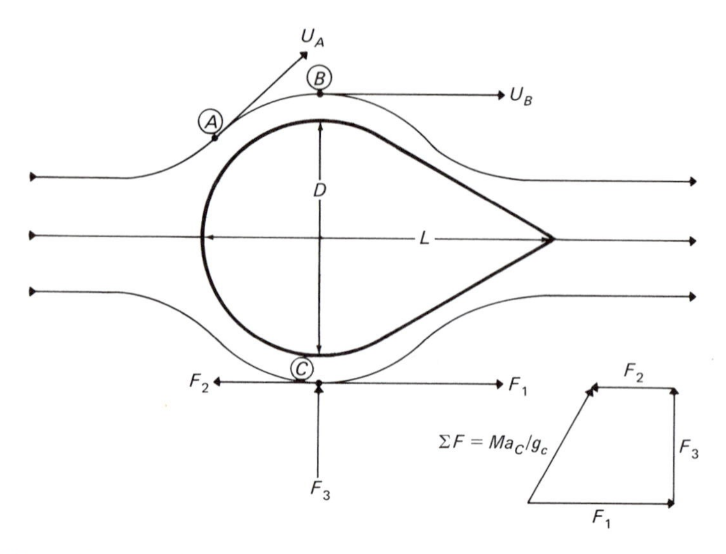

FIGURE 6-3 Notation for dynamic similarity

For kinematic similarity,

$$\frac{U_{Am}}{U_{Ap}} = \frac{U_{Bm}}{U_{Bp}} = V_r \tag{6-4}$$

Consider next the forces acting on point C of Fig. 6-3 without specifying their nature. From the geometric similarity of their vector polygons and Newton's law, which of course applies to both model and prototype:

For dynamic similarity,

$$\frac{F_{1m}}{F_{1p}} = \frac{F_{2m}}{F_{2p}} = \frac{F_{3m}}{F_{3p}} = \frac{M_m a_{Cm}}{M_p a_{Cp}} = F_r \tag{6-6}$$

For dynamic similarity, these force ratios must be maintained on *all* corresponding fluid particles throughout the flow pattern. From the force polygon of Fig. 6-3 it is evident that*

$$F_1 \leftrightarrow F_2 \leftrightarrow F_3 = \frac{Ma_C}{g_C} \tag{6-7}$$

Examination of Eq. (6-7) as well as the force polygon leads to the conclusion that if three of the four terms are known, the other may be determined. This leads to a more general conclusion that dynamic

Note: Forces are vectors. The symbol $\leftrightarrow$ denotes addition of vectors.

similarity may be characterized by an equality of force ratios numbering one less than the total number of forces involved. Any force ratio may be eliminated depending upon the quantities which are desired in the equations. For total model-prototype force ratio, comparisons of force polygons yield

$$F_r = \frac{F_{1m} \leftrightarrow F_{2m} \leftrightarrow F_{3m}}{F_{1p} \leftrightarrow F_{2p} \leftrightarrow F_{3p}} = \frac{M_m a_{Cm}}{M_p a_{Cp}} \tag{6-8}$$

Fluid Forces

The fluid forces that are considered here are those acting on a fluid element whose mass $= \rho L^3$, area $= L^2$, length $= L$, and velocity $= (L/T)$.

Inertia force

$$F_i = (\text{mass})(\text{acceleration}) = (\rho L^3)\left(\frac{L}{T^2}\right)$$

$$= \rho L^2\left(\frac{L^2}{T^2}\right) = \rho L^2 V^2 \tag{6-9}$$

Viscous force

$$F_\mu = (\text{viscous shear stress})(\text{shear area})$$

$$= \tau L^2 = \mu\left(\frac{dU}{dy}\right)L^2 = \mu\left(\frac{V}{L}\right)L^2 = \mu L V \tag{6-10}$$

Gravity force

$$F_g = (\text{mass})(\text{acceleration due to gravity})$$

$$= (\rho L^3)(g) = \rho L^3 g \tag{6-11}$$

Pressure force

$$F_p = (\text{pressure})(\text{area}) = pL^2 \tag{6-12}$$

Centrifugal force

$$F_\omega = (\text{mass})(\text{acceleration}) = (\rho L^3)\left(\frac{L}{T^2}\right)$$

$$= (\rho L^3)(L\omega^2) = \rho L^4 \omega^2 \tag{6-13}$$

Elastic force

$$F_E = (\text{modulus of elasticity})(\text{area}) = EL^2 \tag{6-14}$$

Surface tension force

$$F_\sigma = (\text{surface tension})(\text{length}) = \sigma L \tag{6-15}$$

Vibratory force

$$F_f = (\text{mass})(\text{acceleration}) = (\rho L^3)\left(\frac{L}{T^2}\right)$$

$$= (\rho L^4)(T^{-2}) = \rho L^4 f^2 \tag{6-16}$$

If all fluid forces were acting on a fluid element,

$$F_r = \frac{F_{im} \leftrightarrow F_{\mu m} \leftrightarrow F_{gm} \leftrightarrow F_{pm} \leftrightarrow F_{\omega m} \leftrightarrow F_{Em} \leftrightarrow F_{\sigma m} \leftrightarrow F_{fm}}{F_{ip} \leftrightarrow F_{\mu p} \leftrightarrow F_{gp} \leftrightarrow F_{pp} \leftrightarrow F_{\omega p} \leftrightarrow F_{Ep} \leftrightarrow F_{\sigma p} \leftrightarrow F_{fp}} = \frac{F_{im}}{F_{ip}} \tag{6-17}$$

Fortunately, in most practical engineering problems, not all of the eight forces are involved because they may not be acting, may be of negligible magnitude, or some may be in opposition to each other in such a way as to compensate. In each application of similarity, a good understanding of the fluid phenomena involved is necessary to eliminate the irrelevant, trivial, or compensating forces. When the flow phenomena are too complex to be readily analyzed, or are not known, then only experimental verification with the prototype or results from a model test will determine what forces should be considered in future model testing.

Standard Numbers

With 8 fluid forces that can act in flow situations, the number of dimensionless parameters that can be formed from their ratios is 56. However, conventional practice is to take the ratio of the inertia force to the other fluid forces, because the inertia force is the vector sum of all of the other forces involved in a given flow situation. Results obtained by dividing the inertia force by each of the other forces is shown in Table 6-1 compared with the standard numbers that are used in conventional practice.

TABLE 6-1 *Standard Numbers*

Force Ratio	Equations	Result	Conventional Practice		
			FORM	SYMBOL	NAME
$\frac{\text{Inertia}}{\text{Viscous}}$	$\frac{F_i}{F_\mu} = \frac{\rho L^2 V^2}{\mu L V}$	$\frac{\rho L V}{\mu}$	$\frac{\rho L V}{\mu}$	**R**	Reynolds
$\frac{\text{Inertia}}{\text{Gravity}}$	$\frac{F_i}{F_g} = \frac{\rho L^2 V^2}{\rho L^3 g}$	$\frac{V^2}{Lg}$	$\frac{V}{\sqrt{Lg}}$	**F**	Froude
$\frac{\text{Inertia}}{\text{Pressure}}$	$\frac{F_i}{F_p} = \frac{\rho L^2 V^2}{\rho L^2}$	$\frac{\rho V^2}{p}$	$\frac{\rho V^2}{p}$	**E**	Euler
			$\frac{2\Delta p}{\rho V^2}$	$\mathbf{C}_p$	Pressure coefficient
$\frac{\text{Inertia}}{\text{Centrifugal}}$	$\frac{F_i}{F_\omega} = \frac{\rho L^2 V^2}{\rho L^4 \omega^2}$	$\frac{V^2}{L^2 \omega^2}$	$\frac{V}{DN}$	**V**	Velocity ratio
$\frac{\text{Inertia}}{\text{Elastic}}$	$\frac{F_i}{F_E} = \frac{\rho L^2 V^2}{E L^2}$	$\frac{\rho V^2}{E}$	$\frac{\rho V^2}{E}$	**C**	Cauchy
			$\frac{V}{\sqrt{E/\rho}}$	**M**	Mach
$\frac{\text{Inertia}}{\text{Surface Tension}}$	$\frac{F_i}{F_\sigma} = \frac{\rho L^2 V^2}{\sigma L}$	$\frac{\rho L V^2}{\sigma}$	$\frac{\rho L V^2}{\sigma}$	**W**	Weber
$\frac{\text{Inertia}}{\text{Vibration}}$	$\frac{F_i}{F_f} = \frac{\rho L^2 V^2}{\rho L^4 f^2}$	$\frac{V^2}{L^2 f^2}$	$\frac{Lf}{V}$	**S**	Strouhal

EXAMPLE A seaplane is to take off at 80 mph. At what speed should a 1/60 model be towed to insure similarity of inertia and gravity forces?

Step A *Derive an Equation for this Application*

Inertia force

(6-9) $F_i = \rho L^2 V^2$

Gravity force

(6-11) $F_g = \rho L^3 g$

$$F_r = \frac{F_i}{F_g} = \frac{\rho L^2 V^2}{\rho L^3 g} = \frac{V^2}{L_g} \quad \text{(a)}$$

For similarity, $F_r F_p = F_m$

or

$$\left(\frac{V^2}{L_g}\right)_m = \left(\frac{V^2}{L_g}\right)_p \quad \text{(b)}$$

Solving (b) for V_m:

$$V_m = V_p \sqrt{L_m/L_p} \quad \text{(c)}$$

Step B *Compute Model Speed*

$$V_m = V_p \sqrt{L_m/L_p} = 80 \sqrt{1/60} = 10.32 \text{ mph} \quad \text{(c)}$$

6-7 VIBRATION

In the flow of fluids around objects and in the motion of bodies immersed in fluids, *vibration* may occur because of the formation of a wake caused by alternate shedding of eddies in a periodic fashion or by the vibration of the object or the body. The *Strouhal number* S is the ratio of the velocity of vibration Lf to the velocity of the fluid V. Since the vibration may be fluid induced or structure induced, two frequencies must be considered, the wake frequency, f_ω, and the natural frequency of the structure, f_n. Fluid-induced forces are usually of small magnitude, but as the wake frequency approaches the natural frequency of the structure, the vibratory forces increase very rapidly. When $f_\omega = f_n$, the structure will go into resonance and fail. This imposes on the model designer the requirement of matching to scale the natural frequency characteristics of the prototype. This subject will be treated in Chapter 7. All further discussions of model-prototype relations are made under the assumption either that vibratory forces are absent or that they are taken care of in the design of the model or in the test program.

6-8 SIMILARITY OF INCOMPRESSIBLE FLOW

This section considers the flow of incompressible fluids around an object, the motion of immersed bodies in incompressible fluids, and the flow of incompressible fluids in conduits. It includes for example a submarine traveling under water but not partly submerged. It also includes aircraft moving in atmospheres that may be considered incompressible.

In these situations, the gravity force, although acting on all fluid particles, does not affect the flow pattern. Except for rotating machinery, which will be considered in a later section, centrifugal forces are absent. By definition of an incompressible fluid, elastic forces are zero. Since there is no liquid-gas interface, surface tension forces are absent. This leaves inertia, viscous, and pressure forces acting. With these forces acting, the flow can be characterized by two dimensionless parameters. Using standard numbers, these are *Reynolds Number* and the *Euler Number.*

Reynolds Number

This number was named in honor of Osborne Reynolds (1842–1912), an English engineer who developed it analytically and verified it by experiments. In Sec. 6-6, the Reynolds number was derived as $\rho LV/\mu$. Noting from Eq. (1-62) and the definition of kinematic viscosity that $\nu = \mu/\rho$, we can now write

$$\mathrm{R} = \frac{\text{inertia force}}{\text{viscous force}} = \frac{\rho LV}{\mu} = \frac{LV}{\nu} \tag{6-18}$$

Euler Number

This number was named in honor of Leonhard Euler (1707–1783). Conventional practice is to use the pressure coefficient, which is twice the inverse of the Euler number:

$$\text{Pressure coefficient} = \mathbf{C}_p = \frac{2 \times \text{pressure force}}{\text{inertia force}} = \frac{2\Delta p}{\rho V^2} \tag{6-19}$$

The force created by the pressure loss is

$$\text{force} = \Delta p(\text{area}) \sim F = \Delta p L^2 \tag{6-20}$$

which, substituted in Eq. (6-19), becomes

$$\text{force coefficient} = \mathbf{C}_F = \frac{2(F/L^2)}{\rho V^2} = \frac{2F}{\rho L^2 V^2} \tag{6-21}$$

EXAMPLE A submarine is to move submerged through 0°C sea water at a speed of 5 m/s. (a) At what speed should a 1/20 model be towed in fresh water at 20°C? (b) if the thrust of the model is found to be 200 kN, what power will be required to propel the submarine?

Step A — *Data Reduction*

Fluid properties

(1) *sea water* at 0°C

Table A-1: $\rho_p = 1\,028 \text{ kg/m}^3$

Table A-4: $\mu_p = 18.86 \times 10^{-4} \text{ Pa} \cdot \text{s}$

(2) *fresh water* at 20°C

Table A-1: $\rho_m = 998.3 \text{ kg/m}^3$

Table A-4: $\mu_m = 10.02 \times 10^{-4} \text{ Pa} \cdot \text{s}$

Step B — *Compute Model Speed*

(6-18)
$$\mathbf{R}_m = R_p = \left(\frac{\rho V L}{\mu}\right)_m = \left(\frac{\rho V L}{\mu}\right)_p$$

or

$$V_m = V_p\left(\frac{L_p}{L_m}\right)\left(\frac{\mu_m}{\mu_p}\right)\left(\frac{\rho_p}{\rho_m}\right)$$

$$V_m = 5\left(\frac{20}{1}\right)\left(\frac{10.02 \times 10^{-4}}{18.86 \times 10^{-4}}\right)\left(\frac{1\,028}{998.3}\right)$$

$$V_m = 54.71 \text{ m/s}$$

Step C — *Compute Prototype Power*

(6-21)
$$\mathbf{C}_{Fm} = \mathbf{C}_{Fp} = \left(\frac{2F}{\rho L^2 V^2}\right)_m = \left(\frac{2F}{\rho L^2 V^2}\right)_p$$

or

$$F_p = F_m\left(\frac{\rho_p}{\rho_m}\right)\left(\frac{L_p}{L_m}\right)^2\left(\frac{V_p}{V_m}\right)^2$$

$$F_p = 200\,000\left(\frac{1\,028}{998.3}\right)\left(\frac{20}{1}\right)^2\left(\frac{5}{54.71}\right)^2$$

$$F_p = 688\,064 \text{ N}$$

(4-68)
$$P_p = F_p V_p = 688\,064 \times 5 = 3\,440\,318 = 3\,440 \text{ kW}$$

6-9 SIMILARITY OF COMPRESSIBLE FLOW

This section considers the flow of compressible fluids around an object, the motion of immersed bodies in compressible fluids, and the flow of compressible fluids in conduits. It does not consider the flow of compressible fluids in rotating machinery. It does not include aircraft during takeoff or after landing.

From the discussion given in Sec. 6-8, it is evident that the only difference between compressible and incompressible flow is the elastic force. This means that ratio of the inertia to elastic forces, or Cauchy number, must now be considered in addition to the Reynolds number. The Cauchy number was named to honor Baron Augustin Louis de Cauchy (1789–1857), a French engineer turned mathematician who contributed greatly to the analysis of wave motion. Conventional practice is to use the square root of the Cauchy number or Mach number.

From Eq. (1-60)

(1-60) $$c = \sqrt{E_s/\rho} = \sqrt{kg_cRT}$$

and from the derivation of Sec. 6-6, we have

$$\mathbf{M} = \sqrt{\frac{\text{inertia force}}{\text{elastic force}}} = \sqrt{pV^2/E} = \frac{V}{\sqrt{E/\rho}} = \frac{V}{c} = \frac{V}{\sqrt{kg_cRT}} \tag{6-22}$$

EXAMPLE A 6-in. valve installed in a schedule 80 wrought iron pipe is designed to receive 12,400 lbm/hr of hydrogen at 100 psia and 86°F. This value is tested with air under dynamically similar conditions. When air is supplied at 86°F, the pressure loss is found to be 5 lbf/in². What pressure loss may be expected with the designed flow of hydrogen?

Step A *Data Reduction*

(1) *unit conversions*

(B-28) $$T_m = T_p = 86 + 459.67 = 545.67°\text{R}$$

(B-89) $$p_p = 100 \times 144 = 14{,}400 \text{ lbf/ft}^2$$

(B-89) $$\Delta p_m = 5 \times 144 = 720 \text{ lbf/ft}^2$$

(B-54) $$\dot{M}_p = \frac{12{,}400}{3600} = 3.444 \text{ lbm/sec}$$

(2) *pipe properties,* wrought iron 6-in. schedule 80

Table A-11 $$A_m = A_p = 0.1810 \text{ ft}^2$$

(3) *fluid properties,* hydrogen at 86°F

Table A-7 $$R_p = 766.5 \text{ ft-lbf/lbm-°R}, \; k_p = 1.410$$

Table A-8 $$\mu_p = 18.71 \times 10^{-8} \text{ lbf-sec/ft}^2$$

(1-42) $$v_p = \frac{R_pT_p}{p_p} = \frac{766.5 \times 545.67}{14{,}400} = 29.05 \text{ ft}^3\text{/lbm}$$

Air at 86°F

Table A-7 $R_m = 53.36$ ft-lbf/lbm-°R, $k_m = 1.400$

Table A-8 $\mu_m = 38.94 \times 10^{-8}$ lbf-sec/ft²

Step B *Determine Air Velocity for Mach Number Similarity*

(5-12)
$$V_p = \frac{\dot{M}_p v_p}{A_p} = \frac{3.444 \times 29.05}{0.1810} = 552.8 \text{ ft/sec}$$

(6-22)
$$\mathbf{M}_m = \mathbf{M}_p = \left(\frac{V}{\sqrt{kg_cRT}}\right)_m = \left(\frac{V}{\sqrt{kg_cRT}}\right)_p$$

or

$$V_m = V_p \sqrt{(k_m/k_p)(R_m/R_p)(T_m/T_p)}$$
$$V_m = 552.8 \sqrt{(1.400/1.410)(53.36/766.5)(1)} = 145.3 \text{ ft/sec}$$

Step C *Determine Air Pressure for Reynolds Number Similarity*

(6-18)
$$\mathbf{R}_m = \mathbf{R}_p = \left(\frac{\rho LV}{\mu}\right)_m = \left(\frac{\rho LV}{\mu}\right)_p$$

(1-44) For an ideal gas $\rho = p/g_cRT$, substituting this in Eq. (6-18),

$$\left(\frac{pLV}{\mu g_cRT}\right)_m = \left(\frac{pLV}{\mu g_cRT}\right)_p \tag{a}$$

or

$$p_m = p_p\left(\frac{L_p}{L_m}\right)\left(\frac{V_p}{V_m}\right)\left(\frac{\mu_m}{\mu_p}\right)\left(\frac{R_m}{R_p}\right)\left(\frac{T_m}{T_p}\right) \tag{b}$$

$$p_m = 14{,}400(1)\left(\frac{552.8}{145.3}\right)\left(\frac{38.94 \times 10^{-8}}{18.71 \times 10^{-8}}\right)\left(\frac{53.36}{766.5}\right)(1)$$

(B-89)
$$p_m = 7{,}938 \text{ lbf/ft}^2 = \frac{7{,}938}{144} = 55.12 \text{ psia}$$

Step D *Determine Pressure Loss for Euler Number Similarity*

(6-19)
$$\mathbf{C}_{pm} = \mathbf{C}_{pp} = \left(\frac{2\Delta p}{\rho V^2}\right)_m = \left(\frac{2\Delta p}{\rho V^2}\right)_p$$

(1-44) For an ideal gas $\rho = \dfrac{p}{g_cRT}$, substituting this in Eq. (6-19),

$$\left(\frac{2\Delta p}{(p/g_cRT)V^2}\right)_m = \left(\frac{2\Delta p}{(p/g_cRT)V^2}\right)_p \tag{c}$$

or

$$\Delta p_p = \Delta p_m\left(\frac{p_p}{p_m}\right)\left(\frac{R_m}{R_p}\right)\left(\frac{T_m}{T_p}\right)\left(\frac{V_p}{V_m}\right)^2 \tag{d}$$

$$\Delta p_p = 720\left(\frac{14{,}400}{7{,}938}\right)\left(\frac{53.36}{766.3}\right)(1)\left(\frac{552.8}{145.3}\right)^2$$

$$\Delta p_p = 1316 \text{ lbf/ft}^2 = \frac{1316}{144} = 9.140 \text{ lbf/in}^2$$

6-10 CENTRIFUGAL FORCES

This section covers the flow of fluids in such centrifugal machinery as compressors, fans, and pumps. It is now necessary to consider centrifugal forces in addition to pressure, inertia, viscous, and elastic forces. This means that the ratio of inertia to centrifugal forces must be considered. Since centrifugal force is really a special case of the inertia force, their ratio as shown in Table 6-1 is a velocity ratio.

Consider the fluid machine shown in Fig. 6-4. The absolute velocity of the fluid as it leaves the machine is V, the velocity of the fluid with respect to the runner is v, and $\omega D/2$ is the tangential velocity of the machine. The velocity ratio is defined as the ratio of the fluid velocity to machine tangential velocity. For kinematic similarity, this ratio must be the same at all corresponding points of geometrically similar models and prototypes.

From the derivation of Sec. 6-6, this is also the ratio of the inertia to centrifugal forces. Conventional practice is to state this ratio in the following form:

$$\sqrt{\frac{\text{inertia force}}{\text{centrifugal force}}} = \frac{\text{fluid velocity}}{\text{tangential velocity}} = \frac{V}{\omega D/2} \sim \mathbf{V} = \frac{V}{DN} \tag{6-23}$$

Substituting Eq. (6-23) for V in Eq. (6-22), we obtain for Mach number

$$\mathbf{M} = \frac{V}{\sqrt{kg_cRT}} = \frac{DN}{\sqrt{kg_cRT}} \tag{6-24}$$

FIGURE 6-4 Notation for velocity ratio

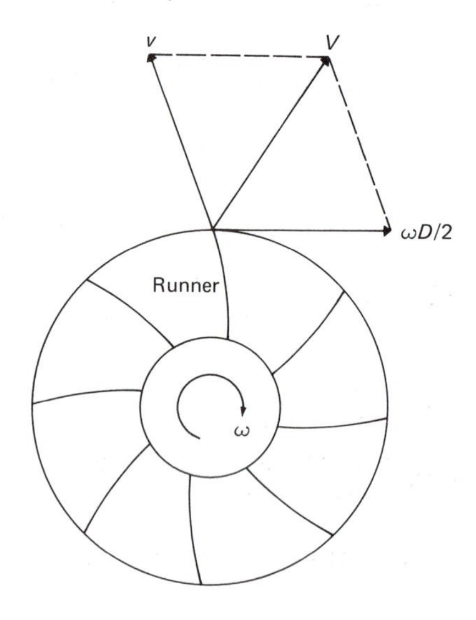

Making the same substitution for Reynolds number in Eq. (6-18),

$$\mathbf{R} = \frac{\rho L V}{\mu} = \frac{\rho L(DN)}{\mu} = \frac{\rho D^2 N}{\mu} \qquad (L \sim D) \tag{6-25}$$

EXAMPLE A centrifugal compressor is to compress methane delivered to it at atmospheric pressure and 30°C. The compressor has an impeller diameter of 300 mm and rotates at 100 rps. It is proposed to test a geometrically similar 75-mm compressor with air. The air source is 20°C and 600 kPa. At what speed should the model be tested for dynamic similarity?

Step A *Data Reduction*

(1) *units*

(B-27) $T_p = 30 + 273.15 = 303.15$ K

(B-27) $T_m = 20 + 273.15 = 293.15$ K

(2) *standard atmosphere*

(B-5) Assume $p_p = 101.3$ kPa

(3) *fluid properties*

methane

Table A-7 $R_p = 518.3$ J/kg · K, $k = 1.310$

Table A-8 $\mu_p = 11.19 \times 10^{-6}$ Pa · s

air

Table A-7 $R_m = 287.1$ J/kg · K, $k = 1.400$

Table A-8 $\mu_m = 18.34 \times 10^{-6}$ Pa · s

Step B *Determine Rotational Speed for Mach Number Similarity*

(6-24)

$$\mathbf{M}_m = \mathbf{M}_p = \left(\frac{DN}{\sqrt{kg_c RT}}\right)_m = \left(\frac{DN}{\sqrt{kg_c RT}}\right)_p$$

or

$$N_m = N_P \left(\frac{D_p}{D_m}\right) \sqrt{(k_m/k_p)\,(R_m/R_p)\,(T_m/T_p)} \tag{a}$$

$$N_m = 100\left(\frac{0.300}{0.075}\right)\sqrt{(1.400/1.310)(287.1/518.3)(293.15/303.15)}$$

$$N_m = 302.6 \text{ rps}$$

Step C *Check Air Pressure for Reynolds Number Similarity at Rotational Speed*

(6-25)

$$\mathbf{R}_m = R_p = \left(\frac{\rho D^2 N}{\mu}\right)_m = \left(\frac{\rho D^2 N}{\mu}\right)_p$$

(1-44) For an ideal gas $\rho = p/g_cRT$ substituting this in Eq. (6-25):

$$\left(\frac{pD^2N}{\mu g_cRT}\right)_m = \left(\frac{pD^2N}{\mu g_cRT}\right)_p \quad \text{(b)}$$

or

$$p_m = p_p\left(\frac{D_p}{D_m}\right)^2\left(\frac{N_p}{N_m}\right)\left(\frac{\mu_m}{\mu_p}\right)\left(\frac{R_m}{R_p}\right)\left(\frac{T_m}{T_p}\right) \quad \text{(c)}$$

$$p_m = (101\,300)\left(\frac{0.300}{0.075}\right)^2\left(\frac{100}{302.6}\right)\left(\frac{18.34\times10^{-6}}{11.19\times10^{-6}}\right)\left(\frac{287.1}{518.3}\right)\left(\frac{293.15}{303.15}\right)$$

$$p_m = 470\,234 \text{ N/m}^2 = 470.2 \text{ kPa}$$

The 600 kPa air source could be throttled down to 470.2 kPa, and thus similarity would be attained at 302.6 rps.

6-11 SIMILARITY OF LIQUID SURFACES

This section considers flow at liquid-gas interfaces. It includes ships, seaplanes during take-off, submarines partly submerged, piers, dams, rivers, spillways, harbors, etc. Resistance at liquid surfaces is due to surface tension and wave action. Since wave action is due to gravity, we will now consider the remaining force ratios discussed in Sec. 6-6.

Surface tension, as was stated in Chapter 1, enables the fluid to support a very small tensile force. It is generally a minor property in fluid mechanics and exerts a negligible effect on wave formation except when the waves are small, say less than 1 in. (25 mm). Thus, the effects of surface tension on a model might be considerable, but not on the prototype. To avoid this type of "scale effect" surface tension should be considered.

The ratio of inertia to surface tension forces is known as the Weber number, in honor of Moritz Weber (1871–1951), a professor of naval mechanics at the Polytechnic Institute of Berlin, who first formulated this number. From the derivation in Sec. 6-6:

$$\mathrm{W} = \frac{\text{inertia force}}{\text{surface tension force}} = \frac{\rho L V^2}{\sigma} \quad \text{(6-26)}$$

The effect of wave resistance is very important in obtaining similarity at liquid surfaces. The ratio of the inertia to gravity forces is usually considered in its square root form. In this form, it is called the Froude number, in honor of William Froude (1810–

1879), an Englishman who developed many towing-tank techniques, particularly the conversion of wave and boundary layer resistance from model to prototype scale. It is one of the ironies of history that Froude's name is inseparably associated with a law of similarity and a number, the first of which he did not originate and the second of which he never used. His very great contribution to boundary layer research is relatively unknown outside the field of naval architecture. From the derivation of Sec. 6-6:

$$\mathbf{F} = \sqrt{\frac{\text{inertia force}}{\text{gravity force}}} = \frac{V}{\sqrt{Lg}} \tag{6-27}$$

As will be seen in the examples to follow, it is impractical to obtain complete dynamic similarity in model-prototype arrangements when liquid surfaces are involved

EXAMPLE 1 Surface Vessel Similarity An ocean vessel 500 ft long is to travel at a speed of 15 knots. A 1/25 model of this ship is to be tested in a towing tank using sea water at design temperature. Determine (a) the model speed required for wave resistance similarity, viscous or skin friction similarity, and surface tension similarity, and (b) the model size required for complete dynamic similarity.

Step A *Determine Speed for Wave Resistance Similarity*

(6-27)
$$\mathbf{F}_m = \mathbf{F}_p = \left(\frac{V}{\sqrt{Lg}}\right)_m = \left(\frac{V}{\sqrt{Lg}}\right)_p$$

or

$$V_m = V_p \sqrt{L_m/L_p} = 15\sqrt{1/25} = 3 \text{ knots} \tag{a}$$

Step B *Determine the Speed for Skin Friction Similarity*

(6-18)
$$\mathbf{R}_m = \mathbf{R}_p = \left(\frac{\rho L V}{\mu}\right)_m = \left(\frac{\rho L V}{\mu}\right)_p$$

or

$$V_m = V_p\left(\frac{\rho_p}{\rho_m}\right)\left(\frac{L_p}{L_m}\right)\left(\frac{\mu_m}{\mu_p}\right) = 15(1)\,(25)\,(1) = 375 \text{ knots} \tag{b}$$

Step C *Determine the Speed for Surface Tension Similarity*

(6-26)
$$\mathbf{W}_m = \mathbf{W}_p = \left(\frac{\rho L V^2}{\sigma}\right)_m = \left(\frac{\rho L V^2}{\sigma}\right)_p$$

or

$$V_m = V_p \sqrt{(\rho_p/\rho_m)(L_p/L_m)(\sigma_m/\sigma_p)} = 15\sqrt{(1)(25)(1)}$$
$$= 75 \text{ knots}$$

Step D *Determine Model Size for Complete Similarity*

First try for Reynolds and Froude number similarity. Setting Eq. (b) equal to Eq. (a),

$$V_m = V_p\left(\frac{\rho_p}{\rho_m}\right)\left(\frac{L_p}{L_m}\right)\left(\frac{\mu_m}{\mu_p}\right) = V_p\sqrt{(L_m/L_p)} \quad \textbf{(c)}$$

which reduces to $\dfrac{L_m}{L_p} = \left(\dfrac{\rho_p}{\rho_m}\right)^{2/3}\left(\dfrac{\mu_m}{\mu_p}\right)^{2/3}$

For the same fluid $L_m/L_p = 1$ or model and prototype must be same size. No practical way has been found to model for complete similarity. Engineering practice is to model for wave resistance and correct for skin friction resistance.

EXAMPLE 2 Time Similarity A 1/256 model of a reservoir is drained in 5 min by opening the sluice gate. How long should it take to empty the prototype?

Step A *Determine Approach*

Since, from prior discussion, complete dynamic similarity cannot be obtained, it is evident in this case that although viscous forces must be present the dominating forces are inertia and gravity and the Froude number should be used for similarity.

Step B *Develop a Specific Equation for This Application and Solve*

(6-27)
$$\mathbf{F}_m = \mathbf{F}_p = \left(\frac{V}{\sqrt{Lg}}\right)_m = \left(\frac{V}{\sqrt{Lg}}\right)_p$$

(6-4)
$$V_r = L_r T_r^{-1}$$

or

$$\left(\frac{L}{TV}\right)_m = \left(\frac{L}{TV}\right)_p$$

Substituting in Eq. (6-27),

$$\mathbf{F} = \frac{L}{T\sqrt{Lg}} = \left(\frac{L^{1/2}}{Tg^{1/2}}\right)_m = \left(\frac{L^{1/2}}{Tg^{1/2}}\right)_p$$

or

$$T_p = T_m\left(\frac{L_p}{L_m}\right)^{1/2} = 5(256)^{1/2} = 80 \text{ min}$$

6-12 DIMENSIONAL ANALYSIS

Dimensional analysis is the mathematics of dimensions and quantities and provides procedural techniques whereby the variables that are assumed to be significant in a problem can be formed into dimensionless parameters, the number of parameters being less than the number of variables. This is a great advantage because fewer experimental runs are then required to establish a relationship between the parameters than between the variables. While the user is not presumed to have any knowledge of the fundamental physical equations, the more knowledgeable the user, the better the results. If any significant variable or variables are omitted, then the relationship obtained from dimensional analysis will not apply to the physical problem. On the other hand, inclusion of all possible variables will result in losing the principal advantage of dimensional analysis, i.e., the reduction of the amount of experimental data required to establish relationships. Two formal methods of dimensional analysis are used, the Method of Lord Rayleigh and Buckingham's Π theorem.

Dimensions used in mechanics are mass, M, length, L, time, T, and force, F. Corresponding *units* for these dimensions are the slug (kilogram), the foot (metre), the second (second), and the pound force (newton). Any system in mechanics can be defined by three fundamental dimensions. Two systems are used, the force (FLT) and the mass (MLT). In the force system, mass is a derived quantity and in the mass system, force is a derived quantity. Force and mass are related by Newton's law: $F = MLT^{-2}$ and $M = FL^{-1}T^2$. Table A-16 shows common variables and their units and dimensions.

6-13 LORD RAYLEIGH'S METHOD

The method developed by Lord Rayleigh uses algebra to determine interrelationships between variables. While this method may be used for any number of variables, it becomes relatively complex and is not generally used for more than four. This method is most easily described by examples.

EXAMPLE 1 In laminar flow, the unit shear stress τ is some function of the fluid dynamic viscosity μ, the velocity gradient dU and the distance between laminae dy. Develop a relationship using the Lord Rayleigh method of dimensional analysis.

Step A *Write a Functional Relationship of the Variables*

$$\tau = f(\mu^a dU^b\, dy^c)$$

Step B *Write a Dimensional Equation in either FLT or MLT Systems*

$$(FL^{-2}) = f(FL^{-2}T)^a (LT^{-1})^b (L)^c$$

Step C *Solve the Dimensional Equation for Exponents*

	τ	μ	dU	dy	Solution
Force, F	$1 =$	a	$+\ 0$	$+\ 0$	$a = 1$
Length, L	$-2 =$	$-2a$	$+\ b$	$+\ c$	
	$-2 =$	$-2(1)$	$+\ b$	$+\ c$	$b = -c$
Time, T	$0 =$	a	$-\ b$	$+\ 0$	$b = a = 1$
					$c = -b = -1$

Step D *Insert Exponents in Functional Equation*

$$\tau = f(\mu^a\, dU^b\, dy^c) = f(\mu^1\, dU^1\, dy^{-1}) = f\frac{\mu\, dU}{dy}$$

The functional relationship cannot be obtained from dimensional analysis. Only physical analysis and/or experiments can determine this. From the physical analysis of Sec. 1-17.

(1-61)
$$\tau = \frac{\mu\, dU}{dy}$$

EXAMPLE 2 The velocity of sound c in a gas depends upon fluid density ρ, pressure p, and dynamic viscosity μ. Develop a relationship using the Lord Rayleigh method of dimensional analysis.

Step A *Write a Functional Relationship of the Variables*

$$c = f(\rho^a p^b \mu^c)$$

Step B *Write a Dimensional Equation in either FLT or MLT Systems*

$$(LT^{-1}) = f(ML^{-3})^a (ML^{-1}T^{-2})^b (ML^{-1}T^{-1})^c$$

Step C *Solve the Dimensional Equation*

	c	ρ	p	μ	Solution
Mass, M	$0 =$	a	$+\ b$	$+\ c$	
Length, L	$1 =$	$-3a$	$-\ b$	$-\ c$	
Time, T	$-1 =$	0	$-\ 2b$	$-\ c$	
$M + L$	$1 =$	$-2a$	$+\ 0$	$+\ 0$	$a = -\frac{1}{2}$
$L - T$	$2 =$	$-3a$	$+\ b$	$+\ 0$	
	$2 =$	$-3(-\frac{1}{2})$	$+\ b$	$+\ 0$	$b = +\frac{1}{2}$
M	$0 =$	$-\frac{1}{2}$	$+\ \frac{1}{2}$	$+\ c$	$c = 0$

Step D *Insert Exponents in Functional Equation*

$$c = f(\rho^{-1/2}p^{1/2}\mu^0) = f\left(\frac{p}{\rho}\right)^{1/2}$$

Note that this analysis rejected viscosity, showing that the velocity of sound is involved in a frictionless process. Again we cannot determine the functional relationship from dimensional analysis alone. From the physical analysis of Sec. 1-16:

$$c = \sqrt{E_s/\rho} \tag{1-59}$$

For an ideal gas from Eq. (1-54), $E_s = kp$, substituting in Eq. (1-59),

$$c = \sqrt{kp/\rho}$$

6-14 THE BUCKINGHAM Π THEOREM

The Buckingham Π theorem serves the same purpose as the method of Lord Rayleigh for deriving equations expressing one variable in terms of its dependent variables. The Π theorem is preferred when the number of variables exceeds four. Application of the Π theorem results in the formation of dimensionless parameters called π ratios. These π ratios have no relation to 3.1416. Application of this theorem is illustrated in the following example.

EXAMPLE Experiments are to be conducted with gas bubbles rising in a still liquid. Consider a gas bubble of diameter D rising in a liquid whose density is ρ, surface tension σ, viscosity μ, rising with a velocity of V in a gravitational field of g. Find a set of parameters for organizing experimental results.

Step A *Data Reduction*

(1) List all the physical variables involved according to type: geometric, kinematic, or dynamic. (See Table A-16.)

(2) Choose either the FLT or MLT system of dimensions.

(3) Select a "basic group" of variables characteristic of the flow as follows:
  (a) B_G, a geometric variable
  (b) B_K, a kinematic variable
  (c) B_D, a dynamic variable
  (if three dimensions are used)

(4) Assign A numbers to the remaining variables, starting with A_1.

Type	Symbol	Description	Dimensions	Number
Geometric	D	Bubble diameter	L	B_G
Kinematic	V	Bubble velocity	LT^{-1}	B_K
	g	Acceleration of gravity	LT^{-2}	A_1
Dynamic	ρ	Liquid density	ML^{-3}	B_D
	σ	Surface tension	MT^{-2}	A_2
	μ	Liquid viscosity	$ML^{-1}T^{-1}$	A_3

Step B *Derive π Ratios*

(1) Write the basic equation for each π ratio as follows:

$$\pi_1 = (B_G)^{x_1}(B_K)^{y_1}(B_D)^{z_1}(A_1),$$
$$\pi_2 = (B_G)^{x_2}(B_K)^{y_2}(B_D)^{z_2}(A_2)$$
$$\cdots \pi_n = (B_G)^{x_n}(B_K)^{y_n}(B_D)^{z_n}(A_n) \qquad \textbf{(6-28)}$$

Note that the number of π ratios is equal to the number of A numbers and thus equal to the number of variables less the number of fundamental dimensions in a problem.

(2) Using the algebraic method of balancing exponents by writing dimensional equations, determine the value of exponents x, y, and z for each π ratio. Note that for all π ratios,

$$\Sigma F = 0,\ \Sigma L = 0, \text{ and } \Sigma T = 0.$$
$$\pi_1 = (B_G)^{x_1}(B_K)^{y_1}(B_D)^{z_1}(A_1) = (D)^{x_1}(V)^{y_1}(\rho)^{z_1}(g)$$
$$(M^0L^0T^0) = (L^{x_1})\,(L^{y_1}T^{-y_1})\,(M^{z_1}L^{-3z_1})\,(LT^{-2})$$

	π_1	D	V	ρ	g	Solution
Mass, M	$0 =$	$0 +$	$0 +$	$z_1 +$	0	$z_1 = 0$
Length, L	$0 =$	$x_1 +$	$y_1 +$	$-3z_1 +$	1	
Time, T	$0 =$	$0 +$	$-y_1 +$	$0 +$	-2	$y_1 = -2$
L	$0 =$	$x_1 +$	$-2 +$	$0 +$	1	$x_1 = 1$

$$\pi_1 = D^1V^{-2}\rho^0 g = \frac{Dg}{V^2}$$

$$\pi_2 = (B_G)^{x_2}(B_K)^{y_2}(B_D)^{z_2}(A_2) = (D)^{x_2}(V)^{y_2}(\rho)^{z_2}(\sigma)$$
$$(M^0L^0T^0) = (L^{x_2})\,(L^{y_2}T^{-y_2})\,(M^{z_2}L^{-3z_2})\,(MT^{-2})$$

	π_2	D	V	ρ	σ	Solution
Mass, M	$0 =$	$0 +$	$0 +$	$z_2 +$	1	$z_2 = -1$
Length, L	$0 =$	$x_2 +$	$y_2 +$	$-3z_2 +$	0	
Time, T	$0 =$	$0 +$	$-y_2 +$	$0 +$	-2	$y_2 = -2$
L	$0 =$	$x_2 +$	$-2 +$	$-3(-1) +$	0	$x_2 = -1$

$$\pi_2 = D^{-1}V^{-2}\rho^{-1}\sigma = \sigma/DV^2\rho$$

$$\pi_3 = (B_G)^{x_3}(B_K)^{y_3}(B_D)^{z_3}(A_3) = (D)^{x_3}(V)^{y_3}(\rho)^{z_3}(\mu)$$

$$(M^0L^0T^0) = (L^{x_3})\,(L^{y_3}T^{-y_3})\,(M^{z_3}L^{-3z_3})\,(ML^{-1}T^{-1})$$

	π_3 D V ρ μ	Solution
Mass, M	$0 = 0 + 0 + z_3 + 1$	$z_3 = -1$
Length, L	$0 = x_3 + y_3 + -3z_3 + -1$	
Time, T	$0 = 0 + -y_3 + 0 + -1$	$y_3 = -1$
L	$0 = x_3 + -1 + -3(-1) + -1$	$x_3 = -1$

$$\pi_3 = D^{-1}V^{-1}\rho^{-1}\mu = \mu/DV\rho$$

Step C *Convert Ratios to Conventional Practice*

One statement of the Buckingham Π theorem is that any π ratio may be taken as a function of all of the others, or

$$f(\pi_1, \pi_2, \pi_3 \cdot\cdot\cdot \pi_n) = 0 \qquad \textbf{(6-29)}$$

Equation (6-29) is mathematical shorthand for a functional statement. It could be written, for example, as

$$\pi_2 = f(\pi_1, \pi_3 \cdot\cdot\cdot \pi_n) \qquad \textbf{(a)}$$

Equation (a) states that π_2 is some function of π_1 and π_3 through π_n, but it is not a statement of *what* function π_2 is of the other π ratios. This can only be determined by physical and/or experimental analysis. Thus we are free to substitute *any* function in Eq. (6-28); for example, π_1 may be replaced with $2\pi_1^{-1}$ or π_n with $a\pi_n^b$.

The procedures set forth in this section are designed to produce π ratios containing the same terms as those resulting from the application of the principles of similarity so that the physical significance may be understood. However, any other combinations might have been used. The only real requirement for a "basic group" is that it contain the same number of terms as there are dimensions in a problem and that each of these dimensions be represented in it.

The maximum number of combinations C or π ratios that can be obtained from V independent variables in B fundamental dimensions is given by

$$C(V, B+1) = \frac{V!}{(B+1)!(V-B-1)!} \qquad \textbf{(6-30)}$$

Solution of Eq. (6-30) for $B = 3$ fundamental dimensions results in the tabulation shown at the top of page 220. This tabulation indicates the importance of selecting the variables that make up the "Basic Group." It is not that the other solutions are incorrect, they are just as valid as the conventional ratios, but their relation to force ratios may not be so easily seen. In the force ratios of Sec. 6-6, the inertia force was always used in combination with the

V, Variables	C, Combinations
5	5
6	15
7	35
8	70
9	126
10	210

other forces. It would have been just as correct to use any of the other forces.

The π ratios derived for this example may be converted into conventional practice as follows:

$\pi_1 = \dfrac{Dg}{V^2}$ is recognized from Sec. 6-6 as the inverse of the square of the Froude number **F**

$\pi_2 = \dfrac{\sigma}{DV^2 \rho}$ is the inverse of the Weber number **W**

$\pi_3 = \dfrac{\mu}{DV \rho}$ is the inverse of the Reynolds number **R**

Let $$\pi_1 = f(\pi_2, \pi_3)$$

Then $$V = K(Dg)^{1/2}$$

where $$K = f(\mathbf{W}, \mathbf{R}) \qquad \text{(b)}$$

Equation (b) tells us that the experimental program must include the variation of the 3 π ratios instead of the six original variables.

To conserve space, the format for dimensional analysis shown in Table 6-2 will be used throughout the balance of this text.

6-15 PARAMETERS FOR JOURNAL BEARINGS

Dimensional Analysis

Consider the journal bearing shown in Fig. 6-5 running at N revolutions per unit time supporting a load pressure per unit of projected area of p. The journal has a diameter of D, the tangential friction force per unit of projected area is p_f, the diametrical clearance is C, and the viscosity of the lubricant μ. Application of the Buckingham Π theorem is shown in Table 6-3.

If we let $$\pi_1 = f(\pi_2, \pi_3)$$

then $$\mathbf{f} = f(m, N\mu/p)$$

TABLE 6-2 *Format for Dimensional Analysis*

Item		Variables considered					
		BUBBLE		FLUID DENSITY	ACCELERATION OF GRAVITY	FLUID	
		Diameter	*Velocity*			*Surface tension*	*Viscosity*
Symbol		D	V	ρ	g	σ	μ
Selection		B_G^x	B_K^y	B_D^z	A_1	A_2	A_3
π Ratios		**Basic group**			π_1	π_2	π_3
Mass Dimension		L	LT^{-1}	ML^{-3}	LT^{-2}	MT^{-2}	$ML^{-1}T^{-1}$
Exponents	M	0	0	z	0	1	1
for	L	x	y	$-3z$	1	0	-1
equations	T	0	$-y$	0	-2	-2	-1

TABLE 6-2 *Solution*

π Ratio	Solved exponents			Derived Ratio	Conventional practice		
	D^x	V^y	ρ^z		FORM	SYMBOL	NAME
π_1	1	-2	0	$\frac{Dg}{V^2}$	$\frac{V}{\sqrt{Dg}}$	**F**	Froude number
π_2	-1	-2	-1	$\frac{\sigma}{DV^2\rho}$	$\frac{DV^2\rho}{\sigma}$	**W**	Weber number
π_3	-1	-1	-1	$\frac{\mu}{DV\rho}$	$\frac{DV\rho}{\mu}$	**R**	Reynolds number

This is as far as we can go with dimensional analysis. To determine *what* function m and $N\mu/p$ are of f, we must turn next to a theoretical analysis and finally to experimental data.

Theoretical Analysis

In example Problem 4 of Sec. 1-17, the velocity profile and friction force produced by the rotation of a cylinder inside a

FIGURE 6-5 Notation for journal bearing study

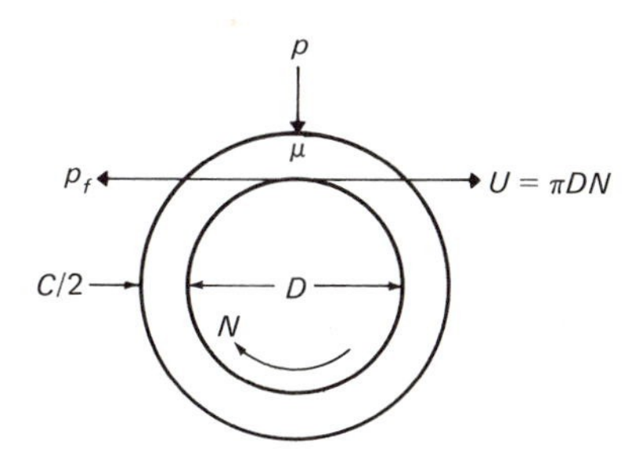

TABLE 6-3 *Journal Bearing Dimensional Analysis*

Item		Variables considered					
		JOURNAL					Fluid Viscosity
		Diameter	Speed	Pressure: Load	Pressure: Friction	Clearance	
Symbol		D	N	p	p_f	C	μ
Selection		B_G^x	B_K^y	B_D^z	A_1	A_2	A_3
π Ratios		Basic group			π_1	π_2	π_3
Force dimensions		L	T^{-1}	FL^{-2}	FL^{-2}	L	$FL^{-2}T$
Exponents for Equations	F	0	0	z	1	0	1
	L	x	0	$-2z$	-2	1	-2
	T	0	$-y$	0	0	0	1

TABLE 6-3 *Solution*

π Ratios	Solved exponents: D^x	N^y	p^z	Derived ratio	Conventional practice: FORM	SYMBOL	NAME
π_1	0	0	-1	p_f/p	p_f/p	f	Coefficient of friction
π_2	-1	0	0	C/D	C/D	m	Clearance ratio
π_3	0	1	-1	$N\mu/p$	$N\mu/p$	P_p	Plotting parameter

fixed tube was developed. Equation (g) developed for the example problem may be converted to the notation of this section as follows, noting that $U_i = \pi DN$, and that

$$\frac{r_o}{r_i} = \frac{D+C}{D} = 1 + m:$$

$$f = \frac{p_f}{p} = \frac{F_s/LD}{p} = \frac{2\pi L(U_i - U_o)\mu}{(LD)\log_e(r_o/r_i)p} = \frac{2\pi(\pi DN - 0)\mu}{pD\log_e(1+m)}$$

$$f = 2\pi^2\left(\frac{N\mu/p}{\log_e(1+m)}\right) \tag{6-31}$$

From Eq. (6-31), the coefficient of friction varies directly with the plotting parameter and inversely with the $\log_e$ of one plus the clearance ratio.

Because bearing clearances are very small and Eq. (6-31) is not easy to manipulate, lubrication engineers assume that the velocity gradient across the small clearance space is linear. With this assumption, noting that Eq. (1-61) integrates to $\tau = \mu U/y$ for a linear velocity profile, Example 3, Sec. 1-17, and that $U = \pi DN$ and $y = C/2$, we can write

$$f = \frac{p_f}{p} = \frac{\tau A_s/LD}{p} = \frac{(\mu U/y)(\pi DL)/(LD)}{p} = \frac{(\mu \pi DN/C/2)\pi}{p}$$

$$f = 2\pi^2\left(\frac{N\mu}{p}\right)\left(\frac{D}{C}\right) = \frac{2\pi^2(N\mu/p)}{m} \tag{6-32}$$

Equation (6-32) is often called Petroff's equation in honor of N. Petroff, a Russian engineer who first published it in 1883. This equation shows that the coefficient of friction varies directly with the plotting parameter and inversely with the clearance ratio.

Experimental Data

Examination of Fig. 6-6 indicates that the plotting parameter serves not only to determine the coefficient of friction but also to set optimum conditions for the design of journal bearings. Our dimensional analysis may be applied to any condition of lubrication, but the theoretical analysis was limited to the thick-film range of bearing operation. Experiments in this range have been correlated by S. A. and T. R. McKee, who proposed the following equation (see References, McKee and McKee).

$$f = 0.002 + \frac{19.56\,(N\mu/p)}{m} \tag{6-33}$$

FIGURE 6-6 Typical experimental results

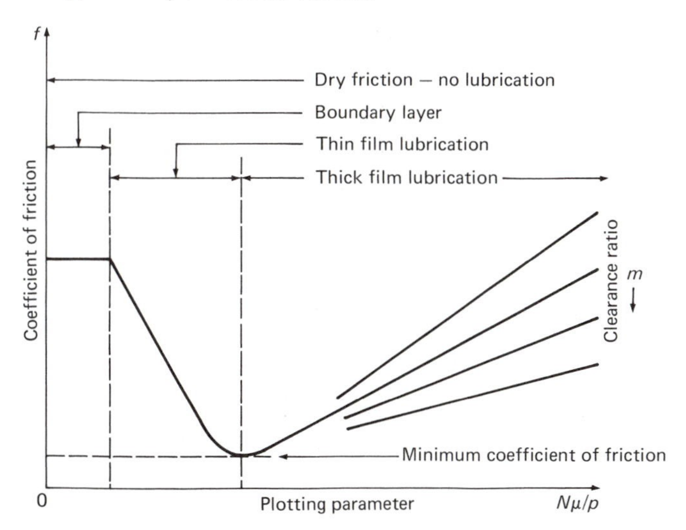

Note that this gives results higher than the Petroff equation (6-32), ($2\pi^2 = 19.74$). McKee found the minimum value of friction factor to be obtained at about a value of $N\mu/p = 0.1 \times 10^{-6}$.

Power and Load

The power lost, P, due to bearing friction is equal to force × distance × time^{-1}, or $(fpDL)(\pi D)N$:

$$P = \pi f p D^2 L N \tag{6-34}$$

The total loading force, F_b, that a bearing may sustain is the load pressure × the load area, or $(p)(DL)$:

$$F_b = p\,DL \tag{6-35}$$

Substituting Eq. (6-35) in Eq. (6-34) results in

$$P = \pi f F_b D N \tag{6-36}$$

EXAMPLE A shaft 3 in. in diameter rotates at 1800 rpm in a bearing 3 in. long with a clearance ratio of 0.001. The bearing is lubricated with linseed oil at 122°F. Estimate the maximum total load that can be imposed on this bearing for thick-film lubrication and the power loss in the bearing at this load.

Step A *Data Reduction*

(1) *unit conversions*

(B-55) $D = 3/12 = 0.25$ ft, $L = 3/12 = 0.25$ ft

(B-68) $$N = \frac{1800}{60} = 30 \text{ sec}^{-1}$$

(2) *fluid properties*, linseed oil at 122°F

$\mu = 367.6 \times 10^{-6}$ lbf-sec/ft^2

Step B *Determine Maximum Load*

Examination of the plotting parameter relationship:

Table 6-3 $$P_p = \frac{N\mu}{p}, \quad \text{or} \quad p = \frac{N\mu}{P_p} \tag{a}$$

Indicates that the maximum load is sustained when the plotting parameter is a minimum. From Section 6-15, $N\mu/p = 0.1 \times 10^{-6}$ is the minimum value for thick-film lubrication.

$$p = \frac{N\mu}{P_p} = \frac{30 \times 367.6 \times 10^{-6}}{0.1 \times 10^{-6}} = 110{,}280 \text{ lbf/ft}^2 \qquad \text{(a)}$$

(6-35) $$F_b = pDL = 110{,}280 \times 0.25 \times 0.25$$

$$F_b = 6{,}893 \text{ lbf}$$

Step C *Compute Power Loss*

(6-33) $$f = \frac{0.002 + 19.56(N\mu/p)}{m}$$

$$f = \frac{0.002 + 19.56(0.1 \times 10^{-6})}{0.001}$$

$$f = 0.003956$$

(6-36) $$P = \pi f F_b DN = \pi(0.003956)(6{,}893)(0.25)(30)$$

$$= 643.0 \text{ ft-lbf/sec}$$

(B-52) $$\text{Horsepower} = \frac{643.0}{550} = 1.169 \text{ hp}$$

6-16 PARAMETERS FOR FLUID MACHINERY

Dimensional Analysis

Consider any fluid machine (turbine, pump, compressor, fan, etc.) handling a fluid at a volume rate of flow of Q, density ρ, viscosity μ, bulk modulus of elasticity E, and with an energy transfer rate (head) per unit mass H. The power developed or supplied is P, the characteristic machine diameter is D, and the machine operates at a rotational speed of N. Determine the parameters that define machine characteristics.

Application of the Buckingham Π theorem is shown in Table 6-4. Let $\pi_1 = f(\pi_2, \pi_3, \pi_4, \pi_5)$

Then $$\frac{Q}{D^3N} = f\left(\frac{H}{D^2N^2}, \mathbf{C}_p, \mathbf{R}, \mathbf{M}\right) \qquad \text{(6-37)}$$

Development of "Pump Laws" from Dimensional Analysis

The so-called pump laws are used by engineers for similarity relations in fluid machinery applications. For incompressible fluids, Mach number is not a parameter, and for geometrically similar machines and for the same fluids, Reynolds number and density are constant, so that

TABLE 6-4 *Fluid Machinery Dimensional Analysis*

Item		Variables considered							
		MACHINE		FLUID		MACHINE		FLUID	
		Diameter	*Speed*	*Density*	*Flow rate*	*Head*	*Power*	*Viscosity*	*Bulk modulus*
Symbol		D	N	ρ	Q	H	P	μ	E
Selection		B_G^x	B_K^y	B_D^z	A_1	A_2	A_3	A_4	A_5
π **Ratios**		**Basic group**			π_1	π_2	π_3	π_4	π_5
Force dimensions		L	T^{-1}	$FL^{-4}T^2$	L^3T^{-1}	L^2T^{-2}	FLT^{-1}	$FL^{-2}T$	FL^{-2}
Exponent for equation	F	0	0	z	0	0	1	1	1
	L	x	0	$-4z$	3	2	1	-2	-2
	T	0	$-y$	$2z$	-1	-2	-1	1	0

TABLE 6-4 *Solution*

π Ratio	Solved exponents			Derived ratio	Conventional practice		
	D^x	N^y	ρ^z		FORM	SYMBOL	NAME
π_1	-3	-1	0	Q/D^3N	Q/D^3N	**V**	Velocity ratio
π_2	-4	-2	0	H/D^2N^2	H/D^2N^2	—	Energy ratio
π_3	-5	-3	-1	$P/D^5N^3\rho$	$P/D^5N^3\rho$	$\mathbf{C}_P$	Power coefficient
π_4	-2	-1	-1	$\mu/D^2N\rho$	$D^2N\rho/\mu$	**R**	Reynolds number
π_5	-2	-2	1	$E/D^2N^2\rho$	$DN/\sqrt{E/\rho}$	**M**	Mach number

$$\pi_1 = \frac{Q}{D^3N} \quad \text{or} \quad Q = f(D^3N) \tag{6-38}$$

First law Discharge varies directly with the speed and as the cube of the diameter, as shown in Eq. (6-38).

$$\pi_2 = \frac{H}{D^2N^2} \quad \text{or} \quad H = f(D^2N^2) \tag{6-39}$$

Second law Head varies directly with the square of the speed and diameter, as shown in Eq. (6-39).

$$\pi_3 = \frac{P}{D^5N^3\rho} \quad \text{for } \rho = \text{a constant} \quad P = f(D^5N^3) \tag{6-40}$$

Third law Power varies directly with the cube of the speed and the fifth power of the diameter, as shown in Eq. (6-40).

EXAMPLE 1 *Pump Law Application* A fan 500 mm in diameter and operating at 12.5 rps requires 3 kW, delivers 3.26 m^3/s of air at a static head of 250 Pa. Estimate the performance of a geometrically similar fan whose diameter is 1100 mm operating at a speed of 7 rps.

Step A *Compute Capacity*

(6-38) $$Q_2 = Q_1\left(\frac{D_2}{D_1}\right)^3\left(\frac{N_2}{N_1}\right) = 3.26\left(\frac{1100}{500}\right)^3\left(\frac{7}{12.5}\right)$$

$$Q_2 = 19.46 \text{ m}^3/\text{s}$$

Step B *Compute Head*

(6-39) $$H_2 = H_1\left(\frac{D_2}{D_1}\right)^2\left(\frac{N_2}{N_1}\right)^2 = 250\left(\frac{1100}{500}\right)^2\left(\frac{7}{12.5}\right)^2$$

$$H_2 = 379.5 \text{ Pa}$$

Step C *Compute Power*

(6-40) $$P_2 = P_1\left(\frac{D_2}{D_1}\right)^5\left(\frac{N_2}{N_1}\right)^3 = 3\left(\frac{1100}{500}\right)^5\left(\frac{7}{12.5}\right)^3$$

$$P_2 = 2\,715 \text{ kW}$$

Concept of Specific Speed

It is desired to develop a parameter for comparing pump or fan performance under the following conditions:

(a) *incompressible fluid* This eliminates the bulk modulus of elasticity E and π_5, the Mach number.
(b) *identical fluid properties* This eliminates density and viscosity as well as π_4, the Reynolds number.
(c) *optimum efficiency* This means identical velocity ratios and eliminates π_1, the velocity ratio.
(d) *in terms of head rather than power input* This eliminates power and π_3, the power coefficient.
(e) *independent of size* This eliminates impeller diameter D. Under these conditions, the only remaining variables are Q, H, and N. Let N_s (specific speed) be some function of these, and use the method of Lord Rayleigh for dimensional analysis:

$$N_s = Q^a H^b N$$

or $$L^\circ T^\circ = (L^3T^{-1})^a(L^2T^{-2})^b(T^{-1})$$

Solved exponents $a = 1/2$, $b = -3/4$

$$N_s = Q^{1/2}H^{-3/4}N = \frac{N\sqrt{Q}}{H^{3/4}} \tag{6-41}$$

For pumps:

$$N_s = \frac{\text{RPM}\sqrt{\text{GPM}}}{H^{3/4}} \tag{6-42}$$

The specific speed of a pump impeller is the speed in revolutions per minute at which a geometrically similar impeller would operate to develop 1 ft of head when displacing 1 gallon per minute of the same fluid.

For fans,

$$N_s = \frac{\text{RPM}\sqrt{\text{CFM}}}{H^{3/4}} \tag{6-43}$$

The specific speed of a fan is the speed in revolutions per minute at which a geometrically similar fan would operate to develop 1 ft of head when displacing 1 ft^3/min of the same fluid.

For Turbines

All of the conditions for pumps and fans apply except that in the case of hydraulic turbines interest is in power developed rather than fluid displacement. The power output of a hydraulic turbine is

$$P_o = \gamma HQe_h \tag{6-44}$$

Where e_h is the hydraulic efficiency.

Substituting from Eq. (6-44) $Q = P_o/\gamma He_h$ in Eq. (6-41):

$$N_s = \frac{N\sqrt{P_o/\gamma He_h}}{H^{3/4}} = \frac{N\sqrt{P_o/\gamma e_h}}{H^{3/4}H^{1/2}} = \frac{N\sqrt{P_o/\gamma e_h}}{H^{5/4}} \tag{6-45}$$

For constant specific weight, density, and efficiency, Eq. (6-45) becomes

$$N_s = \frac{N\sqrt{P_o}}{H^{5/4}} \tag{6-46}$$

For hydraulic turbines, Eq. (6-46) is written as

$$N_s = \frac{\text{RPM}\sqrt{\text{BHP}}}{H^{5/4}} \tag{6-47}$$

The specific speed of a hydraulic turbine is the speed in revolutions per minute at which a geometrically similar hydraulic turbine would operate to develop one brake horsepower under a head of one foot of the same fluid.

Design Specific Speeds

For pumps, the specific speed ranges from 500 to 3,500 for radial flow impellers, from 3,500 to 10,000 for mixed flow and from 10,000 to 15,000 for axial flow pump impellers. Maximum pump efficiency occurs at about a specific speed of 2,500.

For fans, the specific speed ranges from 150 to 1,000 for narrow straight-blade centrifugal fans, from 1,000 to 3,000 for curved-blade centrifugal fans, and from 4,000 to 8,000 for axial-flow fans.

For hydraulic turbines, the specific speed ranges from 0 to 4.5 for impulse wheels, from 10 to 100 for reaction turbines, and from 80 to 200 for axial-flow turbines. Impulse turbines are used for heads over 800 ft with an average efficiency of 82%. Reaction turbines are used from 800 ft to 15 ft, with average efficiencies of 90%. Axial-flow turbines operate at heads below 100 ft with an average efficiency of 90%.

EXAMPLE 2 Pump Specific Speed How many stages are necessary for a pump to deliver 220 gallons per minute against a head of 600 ft-lbf/lbm when operating at 3,600 revolutions per minute if the N_s of 2,500 for maximum efficiency is desired?

Step A *Determine the "Head" per Stage*

(6-42)
$$N_s = \frac{\text{RPM}\sqrt{\text{GPM}}}{H^{3/4}}$$

Solving for H,

$$H = (\text{RPM}\sqrt{\text{GPM}}/N)^{4/3} \tag{a}$$

$$H = (3{,}600\sqrt{220}/2{,}500)^{4/3} = 59.26 \text{ ft-lbf/lbm}$$

Step B *Determine Number of Stages*

$$N = \frac{H_t}{H} = \frac{600}{59.26} = 10.12 = 10 \text{ stages} \tag{b}$$

EXAMPLE 3 Hydraulic Turbine Water supply is available at a height of 15 m. It is desired to develop 37 500 kW with a hydraulic turbine operating at 1 rps. What type of turbine should be selected?

Step A *Data Reduction*

(B-41)
$$H = \frac{15}{0.3048} = 49.21 \text{ ft}$$

(B-53)
$$\text{BHP} = \frac{37\,500 \times 1000}{745.7} = 50{,}288$$

(B-68)
$$N = 1 \times 60 = 60 \text{ RPM}$$

Step B *Select Turbine*

(6-47)
$$N_s = \frac{\text{RPM}\sqrt{\text{BHP}}}{H^{5/4}}$$
$$= \frac{60\sqrt{50{,}288}}{(49.21)^{5/4}} = 103$$

This is in the specific speed range of 80 to 200 for axial-flow turbines, and an axial-flow turbine should be selected.

PROBLEMS

1. The forces present in a journal bearing rotating at a speed of N, under a loading pressure of p, lubricated by a fluid whose viscosity is μ are: inertia, centrifugal, pressure, and viscous. Derive a number that is proportional to the ratio of the total viscous force divided by the total pressure force.
2. Consider a flow in which the types of forces acting are: inertia, elastic, and pressure. Derive a number which is proportional to the total elastic force divided by the total pressure force.
3. In a certain chemical process, a thin sheet of benzene at 86°F is to flow down an inclined plane whose length is 5 ft. The average velocity of the benzene is to be 1 ft/sec. A model of the inclined plane using water at 86°F is to be tested. It is judged that the predominant forces involved are: surface tension, viscous, and gravity. For complete dynamic similarity estimate (a) the length of the model and (b) the water velocity.
4. Mercury at the rate of 0.75 kg/s is to flow in a 6-in. diameter schedule 40 pipe at a temperature of 20°C. It is desired to test this pipe with water at a temperature of 20°C. How many kg/s of water must flow in the pipe for complete dynamic similarity?

5. A submarine is to move through sea water at a speed of 5 knots. A towing tank is available with sea water at the design temperature. If the maximum towing speed is 50 knots, what is the smallest size model that can be used for dynamic similarity?
6. An airplane is to be designed to fly at an altitude of 10,000 m. Model tests are to be conducted in a wind tunnel where the pressure is 1 400 kPa and the temperature 20°C. What size model is required for dynamic similarity?
7. Methane at 20 psia and 86°F flows in a 6-in. schedule 40 steel pipe. It is desired to model this flow using air at 86°F in a 3-in. schedule 40 steel pipe. (a) What pressure should the air supplied attain for complete dynamic similarity and (b) if the pressure loss in the model is 10 psia, what will be the loss in the prototype?
8. A torpedo is designed to travel at a speed of 11.3 m/s in sea water. A model tested under dynamically similar conditions in a towing tank using sea water at design conditions was found to have a drag of 1000 N. Estimate the drag of the torpedo.
9. A centrifugal pump designed to rotate at 900 rpm has an impeller diameter of 3 in. The pump is to handle castor oil at 68°F. A geometrically similar pump whose impeller diameter is 6 in. is to be tested using air at 68°F. Determine (a) the model speed when air is supplied at 14.7 psia, neglecting the effects of elasticity, and (b) the model speed and air pressure required for complete dynamic similarity.
10. The performance of a centrifugal pump of 150-mm diameter impeller designed to handle ethyl alcohol at 30°C is to be determined by model testing. For the model test, air is to be supplied at 30°C and 101 kPa. What size model is required for complete dynamic similarity?
11. An ocean vessel 500 ft long is to travel at 15 knots in 68°F sea water. A model of this ship is to be towed in a tank containing mercury at 122°F. Determine (a) the required length of the model and (b) the towing speed.
12. A $1/9$ model of a seaplane is geometrically similar to its prototype. The model has a take-off speed of 18 m/s. Estimate the take-off speed of the prototype.
13. The two commonly used dimensional systems are the force-length-time, *FLT*, and the mass-length-time, *MLT*, but these are by no means the only systems that could be used. Consider a power-length-time, *PLT*, system. What would be the dimensions of force and mass in the *PLT* system?
14. Using dimensional analysis, derive an expression for stagnation pressure p_s assuming it to be a function of fluid density ρ and the undisturbed velocity V_o.

15. An ideal liquid whose density is ρ flows at a volume rate of Q through a circular restriction of diameter D in a pipeline creating a pressure difference of Δp. Using dimensional analysis derive an expression for the volumetric flow rate Q.
16. In studies of soil erosion due to dust storms, it is judged that the weight W per unit of land area of soil blown away is a function of the wind velocity V, the average diameter of a dust particle D, the particle specific weight γ, the density of air ρ, and the air viscosity μ. Derive an expression using dimensional analysis for the amount of erosion.
17. In a falling-sphere viscometer, the time T for a small sphere of diameter D and weight W to fall at a constant velocity V through a distance Z in a fluid of density ρ and bulk modulus of elasticity E is a measure of the fluid viscosity μ. Using dimensional analysis, determine the parameters involved in viscosity determination.
18. In a certain research project, information is desired concerning a liquid jet of diameter D issuing from a nozzle with a velocity of V. At a distance L from the nozzle, the jet breaks up into a fine spray of liquid droplets. If the liquid has a density of ρ, viscosity of μ, and surface tension of σ, determine a set of coordinates for organizing test data.
19. The ratio zN/p is frequently used in the technical literature on lubrication, where z is in centipoises, N in revolutions per minute, and p in pounds (force) per square inch. What conversion factor is necessary to convert zN/p into a dimensionless ratio in any system of units?
20. The following data were obtained from the test of a 25 mm diameter journal bearing whose length is 25 mm and whose clearance ratio is 0.0016.

Speed	30 rps
Load	3.6 kN
Lubricant-linseed oil	50° C
Friction torque	0.164 mN

Compute the coefficient of friction from (a) test data and (b) the McKee equation.

21. A centrifugal pump has the following characteristics:

Capacity	500 gpm
Speed	1750 rpm
Head	60 ft-lbf/lbm
Horsepower	15 bhp
Impeller diameter	8 in.

The pump is driven by a constant speed motor. In a certain application, the capacity is to be reduced to 211 gpm by changing the impeller. Determine (a) the required impeller diameter and (b) the head developed and (c) the horsepower required with the new impeller.

22. A hydroelectric power plant is to be constructed where water is available at 225 m^3/s at 20°C at an effective head of 16.75 m. The turbine selected is to be an axial-flow type with a specific speed of 150. Estimate the turbine output and the turbine speed.

REFERENCES

Bridgman, P. *Dimensional Analysis,* 2d ed., Yale University Press, New Haven, Conn., 1931.

Buckingham, E. "On Physically Similar Systems: Illustrations of the Use of Dimensional Equations," *Physical Review,* vol VI, 1914.

McKee, S. A., and T. R. McKee. "Friction of Journal Bearings as Influenced by Clearance and Length," *Transactions of the American Society of Mechanical Engineers,* vol. 51, no. 16 (May-Aug., 1929), pp. 161–166.

Lanshарr, R. L. *Dimensional Analysis and Theory of Models,* John Wiley and Sons, New York, 1951.

Guidance Manual on Model Testing. Performance Test Codes Report, American Society of Mechanical Engineers, PTC 37R, (in preparation).

7 Forces on Immersed Objects

7-1 INTRODUCTION

This is the first chapter on applications of fluid mechanics. The approach in the application chapters is to combine the theory developed in Chapters 1 through 5 with the dimensionless parameters derived in Chapter 6. This chapter deals with the flow around bodies.

The number of types of structures that are subject to fluid forces is almost infinite. For this reason, consideration had to be limited to some simple shapes such as the flat plate, sphere, and cylinder. For the purposes of illustration, consideration was given to typical engineering situations in connection with vortex-induced vibration, resistance of ships, properties of lifting vanes, and characteristics of propellers.

Men whose significant contributions underlie the material covered in this chapter are Leonardo da Vinci (1452–1519), who first sketched and described vortex formation, and Ludwig Prandtl (1875–1953), Paul Heinrich Blasius (1883–), Theodor von Kármán (1881–1963), and Hermann Schlichting (1907–), who made contributions to boundary layer theory.

7-2 DRAG AND LIFT

Definition

Figure 7-1 shows the effect of an object placed in a fluid stream. The impingement of the fluid on the object produces a force, F. The horizontal component of this force is the *drag force* and the

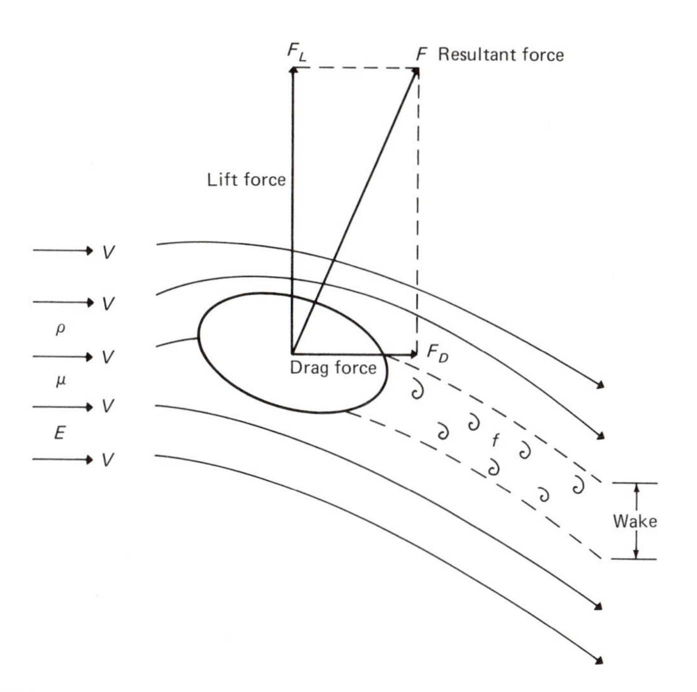

FIGURE 7-1 Notation for drag and lift

vertical component the *lift force.* The presence of this object may produce a *wake* in which eddies are shed regularly from alternate sides at a definite frequency. This shedding pattern is generally known as "Kármán vortex trails" in honor of Theodor von Kármán whose studies formalized this phenomenon.

Dimensional Analysis

Consider a fluid whose density is ρ, viscosity μ, and bulk modulus of elasticity E approaching an object whose characteristic length is L at velocity V. The impingement of the fluid on the object produces a drag force of F_D and a lift force of F_L. The wake sheds eddies with a frequency of f. Derive the dimensionless parameters involved in this phenomenon. Application of the Buckingham Π theorem is shown in Table 7-1.

Physical Analysis

From dimensional analysis we have

$$f(C_D, C_L, \mathbf{R}, \mathbf{M}, \mathbf{S}) = 0 \tag{7-1}$$

TABLE 7-1 *Dimensional Analysis of Drag and Lift*

Item		Variables considered							
		CHARACTERISTIC LENGTH	FLUID		FORCES		FLUID		WAKE FREQUENCY
			Velocity	*Density*	*Drag*	*Lift*	*Viscosity*	*Bulk modulus*	
Symbol		L	V	ρ	F_D	F_L	μ	E	f
Selection		B_G^x	B_K^y	B_D^z	A_1	A_2	A_3	A_4	A_5
π **Ratios**		**Basic group**			π_1	π_2	π_3	π_4	π_5
Force dimensions		L	LT^{-1}	$FL^{-4}T^2$	F	F	$FL^{-2}T$	FL^{-2}	T^{-1}
Exponents for equations	F	0	0	z	1	1	1	1	0
	L	x	y	$-4z$	0	0	-2	-2	0
	T	0	$-2y$	$2z$	0	0	1	0	-1

TABLE 7-1 *Solution*

π Ratios	Solved exponents			Derived ratio	Conventional practice		
	L^x	V^y	ρ^z		FORM	SYMBOL	NAME
π_1	-2	-2	-1	$F_D/L^2V^2\rho$	$\dfrac{2F_D}{\rho V^2(\text{Area})}$	C_D	Drag (Force) coefficient
π_2	-2	-2	-1	$F_L/L^2V^2\rho$	$\dfrac{2F_L}{\rho V^2(\text{Area})}$	C_L	Lift (Force) coefficient
π_3	-1	-1	-1	$\mu/LV\rho$	$\dfrac{\rho VL}{\mu}$	$\mathbf{R}$	Reynolds number
π_4	0	-2	-1	$E/V^2\rho$	$\dfrac{V}{\sqrt{E/\rho}}$	$\mathbf{M}$	Mach number
π_5	1	-1	0	Lf/V	$\dfrac{fL}{V}$	$\mathbf{S}$	Strouhal number

The formation of a wake with shedding eddies depends upon the ratio of the inertia to viscous forces or Reynolds number. It follows that the Strouhal number must be some function of Reynolds number, or

$$S = f(\mathbf{R}) \qquad \textbf{(7-2)}$$

This function will be developed in a later section for cylinders. We may now write Eq. (7-1) as

$$f(C_D, C_L, \mathbf{R}, \mathbf{M}) = 0 \qquad \textbf{(7-3)}$$

Since the lift and drag forces are independent of each other, we can write Eq. (7-3) as follows:

$$F_D = C_D \frac{\rho V^2 A}{2} \tag{7-4}$$

where $C_D = f(\mathbf{R}, \mathbf{M})$ and A is the characteristic area. Also:

$$F_L = C_L \frac{\rho V^2 A}{2} \tag{7-5}$$

The drag force shown in Figs. 7-1 and 7-2 arises from two sources, the interference of the object to the flow, called the pressure or shape drag, and the friction drag due to the wall shear stress τ_o, called skin friction drag. The drag coefficient is made up of two parts, as follows:

$$F_D = F_p + F_f = C_D \frac{\rho A V^2}{2} = C_s \frac{\rho A V^2}{2} + C_f \frac{\rho A_s V^2}{2} \tag{7-6}$$

or

$$C_D = C_s + \frac{C_f A_s}{A} \tag{7-7}$$

where C_s is the shape coefficient, C_f is the skin friction coefficient, and A_s the characteristic area for shear.

The lift force shown in Figs. 7-1 and 7-2 is a function of shape and the angle of attack α as well as Reynolds and Mach numbers.

EXAMPLE A kite (Fig. 7-3) weighs 3 lb and has an area of 10 ft². The tension in the kite string is 8 lb when the string makes

FIGURE 7-2 Notation for physical analysis

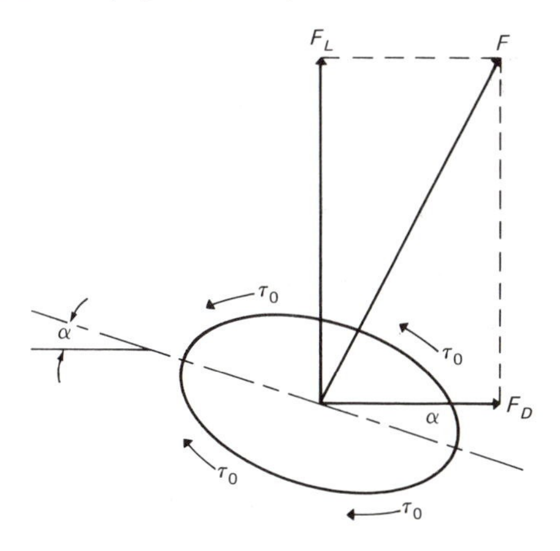

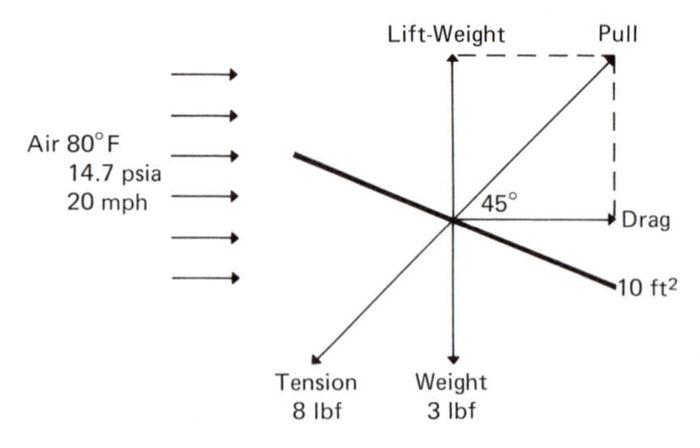

FIGURE 7-3 Notation for example problem

an angle of 45° with the horizontal. For air at 80°F and 14.7 psia with a wind velocity of 20 mph, what are the coefficients of lift and drag?

Step A *Data Reduction*

(1) *unit conversion*

(B-65/54) $$V = 20\left(\frac{5{,}280}{3{,}600}\right) = 29.33 \text{ ft/sec}$$

(B-28) $$T = 80 + 459.67 = 539.67°\text{R}$$

(B-89) $$p = 144 \times 14.7 = 2{,}117 \text{ lbf/ft}^2$$

(2) *proportionality constant*

(1-11) $$g_c = 32.17 \text{ lbm-ft/lbf-sec}^2$$

(3) *fluid properties,* air

Table A-7 $$R = 53.36 \text{ ft-lbf/lbm-°R}$$

(1-44) $$\rho = \frac{p}{g_c RT} = \frac{2{,}117}{(32.17)(53.36)(539.67)}$$

$$\rho = 0.002285 \text{ slugs/ft}^3$$

Step B *Compute Drag Coefficient*

Fig. 7-3 $$F_D = \text{tension} (\cos 45°) = 8 \times \cos 45° = 5.657 \text{ lbf}$$

(7-4) $$C_D = \frac{2F_D}{\rho V^2 A} = \frac{2 \times 5.657}{(0.002285)(29.33)^2(10)}$$

$$C_D = 0.5756$$

Step C *Compute Lift Coefficient*

Fig. 7-3 $$F_L = \text{tension} (\sin 45°) + \text{weight}$$

$$= 8 \times \sin 45° + 3 = 8.657 \text{ lbf}$$

$$C_L = \frac{2F_L}{\rho V^2 A} = \frac{2 \times 8.657}{0.002285(29.33)^2(10)} \tag{7-5}$$

$$C_L = 0.8808$$

7-3 SKIN FRICTION DRAG

Boundary Layer

Fig. 7-4 shows a fluid approaching a flat plate with a uniform velocity profile of V. As the fluid passes over the plate, the velocity at the plate surface is zero and increases to V at some distance from the surface. The region in which the velocity varies from 0 to V is called the *boundary layer*. The thickness of this layer is δ. For some distance along the plate, the flow within the boundary layer is laminar, with the viscous forces predominating, but in the transition zone as the inertia forces begin to exceed the laminar, a turbulent layer begins to form and increases as the laminar layer decreases.

Reynolds Number

The Reynolds number used for skin friction drag is based on the distance X from the leading edge of the flat plate, so that from Eq. (6-18)

FIGURE 7-4 Boundary layer along a smooth, flat plate

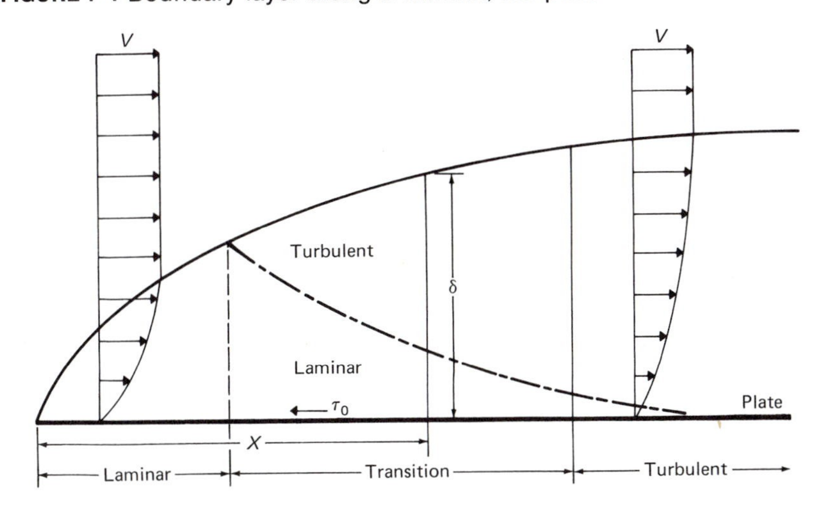

$$\mathbf{R}_X = \frac{\rho X V}{\mu} = \frac{XV}{\nu} \tag{7-8}$$

EQUATIONS Boundary layer thickness and skin friction drag coefficients for incompressible flow over smooth, flat plates may be calculated from the following equations:

Laminar

$$\frac{\delta}{X} = 5.20\ \mathbf{R}_X^{-1/2} \qquad 0 < \mathbf{R}_X < 5 \times 10^5 \tag{7-9}$$

$$C_f = 1.328\ \mathbf{R}_X^{-1/2} \qquad 0 < \mathbf{R}_X < 5 \times 10^5 \tag{7-10}$$

Transition

The Reynolds number at which the boundary layer changes depends upon the roughness of the plate and the degree of turbulence. The generally accepted number is 500,000, but the transition can take place at Reynolds numbers higher or lower. For transition at any Reynolds number $\mathbf{R}_t$:

$$C_f = 0.455\ (\log_{10}\mathbf{R}_X)^{-2.58} - (0.0735\ \mathbf{R}_t^{4/5} - 1.328\ \mathbf{R}_t^{1/2})\mathbf{R}_X^{-1} \tag{7-11}$$

For $\mathbf{R}_t = 5 \times 10^5$,

$$C_f = 0.455\ (\log_{10}\mathbf{R}_X)^{-2.58} - 1725\ \mathbf{R}_X^{-1}. \tag{7-12}$$

Turbulent

$$\frac{\delta}{X} = 0.377\ \mathbf{R}_X^{-1/5} \qquad 5 \times 10^4 < \mathbf{R}_X < 10^6 \tag{7-13}$$

$$\frac{\delta}{X} = 0.220\ \mathbf{R}_X^{-1/6} \qquad 10^6 < \mathbf{R}_X < 5 \times 10^8 \tag{7-14}$$

$$C_f = 0.0735\ \mathbf{R}_X^{-1/5} \qquad 2 \times 10^5 < \mathbf{R}_X < 10^7 \tag{7-15}$$

$$C_f = 0.455\ (\log_{10}\mathbf{R}_X)^{-2.58} \qquad 10^7 < \mathbf{R}_X < 10^8 \tag{7-16}$$

$$C_f = 0.0586\ (\log_{10}C_f\mathbf{R}_X)^{-2} \qquad 10^8 < \mathbf{R}_X < 10^9 \tag{7-17}$$

Figure 7-5 shows the variation of skin friction coefficients with Reynolds number.

EXAMPLE 1 Laminar Flow Castor oil at 20°C flows over a smooth, flat plate with a velocity of 3 m/s. The plate has a length of

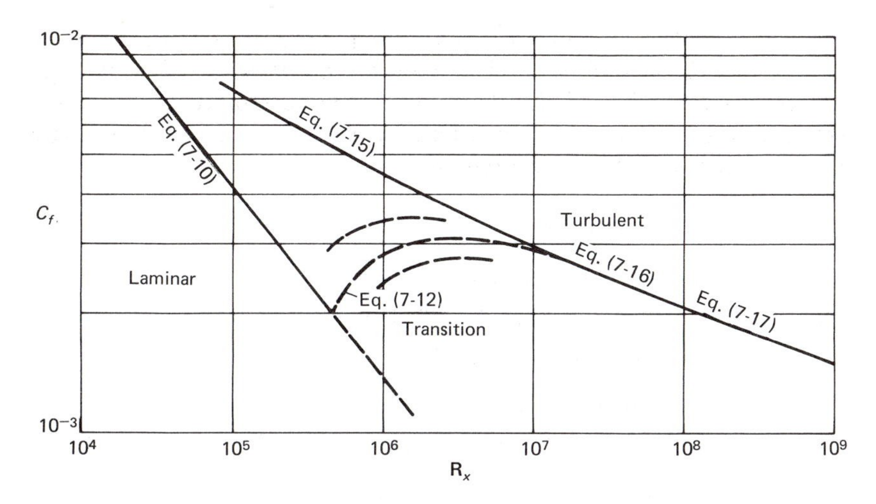

FIGURE 7-5 Skin-friction coefficients for smooth, flat plates

3 m in the direction of the flow and a width of 1.2 m. Estimate the thickness of the boundary layer at the trailing edge of the plate and the skin friction force on the plate.

Step A — *Data Reduction*

Fluid properties, castor oil at 20°C

Table A-1 $\rho = 960.2 \text{ kg/m}^3$

Table A-4 $\mu = 9\,610 \times 10^{-4} \text{ Pa} \cdot \text{s}$

Step B — *Determine Type of Flow*

(7-8) $$\mathbf{R}_X = \frac{\rho X V}{\mu} = \frac{960.2 \times 3 \times 3}{9\,610 \times 10^{-4}}$$

$$\mathbf{R}_X = 8\,993 < 5 \times 10^5 \therefore \text{flow is laminar}$$

Step C — *Compute Boundary Layer Thickness*

(7-9) $$\delta = 5.20 \times \mathbf{R}_X^{-1/2} = 5.20 \times 3(8\,993)^{-1/2}$$

$$\delta = 0.164\,5 \text{ m} = 16.45 \text{ cm}$$

Step D — *Compute Skin Friction Force*

(7-10) $$C_f = 1.328\ \mathbf{R}_X^{-1/2} = 1.328(8\,993)^{-1/2} = 0.014\,00$$

(7-6) $$F_f = \frac{C_f \rho V^2 A}{2} = \frac{0.014\,00 \times 960.2(3)^2(3 \times 1.2)}{2} = 217.8 \text{ N}$$

EXAMPLE 2 Turbulent Flow Air at 68°F and 100 psia flows over a smooth, flat plate with a velocity of 10 ft/sec. The plate is 10 ft in the direction of the flow and 4 ft wide. For turbulent flow, estimate the boundary layer thickness at the trailing edge of the plate and the skin friction force on the plate.

Step A *Data Reduction*

(1) *unit conversions*

(B-28) $T = 68 + 459.67 = 527.67°R$

(B-87) $p = 144 \times 100 = 14{,}400 \text{ lbf/ft}^2$

(2) *proportionality constant*

(1-11) $g_c = 32.17 \text{ lbm-ft/lbf-sec}^2$

(3) *fluid properties,* air at 68°F

Table A-7 $R = 53.36 \text{ ft-lbf/lbm-°R}$

(1-44) $$\rho = \frac{p}{g_c RT} = \frac{14{,}400}{32.17 \times 53.36 \times 527.67}$$

$$= 0.01590 \text{ slugs/ft}^3$$

Table A-8 $\mu = 38.31 \times 10^{-8} \text{ lbf-sec/ft}^2$

Step B *Determine Type of Flow*

(7-8) $$R_X = \frac{\rho XV}{\mu} = \frac{0.01590 \times 10 \times 10}{38.31 \times 10^{-8}} = 4.150 \times 10^6$$

equations (7-14) and (7-15) apply

Step C *Compute Boundary Layer Thickness*

(7-14) $$\delta = 0.220 \times \mathbf{R}_X^{-1/6} = 0.220 \times (4.150 \times 10^6)^{-1/6}$$

$$\delta = 0.1735 \text{ ft}$$

Step D *Compute Drag Force*

(7-15) $$C_f = 0.0735\ \mathbf{R}_X^{-1/5} = 0.0735(4.150 \times 10^6)^{-1/5}$$

$$C_f = 0.003489$$

(7-6) $$F_f = \frac{C_f \rho V^2 A}{2} = \frac{0.003489 \times 0.01590(10)^2(10 \times 4)}{2} = 0.1110 \text{ lbf}$$

7-4 SHAPE DRAG

Experiments with certain objects normal to the flow of a fluid indicate that their drag coefficients are essentially constant

TABLE 7-2 *Shape Coefficients* $\mathbf{R} > 10^3$

Flat plate	Normal to flow	**length/width**	**1**	**5**	**10**	**20**	**30**	**∞**
		C_s	**1.18**	**1.2**	**1.3**	**1.5**	**1.6**	**1.95**
Disk	Normal to Flow		$C_s = 1.18$					
Trains	Locomotive + tender	Streamlined	$C_s = 0.40$					
		Conventional	$C_s = 0.93$					
	Railroad cars	Streamlined	$C_s = 0.15$					
		Conventional	$C_s = 0.40$					
Automotive	Automobiles	Racing-type	$C_s = 0.17$					
		Streamlined	$C_s = 0.23$					
		Fast-back	$C_s = 0.34$					
		Conventional	$C_s = 0.52$					
	Truck	Conventional	$C_s = 0.60$					
Ship	Passenger	Conventional	$C_s = 0.90$ (aerodynamic)					

at Reynolds numbers over 1000. This means that the drag for $\mathbf{R} > 10^3$ is "shape" drag. Table 7-2 shows typical values used by engineers for design and estimating purposes.

EXAMPLE Estimate the total wind force on a billboard 3 m high and 15 m wide when subjected to a 9 m/s wind blowing normal to it. Assume air to be 20°C and 100 kPa.

Step A *Data Reduction*

(1) *unit conversion*

(B-27) $T = 20 + 273.15 = 293.15 \text{ K}$

(2) *proportionality constant*

(1-9) $g_c = 1 \text{ kg} \cdot \text{m/N} \cdot \text{s}^2$

(3) *fluid properties,* air at 20°C

Table A-7 $R = 287.1 \text{ J/kg} \cdot \text{K}$

Table A-8 $\mu = 18.34 \times 10^{-6} \text{ Pa} \cdot \text{s}$

(1-44)
$$\rho = \frac{p}{g_c RT} = \frac{(100\,000)}{(1)\,(287.1)\,(293.15)}$$

$$\rho = 1.188 \text{ kg/m}^3$$

(4) *Geometric*

$$\frac{\text{length}}{\text{width}} = \frac{15}{3} = 5$$

Table A-14 hydraulic radius
$$R_h = \frac{bD}{2}(b + D)$$

$$= \frac{3 \times 15}{2(3 + 15)} = 1.25 \text{ m}$$

(4-11) equivalent diameter

$$D_e = 4R_h = 4 \times 1.25 = 5 \text{ m}$$

Step B *Determine Reynolds Number*

(6-18)
$$\mathrm{R}_e = \frac{\pi D_e V}{\mu} = \frac{1.188 \times 5 \times 9}{18.34 \times 10^{-6}}$$

$$\mathrm{R}_e = 2.915 \times 10^6 > 10^3$$

Table 7-2 may be used

Step C *Compute Force*

Table 7-2
$$\text{For } \frac{\text{length}}{\text{width}} = 5,\ C_s = 1.2$$

(7-6)
$$F_p = \frac{C_s \rho A V^2}{2} = \frac{1.2 \times 1.188\,(3 \times 15)\,(9)^2}{2}$$

$$F_p = 2.598 \text{ kN}$$

7-5 DRAG OF A SPHERE

In calculating the drag of spheres, the sphere diameter D is used as the characteristic length and the projected area, or $\pi D^2/4$, is used as the drag area. The drag of a sphere is obtained by substituting the projected area in Eq. (7-4) (see Fig. 7-6):

$$F_D = \frac{C_D \rho V^2 A}{2} = \frac{C_D \rho V^2 (\pi D^2/4)}{2} = \frac{C_D \pi D^2 \rho V^2}{8} \tag{7-18}$$

FIGURE 7-6 Drag coefficients of spheres

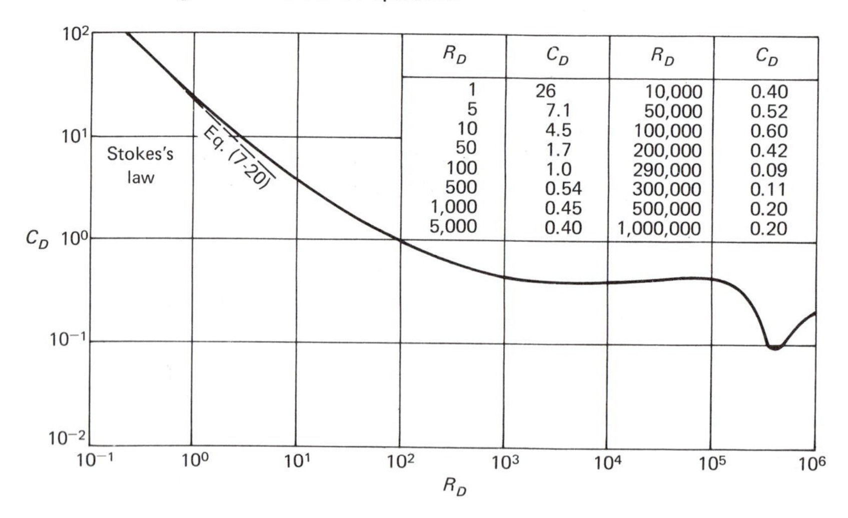

R_D	C_D	R_D	C_D
1	26	10,000	0.40
5	7.1	50,000	0.52
10	4.5	100,000	0.60
50	1.7	200,000	0.42
100	1.0	290,000	0.09
500	0.54	300,000	0.11
1,000	0.45	500,000	0.20
5,000	0.40	1,000,000	0.20

And from Eq. (6-18) the Reynolds number becomes

$$\mathbf{R}_D = \rho \frac{LV}{\mu} = \rho \frac{DV}{\mu} \tag{7-19}$$

Laminar Flow

For laminar flow, Stokes's law is used for Reynolds numbers up to ½. Stokes's law may be stated as follows:

$$C_D = \frac{24}{\mathbf{R}_D} = \frac{24\mu}{\rho DV} \qquad 0 < R_D < 0.5 \tag{7-20}$$

From Eq. (7-18) and (7-20), the drag force of a sphere in laminar flow becomes

$$F_D = C_D \pi D^2 \rho \frac{V^2}{8} = \left(\frac{24\mu}{\rho DV}\right) \pi D^2 \rho \frac{V^2}{8} = 3\pi\mu DV \tag{7-21}$$

EXAMPLE 1 Laminar Drag, Falling Sphere Viscometer A steel ball whose specific weight is 490 lbf/ft³ falls at a steady velocity of 0.16 ft/sec through a mass of oil whose specific weight is 56.78 lbf/ft³. What is the dynamic viscosity of the oil? The sphere diameter is 0.06 in. (see Fig. 7-7).

Step A *Data Reduction*

(1) *unit conversion*

(B-55) $$D = \frac{0.06}{12} = 0.005 \text{ ft}$$

(2) *geometric,* sphere

Table A-10 $$\text{Vol} = \frac{4\pi r^3}{3} = \frac{\pi D^3}{6}$$

FIGURE 7-7 Notation for Example 1

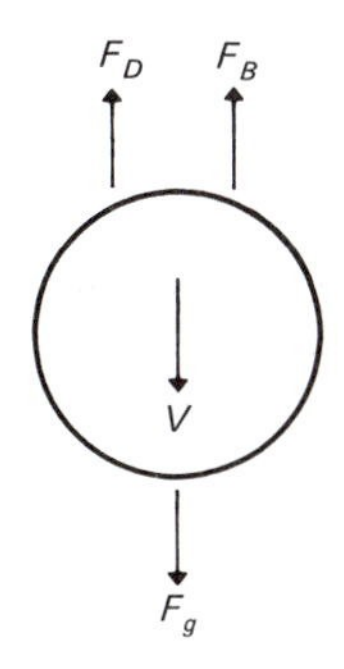

(3) *gravity*

Assume standard

(B-50) $g = 32.17 \text{ ft/sec}^2$

(4) *fluid properties*, oil

(1-35) $\rho = \frac{\gamma}{g} = \frac{56.78}{32.17} = 1.765 \text{ slugs/ft}^3$

Step B *Assume Stokes's Law Applies and Calculate Viscosity*

(7-21) Drag force $F_D = 3\pi\mu DV$

(2-59) Buoyant force $F_B = \frac{\gamma_o \pi D^3}{6}$

(1-35) Gravity force $F_g = \frac{\gamma_s \pi D^3}{6}$

Fig. 7-7 For a "free body"

(2-61) $\Sigma F_z = 0 = F_B - F_g - F_L = F_B - F_g = -F_D$

or

$$F_D = F_g - F_B \quad \text{(a)}$$

Substituting from Eqs. (7-21), (2-59), and (1-35);

$$3\pi\mu\, DV = \frac{\gamma_s \pi D^3}{6} - \frac{\gamma_o \pi D^3}{6} \quad \text{(b)}$$

which reduces to

$$\mu = \frac{(\gamma_s - \gamma_o)D^2}{18V} \quad \text{(c)}$$

$$\mu = \frac{(490 - 56.78)(0.005)^2}{18 \times 0.16}$$

$$= (3.761)(10^{-3}) \text{lbf-sec/ft}^2$$

Step C *Check to Determine if Stokes's Law Applies*

(7-19) $\mathbf{R}_D = \frac{\rho DV}{\mu} = \frac{1.765 \times 0.005 \times 0.16}{3.761 \times 10^{-3}}$

$\mathbf{R}_D = 0.3754 < 0.5 \therefore$ Stokes's law applies

Turbulent Flow

No equations have been developed for the calculation of drag coefficients for turbulent flow around spheres. For estimating purposes, the tabulated drag coefficients given in Fig. 7-6 may be interpolated.

EXAMPLE 2 Turbulent Drag A 75-mm diameter metal ball is towed under fresh water at 20°C by means of a cable fastened to a boat moving steadily in a horizontal direction at 1.8 m/s. Neglect the effect of water on the cable. Estimate the power required to pull the ball.

Step A *Data Reduction*

Fluid properties, water at 20°C

Table A-1 $\rho = 998.3 \text{ kg/m}^3$

Table A-4 $\mu = 10.02 \times 10^{-4} \text{ Pa} \cdot \text{s}$

Step B *Determine Drag of Ball*

(7-19) $$\mathrm{R}_D = \frac{\rho V D}{\mu} = \frac{998.3 \times 1.8 \times 0.075}{10.02 \times 10^{-4}}$$

$$\mathrm{R}_D = 134\ 500$$

Fig. 7-6 $C_D = 0.54$ (linear interpolation of tabulated values)

(7-18) $$F_D = C_D \pi D^2 \frac{\rho V^2}{8}$$

$$\frac{0.54\pi(0.075)^2(998.3)(1.8)^2}{8} = 3.858 \text{ N}$$

Step C *Determine Power*

(4-68) $$P_o = TV = F_D V = (3.858)(1.8) = 6.944 \text{ W}$$

Compressible Flow

For compressible flow, Fig. 7-8 may be used. For estimating, the tabulated values may be linearly interpolated. In each case, it is necessary to check the incompressible drag coefficient of Fig. 7-6 to determine which forces predominate. If the viscous forces predominate, then the drag coefficient of Fig. 7-6 will be higher than that of Fig. 7-8; if the elastic forces predominate, then it will be lower.

EXAMPLE Compressible Flow English artillery tables of the 1750's show that a cannon firing a 4-in. diameter ball weighing 9 lb will have a muzzle velocity of 1,052 ft/sec when charged with 2¼ lb of black powder. Tripling this charge will increase the muzzle velocity by one-half. Gunners are warned not to use the heavier charge not only because the cannon might blow up but also because the increased air resistance will not produce a greater effect. Verify the accuracy of the air resistance statement, assuming air at 68°F and 14.7 psia.

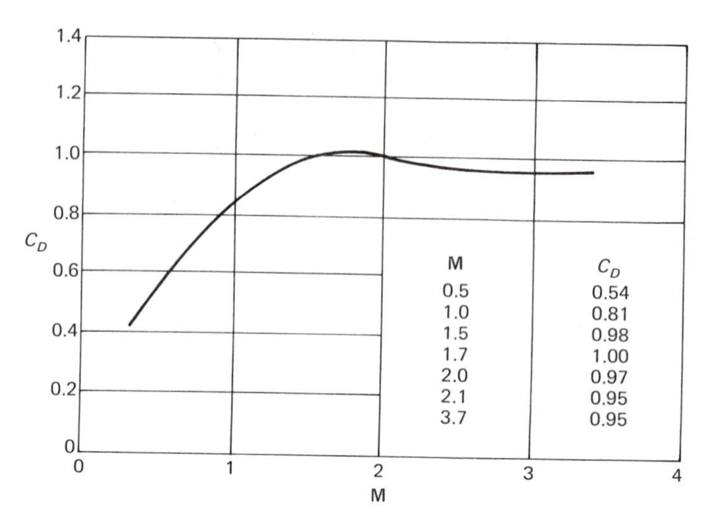

FIGURE 7-8 Drag of a sphere at supersonic velocities

Step A *Data Reduction*

(1) *unit conversions*

(B-55) $D = 4/12 = 1/3$ ft

(B-87) $p = (14.7)(144) = 2{,}117$ lbf/ft^2

(B-28) $T = 68 + 459.67 = 527.67°$R

(2) *proportionality constant*

(1-11) $g_c = 32.17$ lbm-ft/lbf-sec^2

(3) *fluid properties*, air at 68°F

Table A-7 $R = 53.36$ ft-lbf/lbm-°R, $k = 1.400$

Table A-8 $\mu = 38.31 \times 10^{-8}$ lbf-sec/ft^2

(1-44)
$$\rho = \frac{p}{g_c RT} = \frac{2{,}117}{32.17 \times 53.36 \times 527.67}$$

$$\rho = 0.002337 \text{ slugs/ft}^3$$

Step B *Determine Drag with Regular Charge*

(7-19)
$$\mathbf{R}_{D1} = \frac{\rho V_1 D}{\mu} = \frac{0.002337 \times 1{,}052 \times 1/3}{38.31 \times 10^{-8}}$$

$$\mathbf{R}_{D1} = 2.139 \times 10^6$$

Fig. 7-6 $C_{D1} = 0.20$ (assume no change above $\mathbf{R}_D = 10^6$)

(6-22)
$$\mathbf{M}_1 = \frac{V_1}{(kg_c RT)^{1/2}}$$

$$= \frac{1{,}052}{(1.400 \times 32.17 \times 53.36 \times 527.67)^{1/2}}$$

$\mathbf{M}_1 = 0.9342$

Fig. 7-8 $C_{D1} = 0.77$ (linear interpolation of tabulated data)

Since the drag coefficient based on Mach number is higher than the Reynolds number drag coefficient, use Mach number.

(7-18)
$$F_{D1} = \frac{C_{D1}\pi D^2 \rho V^2}{8}$$
$$= \frac{(0.77)(\pi)(1/3)^2(0.002337)(1{,}052)^2}{8}$$
$$= 86.90 \text{ lbf}$$

(1-20)
$$a_{D1} = \frac{g_c F_{D1}}{\mathbf{M}} = \frac{(32.17)(86.90)}{9} = 310.6 \text{ ft/sec}^2$$

(deceleration)

Step C *Determine Drag with Triple Charge*

(7-19)
$$\mathbf{R}_{D2} = \frac{\rho V_2 D}{\mu} = \frac{\rho(1.5V)D}{\mu} = 1.5\mathbf{R}_{D1}$$
$$= (1.5)(2.139)(10^6) = (3.209)(10^6)$$

Fig. 7-6 $C_{D1} = 0.20$ (assume no change above $\mathbf{R}_D = 10^6$)

(6-22)
$$\mathbf{M}_2 = \frac{V_2}{(kg_c RT)^{1/2}} = \frac{(1.5V_1)}{(kg_c RT)^{1/2}}$$
$$= 1.5\ \mathbf{M}_1 = 1.5 \times 0.9342 = 1.401$$

Fig. 7-8 $C_{D2} = 0.95$ (by linear interpolation)

Again drag coefficient based on Mach number is higher.

(7-18)
$$F_{D2} = \frac{C_{D1}\pi D^2 \rho V_2^2}{8}$$
$$= \frac{(0.95)(\pi)(1/3)^2(0.002337)(1.5 \times 1{,}052)^2}{8}$$
$$= 241.2 \text{ lbf}$$

(1-20)
$$a_{D2} = \frac{g_c F_{D2}}{\mathbf{M}} = \frac{(32.17)(241.2)}{9} = 862.2 \text{ ft/sec}^2$$

Step D *Compare Results*

Approximate velocity 1 sec after leaving muzzle:

$$V_{X1} = V_1 - a_{D1}t = 1{,}052 - 310.6(1)$$
$$= 741.4 \text{ ft/sec}$$
$$V_{X2} = V_2 - a_{D2}t = (1.5)(1{,}052) - 862.6(1)$$
$$= 715.4 \text{ ft/sec}$$

Manual is correct. (Or is it? Try a series of shorter time intervals)

7-6 DRAG OF A CYLINDER

In calculating the drag of circular cylinders, the cylinder diameter, D, is used as the characteristic length and the projected area, LD, is used for the drag area. The drag of a cylinder is obtained by substituting the projected area in Eq. (7-4);

$$F_D = \frac{C_D \rho V^2 A}{2} = \frac{C_D \rho V^2 LD}{2} \tag{7-22}$$

Equation (7-19) developed for spheres may be used for Reynolds number calculations. Figure 7-9 gives drag coefficients for cylinders of infinite length and Table 7-3 for those of finite length.

EXAMPLE A cylindrical smokestack is 60 cm in diameter and 900 cm high. A wind of 2.25 m/s blows across the stack. Assume air at 30°C and 101 kPa and estimate the maximum bending moment due to the wind (see Fig. 7-10).

Step A *Data Reduction*

(1) *unit conversion*

(B-27) $T = 30 + 273.15 = 303.15 \text{ K}$

FIGURE 7-9 Drag coefficients of a cylinder of infinite length

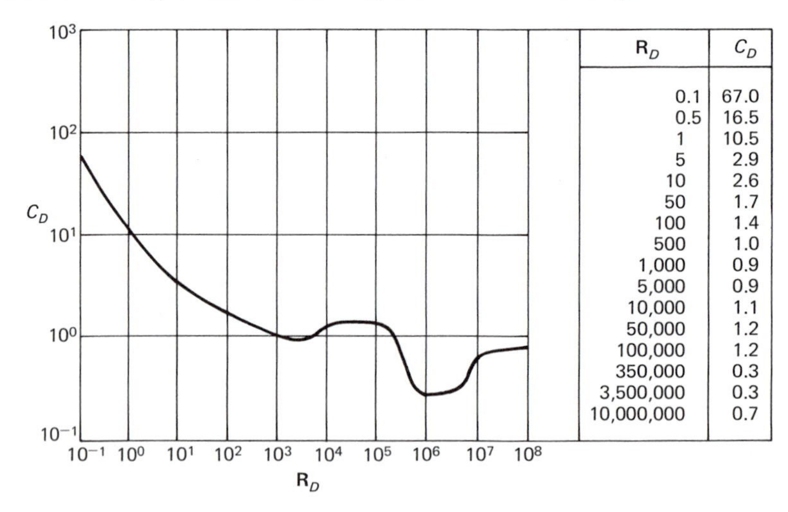

R_D	C_D
0.1	67.0
0.5	16.5
1	10.5
5	2.9
10	2.6
50	1.7
100	1.4
500	1.0
1,000	0.9
5,000	0.9
10,000	1.1
50,000	1.2
100,000	1.2
350,000	0.3
3,500,000	0.3
10,000,000	0.7

TABLE 7-3 *Drag Coefficients for Finite Cylinders*

	L/D	1	5	10	20	30	∞
$10^3 < \mathbf{R}_D < 10^5$	C_D	0.63	0.8	0.83	0.93	1.0	1.2

(2) *proportionality constant*

(1-9) $$g_c = 1 \text{ kg} \cdot \text{m/N} \cdot \text{s}^2$$

(3) *fluid properties,* air at 30°C

Table A-7 $$R = 287.1 \text{ J/kg} \cdot \text{K}$$

Table A-8 $$\mu = 18.64 \times 10^{-6} \text{ Pa} \cdot \text{s}$$

(1-44) $$\rho = \frac{p}{g_c RT} = \frac{101\ 000}{(1)(287.1)(303.15)}$$

$$\rho = 1.161 \text{ kg/m}^3$$

(4) *geometric*

$$\frac{L}{D} = \frac{900}{60} = 15$$

FIGURE 7-10 Notation for example problem

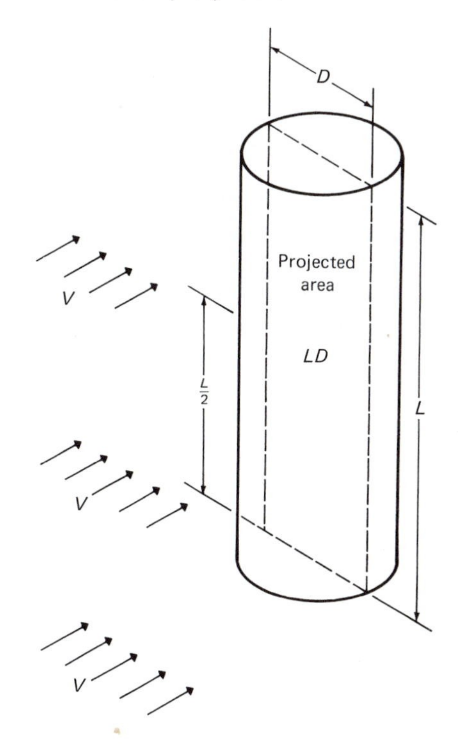

Step B *Determine Drag Coefficient*

(7-19) $$\mathrm{R}_D = \frac{\rho DV}{\mu} = \frac{1.61 \times 0.60 \times 2.25}{18.64 \times 10^{-6}}$$

$\mathrm{R}_D = 8.409 \times 10^4$ lies between 10^3 and 10^5

Table 7-3 $C_D = 0.88$ (by linear interpolation)

Step C *Compute Force and Moment*

(7-22) $$F_D = \frac{C_D \rho V^2 LD}{2} = \frac{(0.88)(1.161)(2.25)^2(0.60)(9)}{2}$$

$F_D = 13.97$ N

Fig. 7-10 Maximum moment will be at the base of the stack or

$$M_{\max} = F_D\left(\frac{L}{2}\right) = 13.97\left(\frac{9}{2}\right) = 62.87 \text{ N}\cdot\text{m}$$

7-7 WAKE FREQUENCY

An object in a fluid stream may be subject to the downstream periodic shedding of vortices from first one side and then the other. The frequency of the resulting transverse (lift) force is a function of the stream Strouhal number. As the wake frequency approaches the natural frequency of the structure, the periodic lift force increases asymptotically in magnitude, and when resonance occurs, the structure fails. Neglecting to take this phenomenon into account in design has been responsible for failures of electric transmission lines, submarine periscopes, smokestacks, bridges, and thermometer wells. The wake frequency characteristics of *cylinders* are shown in Fig. 7-11. At a Reynolds number of about 20, vortices begin to shed alternately. Behind the cylinder is a staggered stable arrangement of vortices known as the Kármán vortex trail. At a Reynolds number of about 10^5, the flow changes from laminar to turbulent. At the end of the transition zone ($\mathrm{R} \approx 3.5 \times 10^5$), the flow becomes turbulent and the alternate shedding stops and the wake is aperiodic. At the end of the supercritical zone ($\mathrm{R} \approx 3.5 \times 10^6$), the wake continues to be turbulent, but the shedding again becomes alternate and periodic.

The alternating *lift force,* is given by

$$F_L(t) = \frac{C_L \rho V^2 A \sin(2\pi f t)}{2} \qquad \text{(7-23)}$$

where t is the time.

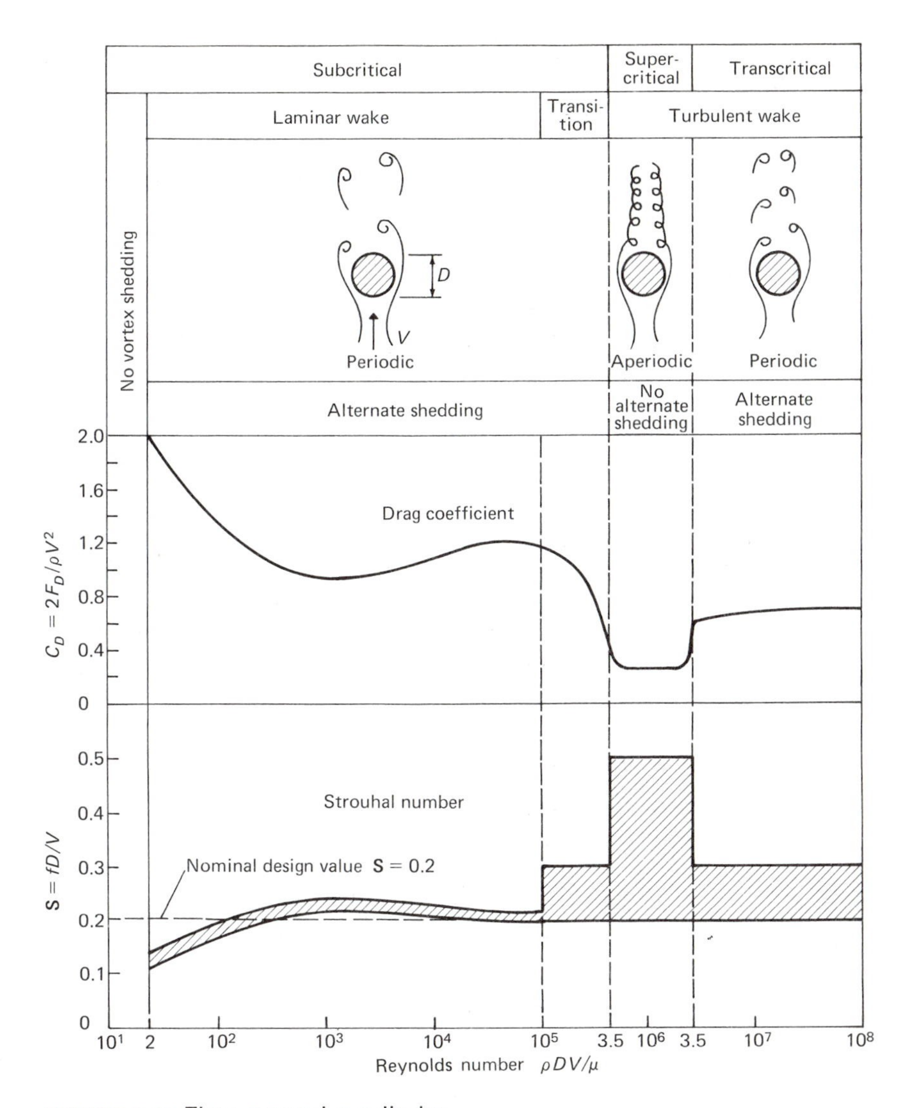

FIGURE 7-11 Flow around a cylinder

The Strouhal number may be computed from Eq. (7-24)

$$S = \frac{fD}{V} \tag{7-24}$$

The Strouhal number is nearly constant to $\mathbf{R} = 10^5$, and a nominal design value of 0.2 is generally used. Above $\mathbf{R} = 10^5$, data from different experimenters vary widely, as indicated by the cross-hatched zone of Fig. 7-11. This wide zone is due to experimental and/or measurement difficulties and the dependence on surface roughness to "trigger" the boundary layer. Examination of Fig. 7-11 reveals an inverse relation of Strouhal number to drag coefficient.

Observation of actual structures shows that they vibrate at their natural frequency and with a mode shape associated with their fundamental (first) mode during vortex excitation.

EXAMPLE A smokestack is to be erected at a location where winter winds up to 50 mph have been observed. The stack design under consideration has a diameter of 16 ft. The natural frequency of this type of stack is given by

$$f_n = \frac{10^5}{H^2}$$

where f_n is the natural frequency in hertz (cycles per second) and H is the height of the stack in feet. To insure that failure will not occur from resonance, good design practice limits the wake frequency to 80% of the natural frequency. What is the maximum height of stack that should be erected? Assume air at 32°F and 14.7 psia.

Step A *Data Reduction*

(1) *unit conversions*

(B-28) $$T = 32 + 459.67 = 491.67°\text{R}$$

(B-89) $$p = 14.7 \times 144 = 2{,}117 \text{ lbf/ft}^2$$

(B-65/54) $$V = 50\left(\frac{5{,}280}{3{,}600}\right) = 73.33 \text{ ft/sec}$$

(2) *proportionality constant*

(1-11) $$g_c = 32.17 \text{ lbm-ft/lbf-sec}^2$$

(3) *fluid properties,* air at 32°F

Table A-7 $$R = 53.36 \text{ ft-lbf/lbm-°R}$$

(1-44) $$\rho = \frac{p}{g_c RT} = \frac{2{,}117}{(32.17)(53.36)(491.67)}$$

$$\rho = 0.002508 \text{ slugs/ft}^3$$

Table A-8 $$\mu = 35.67 \times 10^{-8} \text{ lbf-sec/ft}^2$$

Step B *Compute Wake Frequency*

(7-19) $$\mathbf{R}_D = \frac{\rho D V}{\mu} = \frac{(0.002508)(16)(73.33)}{35.67 \times 10^{-8}}$$

$$\mathbf{R}_D = 8.249 \times 10^6$$

Fig. 7-11 $$S_{\max} = 0.5$$

(7-24) $$f_{\max} = \frac{S_{\max} V}{D} = \frac{0.5 \times 73.33}{16} = 2.292 \text{ Hz}$$

Step C *Compute Height of Stack*

Given $$f_n = \frac{f_{\max}}{0.8} = 2.865 \text{ cps}$$

Given $$f_n = \frac{10^5}{H^2}, \quad \text{or} \quad H = \left(\frac{10^5}{f_n}\right)^{1/2}$$

$$H = \left(\frac{10^5}{2.865}\right)^{1/2} = 186.8 \text{ ft}$$

7-8 RESISTANCE OF SHIPS

Dimensional Analysis

Consider a ship whose length is L moving on the surface of a liquid at a speed of V. The liquid has a density of ρ, viscosity of μ, and a surface tension of σ. The acceleration of gravity is g, and the movement of the ship produces a drag force of F_D. Derive the dimensionless parameter involved in this phenomenon. Results of the application of the Buckingham Π theorem are shown in Table 7-4.

Physical Analysis

The dimensional analysis of ship resistance resulted in the same parameters as were obtained from the similarity analysis of Sec. 6-11. Thus, the total drag of a ship is given by

$$F_D = \frac{C_D \rho V^2(\text{wetted surface})}{2} = \frac{C_D \rho V^2 S}{2} \tag{7-25}$$

where

$$C = f(\mathbf{R}, \mathbf{F}, \mathbf{W})$$

Since, as discussed previously, surface tension forces are very weak, these forces are not considered in computation of ship resistance. Conventional practice is to divide the drag coefficient into three individual coefficients as follows:

$$C_D = C_r + C_f + \Delta C_f \tag{7-26}$$

TABLE 7-4 *Dimensional Analysis of Ship Resistance*

Item		Variables considered						
		SHIP		LIQUID DENSITY	DRAG FORCE	GRAVITY	LIQUID	
		Length	*Speed*				*Viscosity*	*Surface tension*
Symbol		L	V	ρ	F_D	g	μ	σ
Selection		B_G^x	B_K^y	B_D^z	A_1	A_2	A_3	A_4
π **Ratios**		**Basic group**			π_1	π_2	π_3	π_4
Dimensions		L	LT^{-1}	$FL^{-4}T^2$	F	LT^{-2}	$FL^{-2}T$	FL^{-1}
Exponents	F	0	0	z	1	0	1	1
for	L	x	y	$-4z$	0	1	-2	-1
Equations	T	0	$-y$	$2z$	0	-2	1	0

TABLE 7-4 *Solution*

π Ratio	Solved exponents L^x	V^y	ρ^z	Derived ratio	Conventional practice: FORM	SYMBOL	NAME
π_1	-2	-2	-1	$F_D/L^2V^2\rho$	$\frac{2F_D}{\rho V^2(\text{Area})}$	C_D	Drag (Force) coefficient
π_2	-1	-2	0	Lg/V^2	$\frac{V}{\sqrt{Lg}}$	**F**	Froude number
π_3	-1	-1	-1	$\mu/LV\rho$	$\frac{\rho LV}{\mu}$	**R**	Reynolds number
π_4	-1	-2	-1	$\sigma/LV^2\rho$	$\frac{LV^2\rho}{\sigma}$	**W**	Weber number

where

C_r = residual coefficient, a function of Froude number (typical values of C_r for passenger and cargo ships are given in Table 7-5).

C_f = skin friction coefficient for a smooth, flat plate as computed from Eqs. (7-15), (7-16), or (7-17).

ΔC_f = allowance for surface roughness, for actual ships $\Delta C_f = 4 \times 10^{-4}$ for models $\Delta C_f = 0$.

TABLE 7-5 *Typical Values of C_r for Passenger and Cargo Ships*

Froude number	Residual coefficient
0.15	0.53×10^{-3}
0.20	0.72×10^{-3}
0.25	0.87×10^{-3}
0.30	1.53×10^{-3}
0.35	2.00×10^{-3}

The values of the residual coefficient given in Table 7-5 are typical. Actual values of C_r can be obtained only from model or actual ship testing.

Substituting Eq. (7-26) in Eq. (7-25),

$$F_D = \frac{(C_r + C_f + \Delta C_f)\,\rho V^2 S}{2}$$

or

$$F_D = F_r + F_f + F_\Delta \tag{7-27}$$

where F_r is the residual drag, F_f is the skin friction drag, and F_Δ is the drag force allowance for roughness.

EXAMPLE A 1/25 model of a C-4 cargo ship is towed in fresh water at 20°C for Froude number similarity. The C-4 is designed for a speed of 16.5 knots in sea water at 20°C. The ship is 150 m long at its waterline and has a wetted surface of 3 600 m². If it requires 32.38 N to tow the model, estimate the power necessary to drive the ship at design conditions.

Step A *Data Reduction*

(1) *unit conversion*

(B-60) $$V = 16.5\left(\frac{1\,852}{3\,600}\right) = 8.488 \text{ m/s}$$

(2) *fluid properties, fresh water* at 20°C

Table A-1 $\rho_m = 998.3 \text{ kg/m}^3$

Table A-4 $\mu_m = 10.02 \times 10^{-4} \text{ Pa} \cdot \text{s}$

Sea water at 20°C

Table A-1 $\rho_p = 1\,025 \text{ kg/m}^3$

Table A-4 $\mu_p = 10.83 \times 10^{-4} \text{ Pa} \cdot \text{s}$

Step B *Compute Model to Prototype Relationships*

(6-1) $$L_M = \frac{L_p}{25} = \frac{150}{25} = 6 \text{ m}$$

(6-2) $$S_M = S_p\left(\frac{L_M}{L_p}\right)^2 = 3\,600\left(\frac{6}{150}\right)^2 = 5.760 \text{ m}^2$$

(6-27) $$V_M = V_p\left(\frac{L_M}{L_p}\right)^{1/2} \qquad (g_M = g_p)$$

$$V_M = 8.433\left(\frac{6}{150}\right)^{1/2} = 1.687 \text{ m/s}$$

Step C *Determine Skin Friction Drag of Model*

(6-18) $$\mathbf{R}_m = \frac{\rho_m V_m L_m}{\mu_m}$$

$$\mathbf{R}_m = \frac{998.3 \times 1.687 \times 6}{10.02 \times 10^{-4}} = 1.008 \times 10^7$$

(7-16) $$C_{fm} = 0.455(\log_{10} R_m)^{-2.58}$$

$$= 0.455(\log_{10} 1.008 \times 10^7)^{-2.58}$$

$$C_{fm} = 3.00 \times 10^{-3}$$

(7-25) $$F_{fm} = \frac{C_{fm} \rho_m V_m^2 S_m}{2} = \frac{3.000 \times 10^{-3} \times 998.3(1.687)^2 \times 5.760}{2}$$

$$= 24.55 \text{ N}$$

Step D *Determine Residual Coefficient*

(7-27) $$F_{rm} = F_{Dm} - F_{fm} - F_{\Delta m}$$

$$F_{rm} = 32.38 - 24.55 - 0 = 7.83 \text{ N}$$

(7-25) $$C_{rm} = \frac{2F_{rm}}{\rho_m V_m^2 S_m}$$

$$C_{rm} = \frac{2 \times 7.83}{998.3(1.687)^2\ 5.760} = 9.569 \times 10^{-4}$$

$$C_{rm} = C_{rp} = C_r \quad \text{(same Froude number)}$$

Step E *Determine Skin Friction Drag of Ship*

(6-18) $$\mathbf{R}_p = \frac{\rho_p V_p L_p}{\mu_p} = \frac{1\,025 \times 8.488 \times 150}{10.83 \times 10^{-4}}$$

$$\mathbf{R}_p = 1.205 \times 10^9$$

(7-16) $$C_{fp} = 0.455(\log_{10} \mathbf{R}_p)^{-2.58} = 0.455(\log_{10} 1.205 \times 10^9)^{-2.58}$$

$$C_{fp} = 1.535 \times 10^{-3}$$

Step F *Compute Power of Ship*

$$C_{Dp} = C_{rp} + C_{fp} + \Delta C_{fp}$$

$$C_{Dp} = 9.569 \times 10^{-4} + 15.33 \times 10^{-4} + 4 \times 10^{-4}$$

$$C_{Dp} = 2.892 \times 10^{-3}$$

(7-25) $$F_{Dp} = \frac{C_{Dp} \rho_p V_p^2 S_p}{2} = \frac{2.892 \times 10^{-3} \times 1\,025 \times (8.488)^2 \times 3\,600}{2}$$

$$= 384\,419 \text{ N}$$

(4-68) $$P_p = F_{Dp} V_p = 384\,419 \times 8.488 = 3\,263 \text{ kW}$$

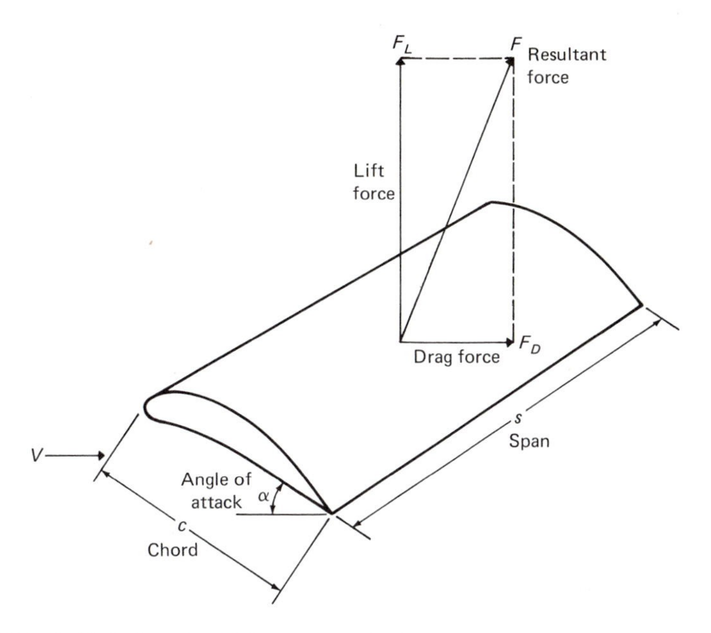

FIGURE 7-12 Notation for lifting vane study

7-9 LIFTING VANES

Definitions

Figure 7-12 shows the notation used in studies of lifting vanes or airfoils

c = Chord length
s = Span of vane
s/c = Aspect ratio
α = Angle of attack (alpha)
cs = Characteristic area

Physical Analysis

When fluid flows about an object such as a lifting vane whose axis is not aligned with the flow, the velocities on either side of the object will have different magnitudes.

With the characteristic area being taken as *cs*, the product of the chord times the span, from Eqs. (7-4) and (7-5) for lifting vanes we can write

$$F_D = \frac{C_D \rho V^2 A}{2} = \frac{C_D \rho V^2 cs}{2} \tag{7-28}$$

$$F_L = \frac{C_L \rho V^2 A}{2} = \frac{C_L \rho V^2 cs}{2} \tag{7-29}$$

Although drag and lift coefficients are a function of Mach and Reynolds numbers, this section will be limited to the consideration of the "shape" effects on a typical lifting vane at constant Reynolds numbers in incompressible fluids.

Examination of Fig. 7-13 indicates that the velocity is higher on the upper side and lower on the underside. From the energy analysis of Sec. 4-4, the increased velocity of the upper side will result in decreased pressures and the decreased velocities on the underside will result in increased pressure.

Figure 7-14 shows the pressure profile around a lifting vane. Lift, of course, is the result of these unbalanced pressures acting on the vane.

While it is possible to determine the lift characteristics by integration of the velocity profile for simple objects, actual practice is to develop the lift and drag coefficients from wind tunnel testing.

Figure 7-15 shows the results of a typical test on a vane. The object is to design a vane to have the minimum drag for the maximum

FIGURE 7-13 Flow around a lifting vane

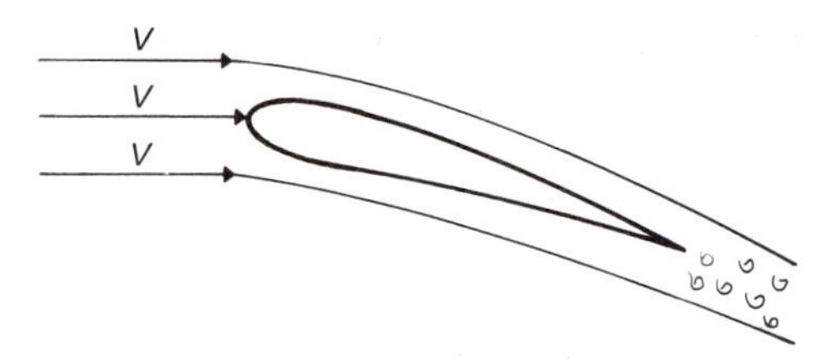

FIGURE 7-14 Pressure profile of a lifting vane

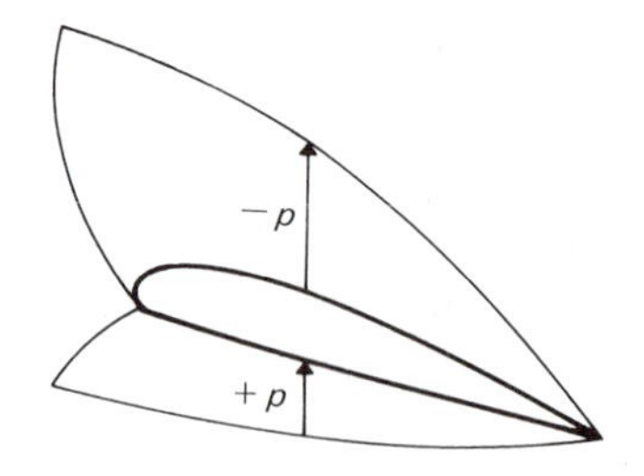

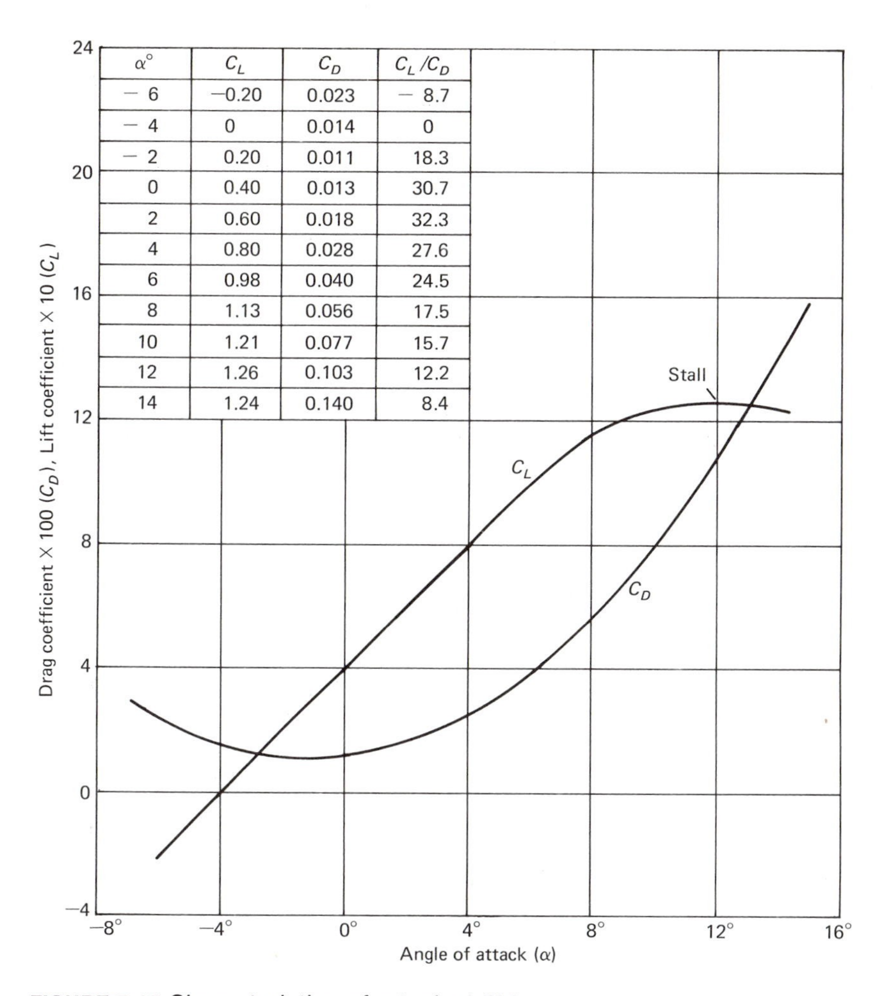

$\alpha°$	C_L	C_D	C_L/C_D
− 6	−0.20	0.023	− 8.7
− 4	0	0.014	0
− 2	0.20	0.011	18.3
0	0.40	0.013	30.7
2	0.60	0.018	32.3
4	0.80	0.028	27.6
6	0.98	0.040	24.5
8	1.13	0.056	17.5
10	1.21	0.077	15.7
12	1.26	0.103	12.2
14	1.24	0.140	8.4

FIGURE 7-15 Characteristics of a typical lifting vane

lift. Maximum C_L/C_D occurs for the vane of Fig. 7-15 at an angle of attack of about 2°.

"Stall" is the point where the coefficient of lift reaches its maximum and starts to decrease in value with increased angle of attack. When stall occurs, there is a marked turbulence in the wake and damage can result to machinery employing lifting-type vanes.

EXAMPLE A Boeing 727 airplane has a wing span of 145.9 ft and a chord length of 19.8 ft. It weighs 328,000 lb when loaded and is designed to cruise at 500 knots at an altitude of 40,000 ft. Estimate the power required to drive the airplane at this speed.

Assume that the wing has the same characteristics as the lifting vane of Fig. 7-15. Neglect lift and drag of the fuselage.

Step A *Data Reduction*

(1) *unit conversion*

(B-59) $V = 500 \times 1.688 = 844.0$ ft/sec

(2) *proportionality constant*

(1-11) $g_c = 32.17$ lbm-ft/lbf-sec^2

(3) *atmospheric properties,* 40,000 ft

Table A-9 $T = 389.97°R$, $p = 393.1$ lbf/ft^2

(4) *fluid properties,* air

Table A-7 $R = 53.36$ ft-lbf/lbm-°R

(1-44) $$\rho = \frac{p}{g_c RT} = \frac{393.1}{32.17 \times 53.36 \times 389.97}$$

$$\rho = 5.872 \times 10^{-4} \text{ slugs/ft}^3$$

Step B *Determine Lift and Drag Coefficients*

For level flight, neglecting the buoyant force of the aircraft,

(2-61) $$\Sigma F_2 = 0 = F_B - F_g - F_L = 0 - F_g - F_L$$

or

$$F_L = -F_g = |F_g| = 328{,}000 \text{ lbf}$$

(7-29) $$C_L = \frac{2F_L}{\rho V^2 cs}$$

$$C_L = \frac{2 \times 328{,}000}{(5.872 \times 10^{-4})(844.0)^2(19.8)(145.9)}$$

$$C_L = 0.5429$$

Fig. 7-15 $C_D = 0.0166$ (linear interpolation of tabulated data)

Step C *Compute Power*

(7-28) $$F_D = \frac{C_D \rho V^2 cs}{2}$$

$$F_D = \frac{(0.0166)(5.872 \times 10^{-4})(844.0)^2(19.8)(145.9)}{2}$$

$$F_D = 10{,}029 \text{ lbf}$$

(4-68) $$P_o = TV = F_D V = (10{,}029)(844.0)$$

$$P_o = 8{,}464{,}476 \text{ ft-lbf/sec}$$

(B-52) $$\text{HP} = \frac{8{,}464{,}476}{550} = 15{,}390 \text{ horsepower}$$

7-10 PROPELLER CHARACTERISTICS

Dimensional Analysis

Consider a propeller rotating in a fixed plane with fluid approaching at a velocity of V. The fluid has a density of ρ, a viscosity of μ, and a bulk modulus of elasticity of E. The propeller rotates at a speed of N, has a diameter of D, and produces a thrust of T for a power input of P. Derive the dimensionless parameters that characterize propeller performance. Results of the application of the Buckingham Π theorem are shown in Table 7-6.

Performance Parameters

An efficient propeller is one in which every cross-section of each blade has the performance characteristics of a well-designed lifting vane. Figure 7-16 shows a cross-section of a

TABLE 7-6 *Dimensional Analysis of Propellers*

Item		**Variables considered**							
		PROPELLER		FLUID				PROPELLER	
		Diameter	*Speed*	*Density*	*Velocity*	*Viscosity*	*Bulk modulus*	*Thrust*	*Power*
Symbol		D	N	ρ	V	μ	E	T	P
Selection		B_G^x	B_K^y	B_D^z	A_1	A_2	A_3	A_4	A_5
π Ratios		**Basic group**			π_1	π_2	π_3	π_4	π_5
Dimensions		L	T^{-1}	$FL^{-4}T^2$	LT^{-1}	$FL^{-2}T$	FL^{-2}	F	FLT^{-1}
Exponents	F	0	0	z	0	1	1	1	1
for	L	x	0	$-4z$	1	−2	−2	0	1
equations	T	0	$-y$	$2z$	−1	1	0	0	−1

TABLE 7-6 *Solution*

π Ratio	**Solved exponents** D^x	N^y	ρ^z	**Derived ratio**	**Conventional practice** FORM	SYMBOL	NAME
π_1	−1	−1	0	V/DN	V/DN	**V**	Advance-diameter ratio (velocity ratio)
π_2	−2	−1	−1	$\mu/D^2N\rho$	$D(ND)\rho/\mu$	**R**	Reynolds number
π_3	−2	−2	−1	$E/D^2N^2\rho$	$DN/\sqrt{E/\rho}$	**M**	Mach number
π_4	−4	−2	−1	$T/D^4N^2\rho$	$T/D^4N^2\rho$	$\mathbf{C}_T$	Thrust coefficient
π_5	−5	−3	−1	$P/D^5N^3\rho$	$P/D^5N^3\rho$	$\mathbf{C}_P$	Power coefficient

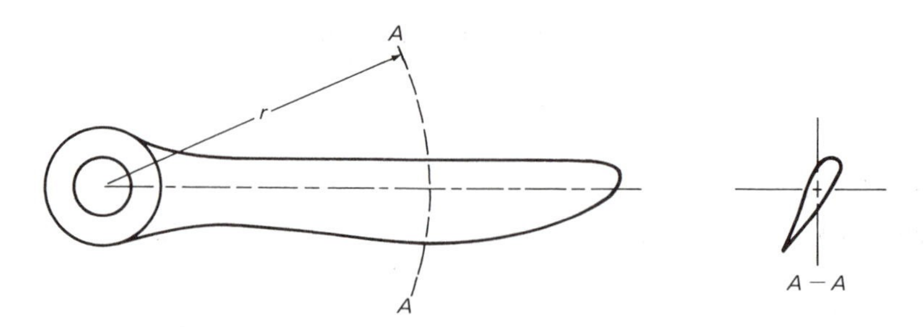

FIGURE 7-16 Propeller blade

propeller blade at radius r. The study and analysis of propeller performance and design is called blade element theory.

As in the case of the lifting vane, propeller performance is based on test data. The first three dimensionless parameters obtained from the dimensional analysis were also obtained from consideration of force similarity in Sec. 6-6. The last parameter was also derived by dimensional analysis of fluid machinery in Sec. 6-16. Only the thrust coefficient was derived for the first time in this text:

$$T = \mathrm{C}_T D^4 N^2 \rho \tag{7-30}$$

The velocity ratio when applied to propellers is called the advance-diameter ratio because it is the distance V/N that the propeller could advance in the fluid per revolution divided by the diameter.

The efficiency η (eta) of a propeller is defined as the useful power divided by the power output. The useful power is defined by Eq. (4-68) as TV, the thrust by Eq. (7-30) as $\mathrm{C}_T D^4 N^2 \rho$, and the power input in Table 6-4 as $\mathrm{C}_P D^5 N^3 \rho$. Thus

$$\eta = \frac{\text{useful power}}{\text{power input}} = \frac{(\mathrm{C}_T D^4 N^2 \rho) V}{\mathrm{C}_P D^5 N^3 \rho} = \frac{\mathrm{C}_T V}{\mathrm{C}_P N D} = (\mathrm{C}_T/\mathrm{C}_P)\mathrm{V} \tag{7-31}$$

EXAMPLE A 3-m diameter propeller has the performance characteristics given in Fig. 7-17, rotates at a speed of 22 rps at an altitude of 5 km. Determine for maximum propeller efficiency (a) air speed, (b) available thrust, and (c) power required to drive propeller.

Step A *Data Reduction*

(1) *proportionality constant*

(1-9) $g_c = 1\ \mathrm{kg \cdot m/N \cdot s^2}$

(2) *atmospheric properties,* 5 km

Table A-9 $T = 255.68$ K, $p = 54.05$ kPa

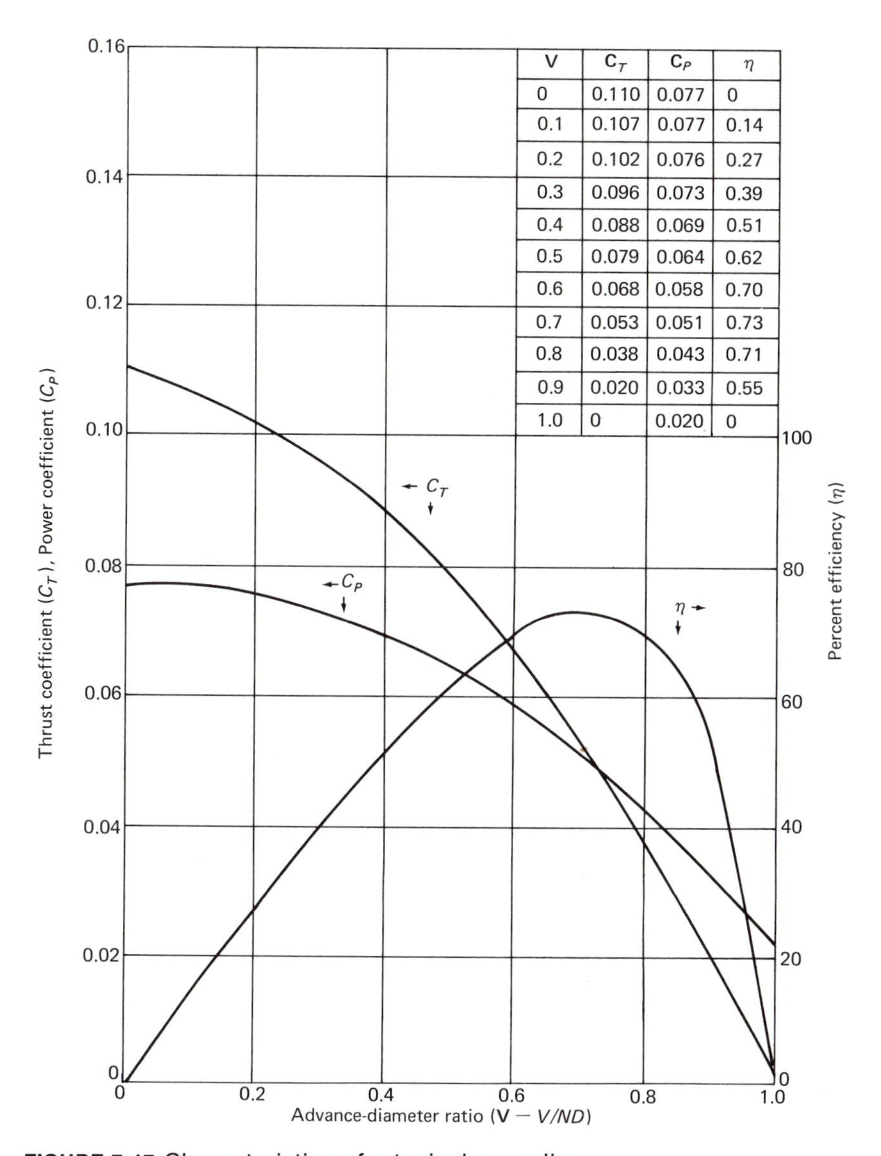

V	C_T	C_P	η
0	0.110	0.077	0
0.1	0.107	0.077	0.14
0.2	0.102	0.076	0.27
0.3	0.096	0.073	0.39
0.4	0.088	0.069	0.51
0.5	0.079	0.064	0.62
0.6	0.068	0.058	0.70
0.7	0.053	0.051	0.73
0.8	0.038	0.043	0.71
0.9	0.020	0.033	0.55
1.0	0	0.020	0

FIGURE 7-17 Characteristics of a typical propeller

(3) *fluid properties,* air

Table A-7 $R = 287.1$ J/kg · K

(1-44)
$$\rho = \frac{p}{g_c RT} = \frac{54\,050}{1 \times 287.1 \times 255.68} = 0.736\,3 \text{ kg/m}^3$$

Step B *Advance-Diameter Ratio for Maximum Efficiency*

Fig. 7-17 Maximum efficiency occurs at $\mathbf{V} = 0.7$; at this point, $\mathbf{C}_T = 0.053$ and $\mathbf{C}_p = 0.051$.

Step C *Compute Air Speed*

(6-23) $V = \mathbf{V}\, DN = 0.7 \times 3 \times 22 = 46.20 \text{ m/s}$

Step D *Compute Thrust*

(7-30) $T = C_T D^4 N^2 \rho$

$T = 0.053(3)^4(22)^2\ 0.736\ 3 = 1\ 530 \text{ N}$

Step E *Compute Power Input*

Table 7-6 $P = C_P D^5 N^3 \rho$

$P = 0.051(3)^5(22)^3\ 0.736\ 3 = 97.16 \text{ kW}$

PROBLEMS

1. A metal ball (specific gravity = 8) is towed under water by means of a cable fastened to a boat moving along the surface in a horizontal direction at 4 mph. The angle between the cable and the vertical is 13°. The ball has a 3-in. diameter and the water temperature is 68°F. Estimate the coefficient of drag. Neglect the effect of water on the cable.
2. A force of 450 N is required to tow a board under water at 20°C and at a velocity of 3 m/s. The board has an area of 0.28 m^2 and an angle of attack of 8°. What is the lift coefficient?
3. Air at 104°F and 30 psia flows over a smooth, flat plate with a velocity of 1 ft/sec. The plate is 40 ft long in the direction of the air flow and 10 ft wide. Estimate (a) the thickness of the boundary layer at the trailing edge and (b) the drag force on the plate.
4. A winter gale sweeps across the flat roof of a building. The length of the building is 120 m in the direction of the wind and 90 m wide. The velocity of the wind is 38 m/s, air temperature is 0°C, and barometric pressure is 101.3 kPa. Estimate the skin friction drag on the roof of the building assuming that the boundary layer is turbulent.
5. Estimate the additional power required to carry a sign facing on-coming traffic on the top of a taxicab moving at 100 km/hr if the sign is 0.6 m × 1.5 m and the atmospheric conditions are 20°C and 101.3 kPa.
6. The General Motors Test Code gives the following equation for estimating the road horsepower of an automobile:

$$RHP = \frac{(K_1 W + K_2 A v^2)v}{375}$$

where RHP = road horsepower, W = weight of the automobile as loaded in pounds, v = the speed in miles per hour, A = the projected area in square feet, $K_1 = 0.015$, and $K_2 = 0.00125$. Standard conditions of atmosphere for code tests are 60°F and 14.7 psia. What coefficients of rolling friction and drag are used in this equation?

7. A typical American passenger automobile has a projected area of 2.42 m^2 and weighs 17.8 kN. If the resistance due to rolling friction is 267 N, estimate the power necessary to drive this car at (a) 30, (b) 60, and (c) 90 km/hr (assume air at 20°C and 101.3 kPa).
8. A rectangular pier in a river is 5 ft wide and 12 ft long. The average velocity of the 68°F water is 10 ft/sec. The depth of water is 9 ft. Estimate the total force imposed by the flow on the pier.
9. A spherical raindrop 0.05 mm in diameter falls through still air. Estimate raindrop velocity if the air is at 40°C and 101.3 kPa.
10. A hollow steel sphere 1 ft in diameter falls through still air. The sphere weighs 8 lb and the air is at 14.7 psia and 86°F. Estimate the terminal velocity of the sphere.
11. A solid steel ball (γ = 77 kN/m^3) 300 mm in diameter falls through still air. The air is at 101.3 kPa and 30°C. Estimate the terminal velocity.
12. A cylindrical chimney 3 ft in diameter and 75 ft high is exposed to a 35 mph wind (ρ = 0.0024 slugs/ft^3). Measurements at the base of the chimney indicate a bending moment of 9,200 lbf-ft. Compute the drag coefficient.
13. A wire 10 mm in diameter and 3 m long is exposed to a flow of carbon dioxide whose velocity is 30 m/s, temperature 30°C, and pressure 101.3 kPa. Estimate the impingement force of the fluid on the wire.
14. A thermometer well is 4 in. long and has a diameter of 13/16 in. If the natural frequency of the well is 1612 Hz, at what minimum fluid velocity will this well go into resonance?
15. A rectangular barge 10 m wide and 30 m long is towed in 20°C fresh water at a speed of 10 knots. The barge and its cargo create a draft of 1 m. Estimate the tension in the towing cable.
16. An ocean liner 600 ft long has a wetted area of 90,000 ft^2. To drive this ship at 16 knots through 68°F sea water 14,000 shaft horsepower are required. Estimate the propulsive efficiency.
17. A Douglass DC-8 airplane has a gross mass of 141 000 kg and a wing area of 226 m^2. Estimate the minimum landing speed of this aircraft without stall in 20°C, 101.3 kPa air. Assume that the wing has the characteristics of the lifting vane of Fig. 7-15. Neglect the lift and drag of the fuselage.
18. A Piper Super-Cub plane has a gross weight of 1,750 lb, a wing area of 178.5 ft^2, and a cruising speed of 100 knots. What is the

overall lift coefficient of this aircraft? What power is required if the ratio of C_L/C_D of the plane is the same as that given in Fig. 7-15. Assume air at 68°F and 14.7 psia.

19. A swamp-buggy is driven by a 1.8-m diameter propeller whose characteristics are identical to those shown in Fig. 7-17. At what speed should this propeller be operated to produce a starting thrust of 250 N in 30°C, 101.3 kPa air?
20. A helicopter weighing 2000 lb is hovering at a certain level where the air density is 0.0024 slugs/ft^3. The propeller diameter is 20 ft and the propeller has the characteristics given in Fig. 7-17. Estimate (a) propeller speed, (b) power input, (c) power added to fluid, and (d) system efficiency.

REFERENCES

Abbot, I. H., and A. E. Von Doenhoff. *Theory of Wing Sections.* Dover Publications, New York, 1959.

Gertler, M., *The Prediction of the Effective Horsepower of Ships by Methods in Use at the David Taylor Model Basin,* David Taylor Model Basin Report No. 576, Carderock, Md., December 1947.

Levins, R. B., "Vortex Induced Vibration of Circular Cylindrical Structures." *American Society of Mechanical Engineers Paper* 72-WA/FE-39, 1972.

Von Mises, R., *Theory of Flight.* Dover Publications, New York, 1959.

Murdock, J. W., "Power Test Code Thermometer Wells." Transactions of the American Society of Mechanical Engineers, *Journal of Engineering for Power,* October 1959, pp. 403–416.

Nanucy, A., "Artillery through the Ages." National Park Service Interpretive Series, History No. 3, U.S. Government Printing Office, Washington, D.C., 1949 (reprint, 1962).

Schlichting, H., *Boundary-Layer Theory,* 6th ed. McGraw-Hill Book Company, New York, 1968.

Staley, C. M., and G. G. Graven, "The Static and Dynamic Wind Design of Steel Stacks." *American Society of Mechanical Engineers Paper* 72-PET-30, 1972.

Automotive Test Code, 5th ed., 3d modification. General Motors Corporation, Detroit, Mich., 1966.

8 Flow in Pipes

8-1 INTRODUCTION

This chapter is a continuation of Chapter 7 in that a pipe may be considered to be a flat plate formed into a channel of a specific shape such as a circular pipe. The approach is the same as in the previous chapter, a combination of the theory developed in the first five chapters with the dimensionless parameters derived in Chapter 6 to produce solutions to engineering problems.

The current approach to pipe friction originates from a paper published in 1944 by Lewis F. Moody (1880–1953). Moody combined the results of Johann Nikuradse on artificially roughened pipe published in 1933 and the analysis of C. F. Colebrook in 1938 that resulted in the equation that bears his name. Colebrook's analysis combined with the boundary layer theory developed by L. Prandtl and T. von Kármán led to the adoption of the Colebrook equation by the major American technical societies and practicing engineers.

8-2 PARAMETERS FOR INCOMPRESSIBLE FLOW IN PIPES

Dimensional Analysis

Consider an incompressible fluid of density ρ and dynamic viscosity μ, flowing in a horizontal pipe with a velocity V, where the pipe has a uniform internal diameter of D for a straight length of L, and an average absolute surface roughness of ϵ. This flow produces a pressure drop of Δp over the length L. Derive the dimensionless parameters involved in this phenomenon. Results of the application of the Buckingham Π theorem are shown in Table 8-1.

Physical Analysis

From dimensional analysis:

$$\pi_4 = f(\pi_1, \pi_2, \pi_3)$$

or

$$\Delta p = \frac{K \rho V^2}{2} \tag{8-1}$$

where

$$K = f\left(\mathbf{R}, \frac{L}{D}, \frac{\epsilon}{D}\right)$$

TABLE 8-1 *Dimensional Analysis of Pipe Flow*

Item		Variables considered						
		PIPE DIAMETER	FLUID			PIPE		
			Velocity	*Density*	*Viscosity*	*Length*	*Roughness*	*Pressure drop*
Symbol		D	V	ρ	μ	L	ϵ	Δp
Selection		B_G^x	B_K^y	B_D^z	A_1	A_2	A_3	A_4
π **Ratios**		**Basic group**			π_1	π_2	π_3	π_4
Dimensions		L	LT^{-1}	$FL^{-4}T^2$	$FL^{-2}T$	L	L	FL^{-2}
Exponents	F	0	0	z	1	0	0	1
for	L	x	y	$-4z$	-2	1	1	-2
equations	T	0	$-y$	$2z$	1	0	0	0

TABLE 8-1 *Solution*

π Ratio	Solved exponents D^x	V^y	ρ^z	Derived ratio	Conventional practice FORM	SYMBOL	NAME
π_1	-1	-1	-1	$\mu/DV\rho$	$\frac{DV\rho}{\mu}$	$\mathbf{R}$	Reynolds number
π_2	-1	0	0	L/D	L/D	—	Relative length
π_3	-1	0	0	ϵ/D	ϵ/D	—	Relative roughness
π_4	-1	-2	0	$\Delta p/\rho V^2$	$\frac{2\Delta p}{\rho V^2}$	K	Resistance coefficient (pressure coefficient)

Equation (8-1) may be written in terms of friction energy or "lost head" by substituting from Eq. (2-6) $\Delta p = \gamma \overline{H}_f$ and from equation (1-35) $\rho = \gamma/g$:

$$\Delta p = \gamma \overline{H}_f = K\rho V^2/2 = K(\gamma/g)V^2/2 \quad \text{or} \quad \overline{H}_f = KV^2/2g \tag{8-2}$$

Conventional practice is to use a friction factor defined as follows:

$$\overline{H}_f = f(L/D)V^2/2g \tag{8-3}$$

where f is the friction factor and $f = f(R, \epsilon/D)$.

For pipe components, the resistance coefficient K is used. The relation of the resistance coefficient K to the friction factor f is obtained by equating Eqs. (8-2) and (8-3):

$$\overline{H}_f = \frac{(fL/D)V^2}{2g} = \frac{KV^2}{2g} \quad \text{or} \quad K = \frac{fL}{D} \tag{8-4}$$

The relation of friction factor to wall shear stress τ_o may be obtained as follows:

$$\Delta pA = \tau_0 A_s \quad \text{for} \quad \Delta p = \frac{\tau_0 A_s}{A}$$

$$\Delta p = \frac{\tau_o(\pi DL)}{(\pi D^2/4)} = \frac{4\tau_o L}{D} \tag{8-5}$$

Substituting Eq. (8-5) in Eq. (8-2),

$$\Delta p = \gamma \overline{H}_f = \frac{4\tau L}{D}$$

or

$$\overline{H}_f = \frac{4\tau_o L}{D\gamma} \tag{8-6}$$

Equating Eqs. (8-3) and (8-6), noting from Eq. (1-35) $\rho = \gamma/g$,

$$\overline{H}_f = \frac{4\tau_o L}{D\gamma} = \frac{(fL/D)V^2}{2g}$$

or

$$\tau_o = \frac{f\gamma V^2}{8g} = \frac{f\rho V^2}{8} \tag{8-7}$$

Dynamic Similarity

Consideration of the forces involved in incompressible flow (inertia, viscous, gravity, buoyant, and pressure) indicates that the gravity forces will be balanced by the buoyant forces. If the inertia and viscous forces are specified, then for equilibrium the pressure force is also specified. Since Reynolds number is the ratio of the inertia to viscous forces, then for a given ϵ/D the types of flow possible are:

Predominant Force	Value of Reynolds Number	Type of Flow
Viscous	Low	Laminar
Neither	Intermediate	Critical Transition
Inertia	High	Turbulent

Experiments are necessary to establish numerical values of Reynolds number for the types of flow shown above. Experimental friction factor data are usually plotted in the form of a Moody diagram, as shown in Fig. 8-1.

8-3 LAMINAR FLOW IN PIPES

Consider the three-dimensional laminar flow shown in Fig. 8-2. The change in pressure is $p_2 - p_1$ for a distance of L along the pipe. For equilibrium of a "free body":

$$\Sigma F = 0 = (p_2 - p_1)A - \tau A_s = (p_2 - p_1)\pi r^2 - \tau 2\pi r L$$

or

$$\tau = \frac{(p_2 - p_1)r}{2L} \tag{8-8}$$

FIGURE 8-1 Moody diagram for pipe friction

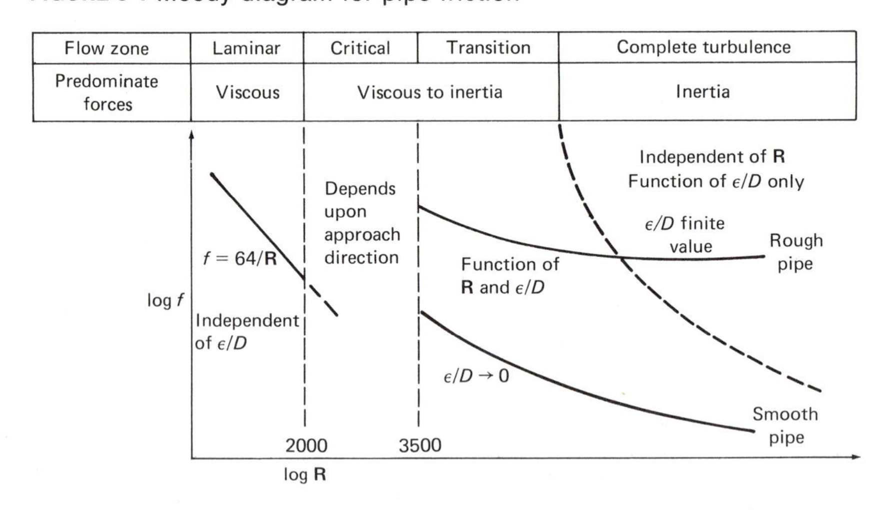

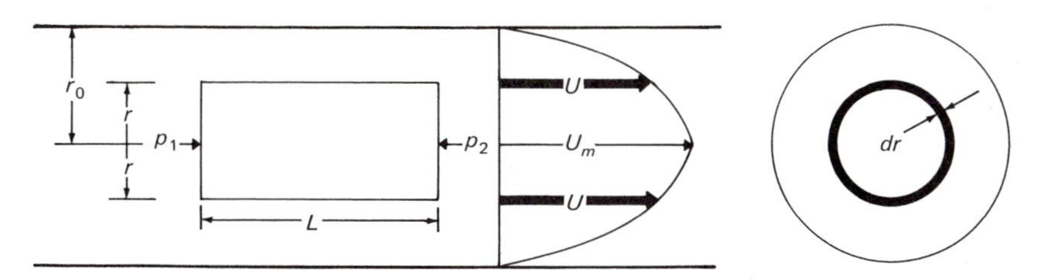

FIGURE 8-2 Notation for laminar flow analysis

If the flow is laminar, only viscous forces are acting, so that Eq. (1-61) $\tau = \mu\, dU/dy$ applies. Noting that $dy = dr$ and substituting in Eq. (8-8) for τ,

$$\mu \frac{dU}{dr} = \frac{(p_2 - p_1)r}{2L}$$

or

$$dU = \left(\frac{p_2 - p_1}{2L\mu}\right) r\, dr \tag{8-9}$$

Integrating Eq. (8-9),

$$\int_{U_m}^{o} dU = \left(\frac{p_2 - p_1}{2L\mu}\right) \int_{o}^{r_o} r\, dr = -U_m = \frac{(p_2 - p_1)r_o^2}{4L\mu}$$

or

$$U_m = \frac{(p_1 - p_2)}{4L\mu}\, r_o^2 = \frac{\Delta p r_o^2}{4L\mu} \tag{8-10}$$

Note that U_m varies as r_o^2, so that the velocity profile is parabolic. From the example of three-dimensional flow reduced to one-dimensional flow for a parabolic velocity profile of Sec. 3-4:

$$U_m = 2V \tag{8-11}$$

Substituting from Eq. (8-11) in Eq. (8-10),

$$V = \frac{\Delta p r_o^2}{8L\mu} \qquad \text{or} \qquad \Delta p = \frac{8L\mu V}{r_o^2} \tag{8-12}$$

From Eq. (8-1) $\Delta p = K\rho V^2/2$ and from Eq. (8-4), $K = fL/D$. Substituting in Eq. (8-12), noting that $r_o = D/2$,

$$\Delta p = \frac{8L\mu V}{r_o^2} = \frac{K\rho V^2}{2} = \frac{(fL/D)\rho V^2}{2} = \frac{8L\mu V}{(D/2)^2}$$

or

$$f = \frac{64\mu}{DV\rho}$$

and from Eq. (6-18) $\mathbf{R} = DV\rho/\mu$ or

$$f = \frac{64}{\mathbf{R}} \tag{8-13}$$

Critical Reynolds Numbers

Experiments show that when the Reynolds number is 2000 or less, the flow is normally laminar. If the flow is initially turbulent, it cannot be maintained indefinitely. For *stable* flow, the Reynolds number of 2000 is the lower limit. It is possible to maintain a laminar flow to very high Reynolds numbers if care is taken to increase the flow gradually, but normally the slightest disturbance will destroy the laminar flow pattern when the Reynolds number is greater than 3000. For stable flow, the Reynolds number of 3000 is the upper limit. Between 2000 and 3000, the flow depends upon many factors, such as external vibration, direction of flow (increasing or decreasing), internal upstream obstructions, etc. Flow between Reynolds numbers of 2000 to 3000 is generally unstable and designers of piping systems must take this into account.

EXAMPLE Castor oil at 68°F flows in a 1-in. schedule 40 horizontal steel pipe 20 ft long at the rate of 0.090 lbm/sec. What is the pressure drop?

Step A *Data Reduction*

(1) *proportionality constant*

(1-11) $g_c = 32.17$ lbm-ft/lbf-sec^2

(2) *unit conversion*

(1-19) $$\dot{m} = \frac{\dot{M}}{g_c} = \frac{0.090}{32.17} = 2.798 \times 10^{-3} \text{ slugs/sec}$$

(3) *fluid properties,* castor oil at 68°F

Table A-1 $\rho = 1.863$ slugs/ft^3

Table A-4 $\mu = 20{,}070 \times 10^{-6}$ lbf-sec/ft^2

(4) *pipe properties,* 1-in. schedule 40

Table A-11 $D = 0.08742$ ft, $A = 0.006002$ ft^2

(5) *kinematic*

(3-8) $$V = \frac{\dot{m}}{\rho A} = \frac{2.798 \times 10^{-3}}{1.863 \times 0.006002}$$

$$V = 0.2502 \text{ ft/sec}$$

Step B *Determine Type of Flow*

(6-18) $$\mathbf{R} = \frac{DV\rho}{\mu} = \frac{0.08742 \times 0.2502 \times 1.863}{20{,}070 \times 10^{-6}}$$

$$\mathbf{R} = 2.030 < 2000 \therefore \text{flow is laminar}$$

Step C *Determine Pressure Drop*

(8-13) $$f = \frac{64}{\mathbf{R}} = \frac{64}{2.030} = 31.53$$

(8-4) $$K = \left(\frac{fL}{D}\right) = \frac{31.53 \times 20}{0.08742} = 7{,}213$$

(8-1) $$\Delta p = \frac{K\rho V^2}{2} = \frac{7{,}213 \times 1.863 \times (0.2502)^2}{2}$$

$$\Delta p = 420.6 \text{ lbf/ft}^2$$

(B-87) $$\Delta p = \frac{420.6}{144} = 2.921 \text{ lbf/in}^2$$

8-4 TURBULENT FLOW IN PIPES

By dimensional, physical, and similarity analyses, it has been determined that the friction factor is some function of Reynolds number and relative roughness. Experimental data indicates that at Reynolds numbers below 3000, the flow is laminar and independent of relative roughness. For Reynolds numbers above 3000, the friction factor is some function of both Reynolds number and relative roughness. Conventional American practice is to use the Colebrook equation, which was the result of an analysis of a large quantity of experimental data. This equation is

$$\frac{1}{\sqrt{f}} = -2 \log_{10}\left(\frac{\epsilon/D}{3.7} + \frac{2.51}{\mathbf{R}\sqrt{f}}\right) \tag{8-14}$$

Examination of Eq. (8-14) indicates that if the value of surface roughness is small compared with the pipe diameter ($\epsilon/D \rightarrow 0$) the friction factor is a function of Reynolds number only. A *smooth pipe* is one in which the ratio $\epsilon/3.7D$ is small compared with $2.51/\mathbf{R}\sqrt{f}$. On the other hand, as the Reynolds number increases so that

$2.51/\mathbf{R}\sqrt{f} \rightarrow 0$ then the friction factor becomes a function of relative roughness only and is called a *rough pipe*. Thus, the same pipe may be smooth under one flow condition and rough under another. The reason for this is that, as the Reynolds number increases, the thickness of the laminar sub-layer decreases, as shown in Fig. 7-4, exposing the surface roughness to the flow.

Equation (8-14) requires a trial and error solution of friction factor. For rough and smooth pipes, Moody derived the following approximation, which will match Eq. (8-14) within ±5%:

$$f_a = 0.0055\left\{1 + \left[20{,}000\left(\frac{\epsilon}{D}\right) + \frac{10^6}{\mathbf{R}}\right]^{1/3}\right\} \tag{8-15}$$

To avoid the trial and error solution required for the Colebrook equation, it may be modified as follows:

$$\frac{1}{\sqrt{f}} = -2\log_{10}\left(\frac{\epsilon/D}{3.7} + \frac{2.51}{\mathbf{R}\sqrt{f_a}}\right) \tag{8-16}$$

where f_a is computed using Eq. (8-15).

Figure 8-3 shows a plot of the Colebrook equation, and Table 8-2 shows recommended values of roughness of new, clean commercial pipes.

FIGURE 8-3 Plot of the Colebrook equation

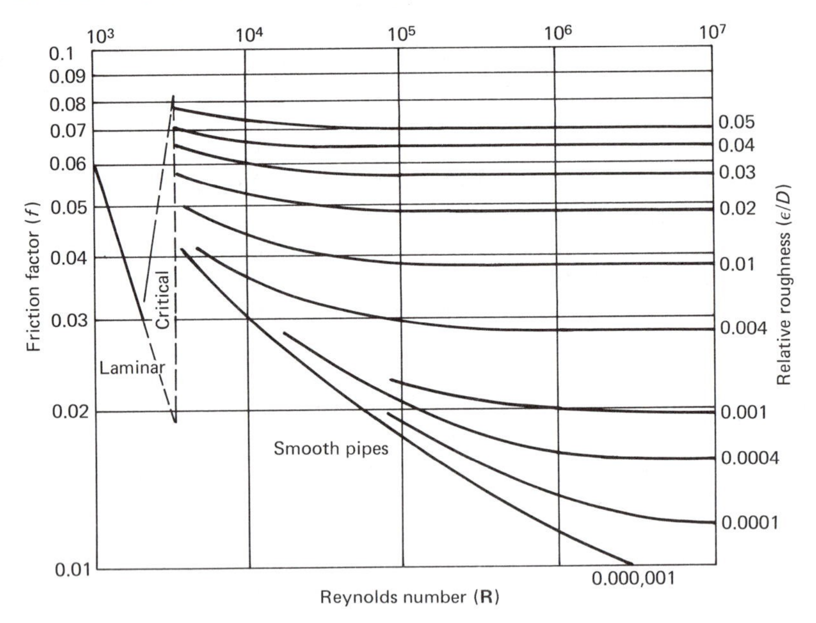

TABLE 8-2 *Values of Absolute Roughness of New Clean Commercial Pipes*

Type of Pipe or Tubing	ϵ ft (0.3048 m) × 10^6		Probable maximum variation of f from design (%)
	RANGE	DESIGN	
Asphalted cast iron	400	400	−5 to +5
Brass and copper	5	5	−5 to +5
Concrete	1,000 to 10,000	4,000	−35 to 50
Cast iron	850	850	−10 to +15
Galvanized iron	500	500	0 to +10
Wrought iron	150	150	−5 to 10
Steel	150	150	−5 to 10
Riveted steel	3,000 to 30,000	6,000	−25 to 75
Wood stave	600 to 3,000	2,000	−35 to 20

Source: *Pipe Friction Manual,* 3d ed. Hydraulic Institute, 1961.

Engineering Calculations

Engineering pipe flow computations usually fall into one of the following classes:

(a) Determination of pressure loss Δp when Q, L, and D are known.
(b) Determination of flow rate Q when L, D, and Δp are known.
(c) Determination of pipe diameter D when Q, L, and Δp are known.

Pressure loss computations can be made to engineering accuracy using the approximation of Eq. (8-16). Flow rate may be determined by direct use of Eq. (8-14). Pipe diameter computation requires a trial and error or graphical method of solution.

EXAMPLE 1 Pressure Loss Calculation Water at 20°C flows in a 3-in. 250 psi cast iron pipe with a velocity of 6 m/s. What is the pressure loss in 30 m of this pipe?

Step A *Data Reduction*

(1) *fluid properties,* water at 20°C

Table A-1 $\rho = 998.3$ kg/m^3

Table A-4 $\mu = 10.02 \times 10^{-4}$ Pa · s

(2) *pipe properties,* 3-in. 250 psi cast iron

Table A-12 $D = 84.8$ mm, $\dfrac{\epsilon}{D} = 3\ 072 \times 10^{-6}$

Step B *Determine Type of Flow*

(6-18)
$$\mathbf{R} = \frac{DV\rho}{\mu} = \frac{(0.0848)(6)(998.3)}{10.02 \times 10^{-4}}$$

$$\mathbf{R} = 506\ 921 > 3\ 000 \therefore \text{flow is turbulent}$$

Step C *Compute Pressure Loss*

(8-15)
$$f_a = 0.0055\left\{1 + \left[20{,}000\left(\frac{\epsilon}{D}\right) + 10^6/\mathbf{R}\right]^{1/3}\right\}$$

$$f_a = 0.0055\{1 + [20{,}000(3\ 072 \times 10^{-6}) + 10^6/506\ 921]^{1/3}\}$$

$$f_a = 0.02743$$

(8-16)
$$\frac{1}{\sqrt{f}} = -2\log_{10}\left(\frac{\epsilon/D}{3.7} + \frac{2.51}{\mathbf{R}\sqrt{f_a}}\right)$$

$$\frac{1}{\sqrt{f}} = -2\log_{10}\left(\frac{3\ 072 \times 10^{-6}}{3.7} + \frac{2.51}{506\ 921\sqrt{0.02743}}\right)$$

$$\frac{1}{\sqrt{f}} = 6.1308,\ f = 0.02660$$

(8-4)
$$K = \frac{fL}{D} = \frac{0.02660 \times 30}{0.0848} = 9.410$$

(8-1)
$$\Delta p = \frac{K\rho V^2}{2} = \frac{9.410 \times 998.3(6)^2}{2}$$

$$\Delta p = 169\ 092\ \text{N/m}^2 = 169.1\ \text{kPa}$$

Computation of Flow Rate

An equation that does not contain flow rate may be derived by solving the friction factor equation (8-3) for V:

(8-3)
$$V = \frac{(2g\overline{H}_f D)^{1/2}}{(fL)^{1/2}}$$

In a like manner, the Reynolds number equation (6-18) may be written as

(6-18)
$$V = \frac{\mathbf{R}\mu}{\rho D}$$

Equating these forms of Eqs. (8-3) and (6-18),

$$V = \frac{\mathbf{R}\mu}{\rho D} = \frac{(2g\overline{H}_f D)^{1/2}}{(fL)^{1/2}}$$

or

$$\mathbf{R}\sqrt{f} = \left(\frac{\rho D}{\mu}\right)\left(\frac{2g\overline{H}_f D}{L}\right)^{1/2} \qquad \textbf{(8-17)}$$

Equation (8-17) is in a form that may be used directly in Eq. (8-14).

EXAMPLE 2 Flow Rate Calculation Air at 14.7 psia and 68°F flows in an asphalted cast iron pipe whose internal diameter is 12.16 in. The pressure loss in this pipe due to friction is 0.4 psi for a straight horizontal run of 1000 ft. Assume that the air is incompressible and determine the volumetric flow rate.

Step A *Data Reduction*

(1) *unit conversion*

(B-28) $T = 68 + 459.67 = 527.67°\text{R}$

(B-87) $p = 144 \times 14.7 = 2{,}117 \text{ lbf/ft}^2$

(B-87) $\Delta p = 144 \times 0.4 = 57.6 \text{ lbf/ft}^2$

(B-55) $D = \dfrac{12.16}{12} = 1.013 \text{ ft}$

(2) *proportionality constant*

(1-11) $g_c = 32.17 \text{ lbm-ft/lbf-sec}^2$

(3) *gravity*
assume standard

(B-50) $g = 32.17 \text{ ft/sec}^2$

(4) *fluid properties,* air at 68°F

Table A-7 $R = 53.36 \text{ ft-lbf/lbm-°R}$

Table A-8 $\mu = 38.31 \times 10^{-8} \text{ lbf-sec/ft}^2$

(1-44) $\rho = \dfrac{p}{g_c RT} = \dfrac{2{,}117}{32.17 \times 53.36 \times 527.67}$

$\rho = 0.002337 \text{ slugs/ft}^3$

(1-35) $\gamma = \rho g = 0.002337 \times 32.17 = 0.07518 \text{ lbf/ft}^3$

(8-2) $\overline{H}_f = \dfrac{\Delta p}{\gamma} = \dfrac{57.6}{0.07518} = 766.2 \text{ ft of air}$

(5) *pipe properties,* asphalted cast iron

Table 8-2 $\epsilon = 400 \times 10^{-6} \text{ ft}$

$\dfrac{\epsilon}{D} = \dfrac{400 \times 10^{-6}}{1.013}$

$= 394.9 \times 10^{-6}$

Table A-10 $A = \dfrac{\pi D^2}{4} = \dfrac{\pi(1.013)^2}{4} = 0.8060 \text{ ft}^2$

Step B *Determine Type of Flow*

(8-17)
$$\mathbf{R}\sqrt{f} = \left(\frac{\rho D}{\mu}\right)\left(\frac{2g\overline{H}_f D}{L}\right)^{1/2}$$

$$\mathbf{R}\sqrt{f} = \left(\frac{0.002337 \times 1.013}{38.31 \times 10^{-8}}\right)\left(\frac{2 \times 32.17 \times 766.2 \times 1.013}{1{,}000}\right)^{1/2}$$

$$\mathbf{R}\sqrt{f} = 43{,}669$$

Assume flow is turbulent, apply Eq. (8-14), and then check.

(8-14)
$$\frac{1}{\sqrt{f}} = -2\log_{10}\left(\frac{\epsilon/D}{3.7} + \frac{2.51}{\mathbf{R}\sqrt{f}}\right)$$

$$\frac{1}{\sqrt{f}} = -2\log_{10}\left(\frac{394.9 \times 10^{-6}}{3.7} + \frac{2.51}{43{,}669}\right)$$

$$\sqrt{f} = 0.1321$$

(8-17)
$$\mathbf{R} = \frac{43{,}669}{\sqrt{f}} = 330{,}575 > 3{,}500$$

$\therefore$ flow is turbulent

Step C *Compute Volume Rate of Flow*

(6-18)
$$V = \frac{\mathbf{R}\mu}{\rho D} = \frac{330{,}575 \times 38.31 \times 10^{-8}}{0.002337 \times 1.013}$$

$$V = 53.49 \text{ ft/sec}$$

(3-4)
$$Q = AV = 0.8060 \times 53.49 = 43.11 \text{ ft}^3\text{/sec}$$

Computation of Pipe Sizes

An expression that does not contain a term for pipe diameter may be derived as follows:

(1) Substitute from the continuity equation (3-4) $Q = AV = (\pi D^2/4)V$ in the friction factor equation (8-3) and solve for D:

$$\overline{H}_f = \frac{(fL/D)V^2}{2g} = \frac{(fL/D)(Q/A)^2}{2g} = \frac{(fL/D)(4Q/\pi D^2)^2}{2g}$$

$$\overline{H}_f = (fL/2g)(4Q/\pi)^2(1/D)^5$$

or

$$D = (fL/2g\overline{H}_f)^{1/5}(4Q/\pi)^{2/5}$$

(2) Substitute from the continuity equation (3-4) in the Reynolds number equation (6-18) and solve for D:

$$\mathbf{R} = \frac{\rho VD}{\mu} = \frac{\rho(Q/A)D}{\mu} = \frac{\rho(4Q/\pi D^2)D}{\mu} = \frac{\rho(4Q/\pi)}{D\mu}$$

or

$$D = \left(\frac{\rho}{\mu \mathbf{R}}\right)\left(\frac{4Q}{\pi}\right) \tag{8-18}$$

(3) Equating these forms of Eqs. (6-18) and (8-3),

$$D = \left(\frac{\rho}{\mu \mathbf{R}}\right)\left(\frac{4Q}{\pi}\right) = \left(\frac{fL}{2g\overline{H}_f}\right)^{1/5}\left(\frac{4Q}{\pi}\right)^{2/5} \tag{8-19}$$

or

$$\mathbf{R}f^{1/5} = \left(\frac{\rho}{\mu}\right)\left(\frac{2g\overline{H}_f}{L}\right)^{1/5}\left(\frac{4Q}{\pi}\right)^{3/5}$$

EXAMPLE 3 Pipe Size Calculations It is required that 225 m^3 of water per hour flow at 350 kPa and 30°C through 300 m of horizontal wrought iron pipe. If the pressure loss may not exceed 100 kPa, what diameter of pipe should be used?

Step A *Data Reduction*

(1) *unit conversion*

(B-54) $Q = \dfrac{225}{3\ 600} = 0.0625\ \text{m}^3/\text{s}$

(2) *gravity*

Assume standard

(B-51) $g = 9.807\ \text{m/s}^2$

(3) *pipe properties,* wrought iron

Table 8-2 $\epsilon = 150 \times 0.3048 \times 10^{-6} = 45.72 \times 10^{-6}\ \text{m}$

(4) *fluid properties,* water at 30°C

Table A-1 $\rho = 995.7\ \text{kg/m}^3$

Table A-4 $\mu = 7.972 \times 10^{-4}\ \text{Pa} \cdot \text{s}$

(1-35) $\gamma = \rho g = 995.7 \times 9.807 = 9\ 765\ \text{N/m}^3$

(8-2) $\overline{H}_f = \dfrac{\Delta p}{\gamma} = \dfrac{100\ 000}{9\ 765} = 10.24\ \text{m}$

Step B *Compute* $\mathbf{R}f^{1/5}$

(8-19)
$$\mathbf{R}f^{1/5} = \left(\frac{\rho}{\mu}\right)\left(\frac{2g\overline{H}_f}{L}\right)^{1/5}\left(\frac{4Q}{\mu}\right)^{3/5}$$

$$= (995.7/7.972 \times 10^{-4})\left(\frac{2 \times 9.807 \times 10.24}{300}\right)^{1/5}$$

$$\times\ (4 \times 0.0625/\pi)^{3/5}$$

$$\mathbf{R}f^{1/5} = 252\ 455 \tag{a}$$

Step C *First Trial*

Assume $\mathbf{R}f^{1/5} \approx \mathbf{R}\sqrt{f}$ and that pipe is smooth $\epsilon/D \to 0$

(8-14) $$\frac{1}{\sqrt{f}} = -2\log_{10}\left(\frac{\epsilon/D}{3.7} + \frac{2.51}{\mathbf{R}\sqrt{f}}\right)$$

$$\frac{1}{\sqrt{f_1}} = -2\log_{10}\left(0 + \frac{2.51}{252\ 455}\right)$$

$$\sqrt{f_1} = 0.099\ 95,\ f_1 = 0.009\ 990$$

$$\mathbf{R}_1 = 252\ 455/f^{1/5} = \frac{252\ 455}{(0.00999)^{1/5}} \quad \text{(a)}$$

$$\mathbf{R}_1 = 634\ 266$$

$$\mathbf{R}_1\sqrt{f_1} = 634\ 266 \times 0.099\ 95 = 63\ 395 \quad \text{(b)}$$

(8-18) $$D_1 = \left(\frac{\rho}{\mu\mathbf{R}_1}\right)\left(\frac{4Q}{\pi}\right)$$

$$D_1 = (995.7/7.972 \times 10^{-4} \times 634\ 266)\,(4 \times 0.062\ 5/\pi)$$

$$D_1 = 0.156\ 7\ \text{m}$$

$$\frac{\epsilon}{D_1} = \frac{45.72 \times 10^{-6}}{0.156\ 7} = 291.8 \times 10^{-6}$$

Step D *Second Trial*

Use first trial values

(8-16) $$\frac{1}{\sqrt{f_2}} = -2\log_{10}\left(\frac{\epsilon/D_1}{3.7} + \frac{2.51}{\mathbf{R}_1\sqrt{f_1}}\right)$$

$$\frac{1}{\sqrt{f_2}} = -2\log_{10}\left(\frac{291.8 \times 10^{-6}}{3.7} + \frac{2.51}{63\ 395}\right)$$

$$\sqrt{f_2} = 0.127\ 3,\ f_2 = 0.016\ 22$$

$$\mathbf{R}_2 = 252\ 455/f_2^{1/5} = 252\ 455/(0.016\ 22)^{1/5} \quad \text{(a)}$$

$$\mathbf{R}_2 = 575\ 670$$

$$\mathbf{R}_2\sqrt{f_2} = 575\ 670 \times 0.127\ 3 = 73\ 283 \quad \text{(b)}$$

(8-18) $$D_2 = \left(\frac{\rho}{\mu\mathbf{R}_2}\right)\left(\frac{4Q}{\pi}\right)$$

$$D_2 = \left(\frac{995.7}{7.972 \times 10^{-4} \times 575\ 670}\right)\left(\frac{4 \times 0.062\ 5}{\pi}\right)$$

$$D_2 = 0.172\ 6\ \text{m}$$

$$\frac{\epsilon}{D_2} = \frac{45.72 \times 10^{-6}}{0.172\ 6} = 264.8 \times 10^{-6}$$

Step E *Third Trial*

Use second trial values

(8-16) $$\frac{1}{\sqrt{f_3}} = -2\log_{10}\left(\frac{\epsilon/D_2}{3.7} + \frac{2.51}{\mathbf{R}_2\sqrt{f_2}}\right)$$

$$\frac{1}{\sqrt{f_3}} = -2 \log_{10}\left(\frac{264.8 \times 10^{-6}}{3.7} + \frac{2.51}{73\ 283}\right)$$

$$\sqrt{f_3} = 0.125\ 8,\ f_3 = 0.015\ 83$$

$$\mathbf{R}_3 = \frac{252\ 455}{(0.015\ 83)^{1/5}} = 578\ 479 \qquad \textbf{(a)}$$

$$\textbf{(8-17)} \qquad D_3 = \left(\frac{995.7}{7.972 \times 10^{-4} \times 578\ 479}\right)\left(\frac{4 \times 0.062\ 5}{\pi}\right) \qquad \textbf{(a)}$$

$$D_3 = 0.171\ 9 \text{ m}$$

Note: In actual practice, the trial and error solution beyond the first trial would not be justified. For this pressure schedule 40 pipe would be used. From Table A-11, 6-in. schedule 40 pipe has an internal diameter of 0.154 1 m, which is too small. In this case, the next larger size, 8 in., which has an internal diameter of 0.202 7 m, would be used.

8-5 VELOCITY PROFILE IN PIPES

Laminar Flow

Figure 8-4 shows the formation of the laminar velocity profile in a pipe. It is evident that some distance is required to develop this profile. This distance, L, is given by

$$\frac{L}{D} \approx 0.058\ \mathbf{R} \qquad \textbf{(8-20)}$$

FIGURE 8-4 Formation of laminar velocity profile

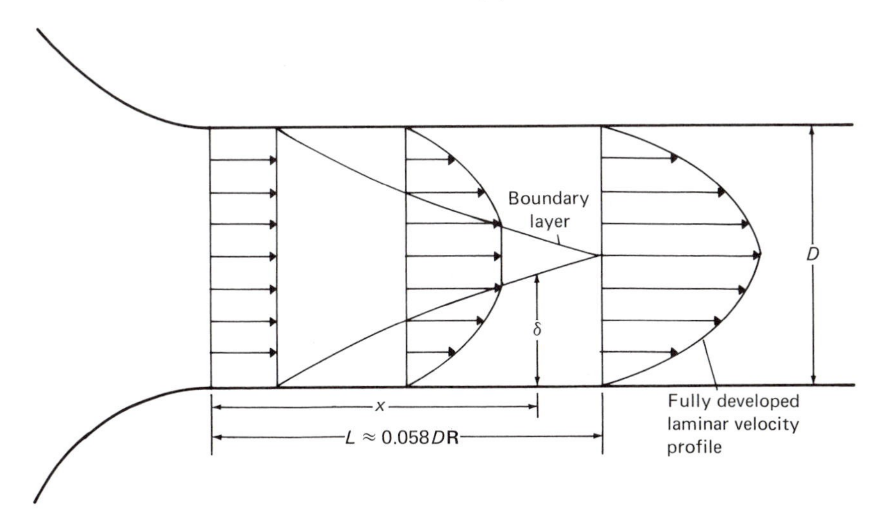

In Sec. 8-3 it was established that the velocity profile for laminar flow is parabolic. Equations for three-dimensional parabolic flow were derived in the example problem of Sec. 3-4 as follows:

$$U = U_m\left[1 - \left(\frac{r}{r_o}\right)^2\right] = 2V\left[1 - \left(\frac{r}{r_o}\right)^2\right] \tag{8-21}$$

EXAMPLE 1 Laminar Velocity Profile At what radius is the local velocity equal to the average velocity?

$$U = V = 2V\left[1 - \left(\frac{r}{r_o}\right)^2\right], \quad 1 = 2 - 2\left(\frac{r}{r_o}\right)^2$$

$$r = (1/2)^{1/2} r_o = 0.7071\ r_o$$

Turbulent Flow

Velocity profiles in turbulent flow are much "flatter" than those of laminar flow, as shown in Fig. 8-5. As the Reynolds number increases, the profile approaches one-dimensionality. For this reason, the one-dimensional analysis gives satisfactory solutions of many practical engineering problems.

Equations for velocity profile in turbulent flow cannot at the present state of the art be developed from theoretical considerations as was done in the case of laminar flow. The following equation is based on boundary layer "theory" and experimental data and is valid except at the very thin laminar sub-layer:

$$U = \left[(1 + 1.326\sqrt{f}) - 2.04\sqrt{f}\log_{10}\left(\frac{r_o}{r_o - r}\right)\right]V \tag{8-22}$$

At the pipe center line, where $r = 0$ and $U = U_m$, Eq. (8-22) reduces to

FIGURE 8-5 Comparison of laminar and turbulent velocity profiles

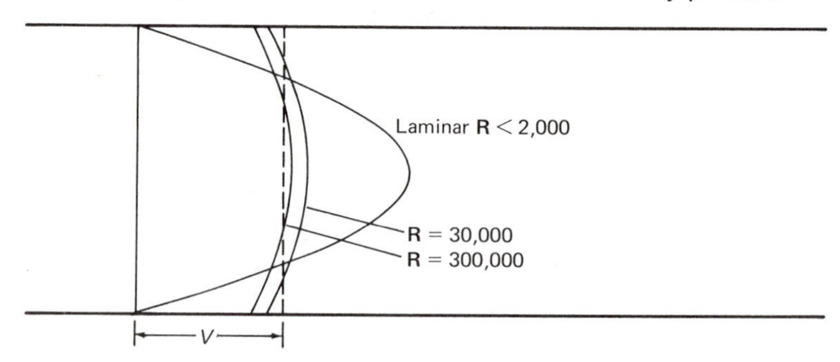

$$U_m = (1 + 1.326\sqrt{f})V \tag{8-23}$$

Equations (8-22) and (8-23) may be used for either smooth or rough pipes, and the friction factor is determined by the methods given in Sec. 8-4.

Because of the difficulty of employing Eqs. (8-22) and (8-23), an alternative equation is sometimes used:

$$U = U_m\left(\frac{r_o - r}{r_o}\right)^b = U_m\left(1 - \frac{r}{r_o}\right)^b \tag{8-24}$$

The value of b for smooth pipe up to values of $\mathbf{R} = 10^5$ is $1/7$; for higher values of $\mathbf{R}$, $b = 1/8$. For rough pipes, $b \approx 1/5$.

EXAMPLE 2 Turbulent Velocity Profile Carbon tetrachloride flows at 86°F through a 3-in. schedule 80 steel pipe with a velocity of 4 ft/sec. Compute (a) the velocity at the center line and (b) the velocity 1 in. from the center line.

Step A *Data Reduction*

(1) *unit conversion*

(B-55) $r = 1/12$ ft

(2) *fluid properties,* carbon tetrachloride at 86°F

Table A-1 $\rho = 3.055$ slugs/ft^3

Table A-4 $\mu = 17.58 \times 10^{-6}$ lbf-sec/ft^2

(3) *pipe properties,* 3 in. schedule 80 steel

Table A-11 $\frac{\epsilon}{D} = 620.6 \times 10^{-6}$

$D = 0.2417$ ft

$r_o = \frac{0.2417}{2} = 0.12085$ ft

Step B *Determine Friction Factor*

(6-18) $\mathbf{R} = \frac{DV\rho}{\mu} = \frac{0.2417 \times 4 \times 3.055}{17.58 \times 10^{-6}}$

$\mathbf{R} = 168{,}008 > 3{,}500 \therefore$ flow is turbulent

(8-15) $f_a = 0.0055\left[1 + \left(\frac{20{,}000\,\epsilon}{D} + \frac{10^6}{\mathbf{R}}\right)^{1/3}\right]$

$f_a = 0.0055\left[1 + \left(20{,}000 \times 620.6 \times 10^{-6} + \frac{10^6}{168{,}008}\right)^{1/3}\right]$

$f_a = 0.02001$

(8-16)
$$\frac{1}{\sqrt{f}} = -2\log_{10}\left(\frac{\epsilon/D}{3.7} + \frac{2.51}{\mathbf{R}\sqrt{f_a}}\right)$$
$$\frac{1}{\sqrt{f}} = -2\log_{10}\left(\frac{620.6\times 10^{-6}}{3.7} + \frac{2.51}{168{,}008\sqrt{0.02001}}\right)$$
$$\sqrt{f} = 0.1403,\ f = 0.01968$$

Step C *Compute Centerline Velocity*

(8-23)
$$U_m = (1 + 1.326\sqrt{f})V = (1 + 1.326\times 0.1403)4$$
$$U_m = 4.744 \text{ ft/sec}$$

Step D *Compute Velocity One Inch from Centerline*

(8-22)
$$U = \left[(1 + 1.326\sqrt{f}) - 2.04\sqrt{f}\log_{10}\left(\frac{r_o}{r_o - r}\right)\right]V$$
$$= \left[(1 + 1.326\times 0.1403) - 2.04\times 0.1430\log_{10}\left(\frac{0.12085}{0.12085 - (1/12)}\right)\right]\times 4$$
$$U = 4.151 \text{ ft/sec}$$

Kinetic Energy Correction Factor α

In Sec. 3.6 in the three-dimensional laminar flow example it was demonstrated that $\alpha = 2$ for laminar flow. It can be shown that substitution of Eq. (8-22) in Eq. (3-13) results in

$$\alpha = 1 + 2.9297 f - 1.5537 f^{3/2} \quad \text{(8-25)}$$

EXAMPLE 3 Turbulent Kinetic Energy Correction Factor Determine the kinetic energy correction factor for Example 2.

(8-25)
$$\alpha = 1 + 2.9297 f - 1.5537 f^{3/2}$$
$$= 1 + 2.927(0.01968) - 1.5537(0.01968)^{3/2}$$
$$\alpha = 1.053$$

8-6 COMPRESSIBLE FLOW IN PIPES

The calculation of pressure loss due to friction when gases and vapors flow in pipelines is in general a rather complex thermodynamic problem and will not be treated exhaustively in this section. The isothermal flow of an ideal gas in a pipeline is a simple but practical case that has wide application in engineering. We will consider the isothermal case of compressible flow only.

Reynolds Number

Application of the continuity equation (3-8) $\dot{m} = \rho AV$ to the equation for Reynolds number, Eq. (6-18):

$$\mathbf{R} = \frac{VD\rho}{\mu} = \frac{(\dot{m}/A)D}{\mu} = \frac{\dot{m}D}{A\mu} \tag{8-26}$$

Equation (8-26) indicates that the Reynolds number for compressible flow is a constant if the viscosity is a constant and does not change with changes in velocity and density. If the flow is isothermal, then because the temperature is constant the viscosity is constant and for a given flow the Reynolds number is the same at all sections of the pipe. This means that, since friction factor is a function of Reynolds number and relative roughness, then the friction factor will also be the same at all sections of the pipe for isothermal flow.

Compressible Flow Equations

Writing the equation of motion (4-12) for a horizontal pipe $(dz = 0)$:

$$\frac{V\,dV}{g_c} + v\,dp + \frac{v\tau\,dL}{R_h} = 0 \tag{8-27}$$

Substituting $f\rho V^2/8$ for τ from Eq. (8-7), $D/4$ for R_h from Eq. (4-10), multiplying by $2g_c/V^2$ and noting from Eq. (1-36) that $\rho v = 1/g_c$,

$$\frac{V\,dV}{g_c}\left(\frac{2g_c}{V^2}\right) + v\,dp\left(\frac{2g_c}{V^2}\right) + \frac{v}{(D/4)}\left(\frac{f\rho V^2}{8}\right)\left(\frac{2g_c}{V^2}\right) dL = 0$$

$$\frac{2\,dV}{V} + \frac{2\,dp}{V^2\rho} + \left(\frac{f}{D}\right) dL = 0 \tag{8-28}$$

Substituting in Eq. (8-28) $V^2 = \dot{m}^2/\rho^2A^2$ from Eq. (3-8),

$$\frac{2\,dV}{V} + \frac{2\rho\,dp}{(\dot{m}/A)^2} + \left(\frac{f}{D}\right) dL = 0 \tag{8-29}$$

Noting from Eq. (1-44) $\rho = p/g_cRT$ and substituting in Eq. (8-29),

$$\frac{2\,dV}{V} + \frac{2p\,dp}{(\dot{m}/A)^2 g_cRT} + \left(\frac{f}{D}\right) dL = 0 \tag{8-30}$$

Equation (8-30) may be used for any ideal gas process for which the relationship between temperature and pressure has been established.

Isothermal Equations

Equation (8-30) may be integrated for the special case of $T = C$ as follows:

$$2\int_1^2 \frac{dV}{V} + \frac{2}{(\dot{m}/A)^2 g_c RT}\int_1^2 p\,dp + \left(\frac{f}{D}\right)\int_0^L dL = 0$$

$$2\log_e\left(\frac{V_2}{V_1}\right) + \frac{p_2^2 - p_1^2}{(\dot{m}/A)^2 g_c RT} + \frac{fL}{D} = 0 \tag{8-31}$$

If the equation of continuity is written for flow in a constant area duct, Eq. (3-8), becomes

$$\dot{m} = \rho_1 A V_1 = \rho_2 A V_2 \qquad \text{or} \qquad \frac{V_2}{V_1} = \frac{\rho_1}{\rho_2} \tag{8-32}$$

From Eq. (1-44), $\rho = p/g_c RT$, substituting in Eq. (8-32),

$$\frac{V_2}{V_1} = \frac{p_1/g_c RT}{p_2/g_c RT} = \frac{p_1}{p_2} \tag{8-33}$$

Substituting p_1/p_2 for V_2/V_1 and $K = fL/D$ from Eq. (8-4) into (8-31),

$$p_1^2 - p_2^2 = (\dot{m}/A)^2 g_c RT[2\log_e(p_1/p_2) + K] \tag{8-34}$$

Equation (8-34) may be solved only by trial and error.

Limiting Isothermal Flow in Pipes

In the derivation of equations for acoustic velocity in Sec. 1-16, the velocity of a pressure wave was developed as $c = \sqrt{E/\rho}$ in Eq. (1-58) where the numerical value of E depended upon the process. In Sec. 1-15, the value of $E_T = p$ was derived as Eq. (1-53). Substituting Eq. (1-53) in Eq. (1-58), noting from Eq. (1-44) $\rho = p/g_c RT$,

$$V_{\max} = c_T = \sqrt{E_T/\rho} = \sqrt{p/\rho} = \sqrt{g_c RT} \tag{8-35}$$

EXAMPLE Natural gas at 20°C and 2400 kPa enters an 18-in. standard steel pipe with a velocity of 3 m/s. The pipe is a straight horizontal run 40 km long. Assume that the natural gas has the same fluid properties as methane. Estimate the pressure at the end of the pipe for (a) incompressible flow, (b) isothermal flow, and (c) the minimum pressure that can exist in the system for isothermal flow.

Step A *Data Reduction*

(1) *unit conversion*

(B-27) $T = 20 + 273.15 = 293.15$ K

(2) *proportionality constant*

$g_c = 1 \text{ kg} \cdot \text{m/N} \cdot \text{s}^2$

(3) *fluid properties,* methane at 20°C

Table A-7 $R = 518.3 \text{ J/kg} \cdot \text{K}$

Table A-8 $\mu = 10.87 \times 10^{-6} \text{ Pa} \cdot \text{s}$

(1-44)
$$\rho = \frac{p}{g_c R T} = \frac{2\ 400\ 000}{(1)(518.3)(293.15)}$$
$$= 15.80 \text{ kg/m}^3$$

(4) *pipe properties,* 18-inch standard steel

Table A-11 $D = 0.438\ 1$ m

Table A-11 $\epsilon/D = 104.3 \times 10^{-6}$

Step B *Determine Friction Factor*

(6-18)
$$\mathbf{R} = \frac{V_1 D \rho_1}{\mu_1} = \frac{3(0.438\ 1)(15.80)}{10.87 \times 10^{-6}}$$

$\mathbf{R} = 1\ 910\ 390 > 3\ 000$ ∴ flow is turbulent

(8-15)
$$f_a = 0.0055\left[1 + \left[20{,}000\left(\frac{\epsilon}{D}\right) + \frac{10^6}{\mathbf{R}}\right]^{1/3}\right]$$

$$f_a = 0.0055\left[1 + \left(20{,}000 \times 104.3 \times 10^{-6} + \frac{10^6}{1\ 910\ 390}\right)^{1/3}\right]$$

$f_a = 0.013\ 07$

(8-16)
$$\frac{1}{\sqrt{f}} = -2\log_{10}\left(\frac{\epsilon/D}{3.7} + \frac{2.51}{\mathbf{R}\sqrt{f_a}}\right)$$

$$\frac{1}{\sqrt{f}} = -2\log_{10}\left(\frac{104.3 \times 10^{-6}}{3.7} + \frac{2.51}{1\ 910\ 390\sqrt{0.013\ 07}}\right)$$

$\sqrt{f} = 0.113\ 6,\ f = 0.012\ 90$

Step C *Compute Loss for Incompressible Flow*

(8-4) $$K = \frac{fL}{D} = \frac{0.0129\,0 \times 40\,000}{0.438\,1} = 1\,178$$

(8-1) $$\Delta p = \frac{K\rho V^2}{2} = \frac{1\,178 \times 15.80(3)^2}{2}$$

$$\Delta p = 83\,756 \text{ Pa} = 83.76 \text{ kPa}$$

Step D *Compute Loss for Isothermal Flow*

(3-8) $$\frac{\dot{m}}{A} = \rho_1 V_1 = 15.80 \times 3 = 47.40 \text{ kg/m}^2 \cdot \text{s}$$

First trial

Assume Δp is the same as for incompressible flow, or

$$p_2 = p_1 - \Delta p = 2\,400\,000 - 83\,756 = 2\,316\,244 \text{ Pa}$$

(8-34) $$p_1^2 - p_2^2 = \left(\frac{\dot{m}}{A}\right)^2 g_c RT \left[2 \log_e\left(\frac{p_1}{p_2}\right) + K\right]$$

$$(2\,400\,000)^2 - p_2^2 = (47.40)^2(1)(518.3) \times (292.15)$$

$$\times \left(2 \log_e \frac{2\,400\,000}{2\,316\,244} + 1178\right)$$

$$p_2 = 2\,314\,701 \text{ Pa}$$

$$\Delta p = p_1 - p_2 = 2\,400\,000 - 2\,314\,701$$

$$\Delta p = 85\,299 \text{ Pa} = 85.30 \text{ kPa}$$

Further trials not necessary, since

$$2 \log_e\left(\frac{p_1}{p_2}\right) = 0.072\,38$$

which is small compared with a K of 1 178

Step E *Minimum Pressure for Isothermal Flow*

(8-35) $$V_{max} = \sqrt{g_c RT} = \sqrt{1 \times 518.3 \times 293.15} \qquad V_{max} = 389.8 \text{ m/s}$$

Minimum pressure will occur at V_{max}

(8-33) $$p_{min} = \frac{p_1 V_1}{V_{max}} = \frac{(2\,400\,000)3}{389.8} \qquad p_{min} = 18\,471 = 18.47 \text{ kPa}$$

8-7 NONCIRCULAR PIPES

The Colebrook equation (8-14) may be used for noncircular pipes. One substitutes the equivalent diameter obtained from the hydraulic radius for the diameter in the Reynolds number equation (6-18), the resistance coefficient equation (8-4), and in computing the relative roughness from the absolute roughness. The following example will illustrate the calculation of friction loss in noncircular pipes.

EXAMPLE OF LOSS IN A RECTANGULAR DUCT Air at 68°F and 14.7 psia flows in a rectangular duct 1 ft by 3 ft at a rate of 150 ft³/sec. The duct is horizontal, 1000 ft long, and made of galvanized iron. Assuming the air to be incompressible, estimate the pressure loss due to friction in this duct.

Step A *Data Reduction*

(1) *unit conversion*

(B-28) $T = 68 + 459.67 = 527.67°\text{R}$

(B-87) $p = 144 \times 14.7 = 2{,}117 \text{ lbf/ft}^2$

(2) *proportionality constant*

(1-11) $g_c = 32.17 \text{ lbm-ft/lbf-sec}^2$

(3) *fluid properties,* air at 68°F

Table A-7 $R = 53.36 \text{ ft-lbf/lbm-°R}$

Table A-8 $\mu = 38.31 \times 10^{-8} \text{ lbf-sec/ft}^2$

(1-44) $$\rho = \frac{p}{g_c RT} = \frac{2{,}117}{32.17 \times 53.36 \times 527.67}$$

$$\rho = 0.002337 \text{ slugs/ft}^3$$

(4) *duct properties,* galvanized iron

Table A-14 $$R_h = \frac{1 \times 3}{2(1+3)} = 3/8 \text{ ft}, \; A = 1 \times 3 = 3 \text{ ft}^2$$

(4-11) $D_e = 4R_h = 4(3/8) = 1.5 \text{ ft}$

Table 8-2 $\epsilon = 500 \times 10^{-6} \text{ ft}$

$$\frac{\epsilon}{D} = \frac{500 \times 10^{-6}}{1.5} = 333.3 \times 10^{-6}$$

(5) *kinematic*

$$V = \frac{Q}{A} = \frac{150}{3} = 50 \text{ ft/sec}$$

Step B *Compute Friction Factor*

(6-18) $$\mathbf{R} = \frac{VD_e\rho}{\mu} = \frac{50 \times 1.5 \times 0.002337}{38.31 \times 10^{-8}}$$

$\mathbf{R} = 457{,}518 > 3{,}500 \therefore$ flow is turbulent

(8-15) $$f_a = 0.0055\left[1 + \left(20{,}000\left(\frac{\epsilon}{D}\right) + \frac{10^6}{\mathbf{R}}\right)^{1/3}\right]$$

$$f_a = 0.0055\left[1 + \left(20{,}000 \times 333.3 \times 10^{-6} + \frac{10^6}{457{,}518}\right)^{1/3}\right]$$

$$f_a = 0.01688$$

(8-16) $$\frac{1}{\sqrt{f}} = -2\log_{10}\left(\frac{\epsilon/D_e}{3.7} + \frac{2.51}{\mathbf{R}\sqrt{f}}\right)$$

$$\frac{1}{\sqrt{f}} = -2\log_{10}\left(\frac{333.3 \times 10^{-6}}{3.7} + \frac{2.51}{457{,}518\sqrt{0.01688}}\right)$$

$$\sqrt{f} = 0.1289 \qquad f = 0.01662$$

Step C *Compute Friction Loss*

(8-4) $$K = \frac{fL}{D_e} = \frac{0.01662 \times 1{,}000}{1.5} = 11.08$$

(8-1) $$\Delta p = \frac{K\rho V^2}{2} = \frac{11.08 \times 0.002337(50)^2}{2}$$

$$\Delta p = 32.37 \text{ lbf/ft}^2$$

(B-87) $$\Delta p = \frac{32.37}{144} = 0.2248 \text{ lbf/in}^2$$

8-8 PIPING SYSTEMS

In piping systems, changes in pressure may occur because of energy conversions as well as frictional losses. In computing the flow through a piping system, it is necessary to consider the equation of motion. In Sec. 4-14, the equation of motion with shaft work was found to be

(4-42) $$W_{sf} + \frac{g}{g_c}(z_2 - z_1) + \frac{V_2^2 - V_1^2}{2g_c} + \int_1^2 v\,dp + H_f = 0$$

For an incompressible fluid in which v is a constant, integrating the fourth term of the equation, multiplying the whole equation by g_c/g, and noting from Eq. (1-36) that $v = 1/\rho g_c$, results in

$$\frac{g_c}{g}W_{sf} + (z_2 - z_1) + \frac{V_2^2 - V_1^2}{2g} + \frac{p_2 - p_1}{\rho g} + \frac{g_c}{g}H_f = 0 \qquad \text{(8-36)}$$

Note that the dimensions of all the terms are length.

Let $$\frac{g_c}{g}W_{sf} = \overline{W} \qquad \text{(8-37)}$$

and

$$\frac{g_c}{g} H_f = \overline{H}_f \tag{8-38}$$

where $\overline{W}$ and $\overline{H}_f$ are the steady-flow shaft work head and friction loss head respectively in dimensions of length. Substitution in Eq. (8-36) results in

$$\overline{W} + (z_2 - z_1) + \frac{V_2^2 - V_1^2}{2g} + \frac{p_2 - p_1}{\rho g} + \overline{H}_f = 0 \tag{8-39}$$

Resistance Coefficients

For valves, branch flow through tees, and the type of components listed in Part 2 of Table 8-3, the pressure drag is predominant and is "rougher" than the pipe to which it is attached, and it will extend the completely turbulent region to lower values of Reynolds number. For bends and elbows, the loss consists of pressure drag due to the change of direction and the consequent secondary flows which are dissipated in 50 diameters or more downstream piping. For this reason, loss through adjacent bends will not be twice that of a single bend.

In long pipelines, the effect of bends, valves, and fittings is usually negligible, but in systems where there is little straight pipe, it is the controlling factor. Under-design will result in the failure of the system to deliver the required capacity. Over-design will result in inefficient operation, because it will be necessary to "throttle" one or more of the valves. For estimating purposes, Table 8-2 may be used as shown in the example. When available, the manufacturer's data should be used, particularly for valves, because of the wide variety of designs for the same type.

EXAMPLE It is desired to pump water at 20°C from Tank 1 to Tank 2 of the piping system shown in Fig. 8-6 at a rate of 225 m^3/hr. All connecting piping is 6-in. schedule 40 wrought iron pipe. Total straight length of pipe is 180 m. All fittings are 6 in. and are welded or flanged. Estimate the power that the pump must add to the fluid.

Step A *Data Reduction*

(1) *unit conversion*

(B-54) $$Q = \frac{225}{3,600} = 0.062\ 5\ \text{m}^3/\text{s}$$

(2) *gravity*

Assume standard

(B-51) $$g = 9.807\ \text{m/s}^2$$

TABLE 8-3 *Resistance Coefficients—Part 1*

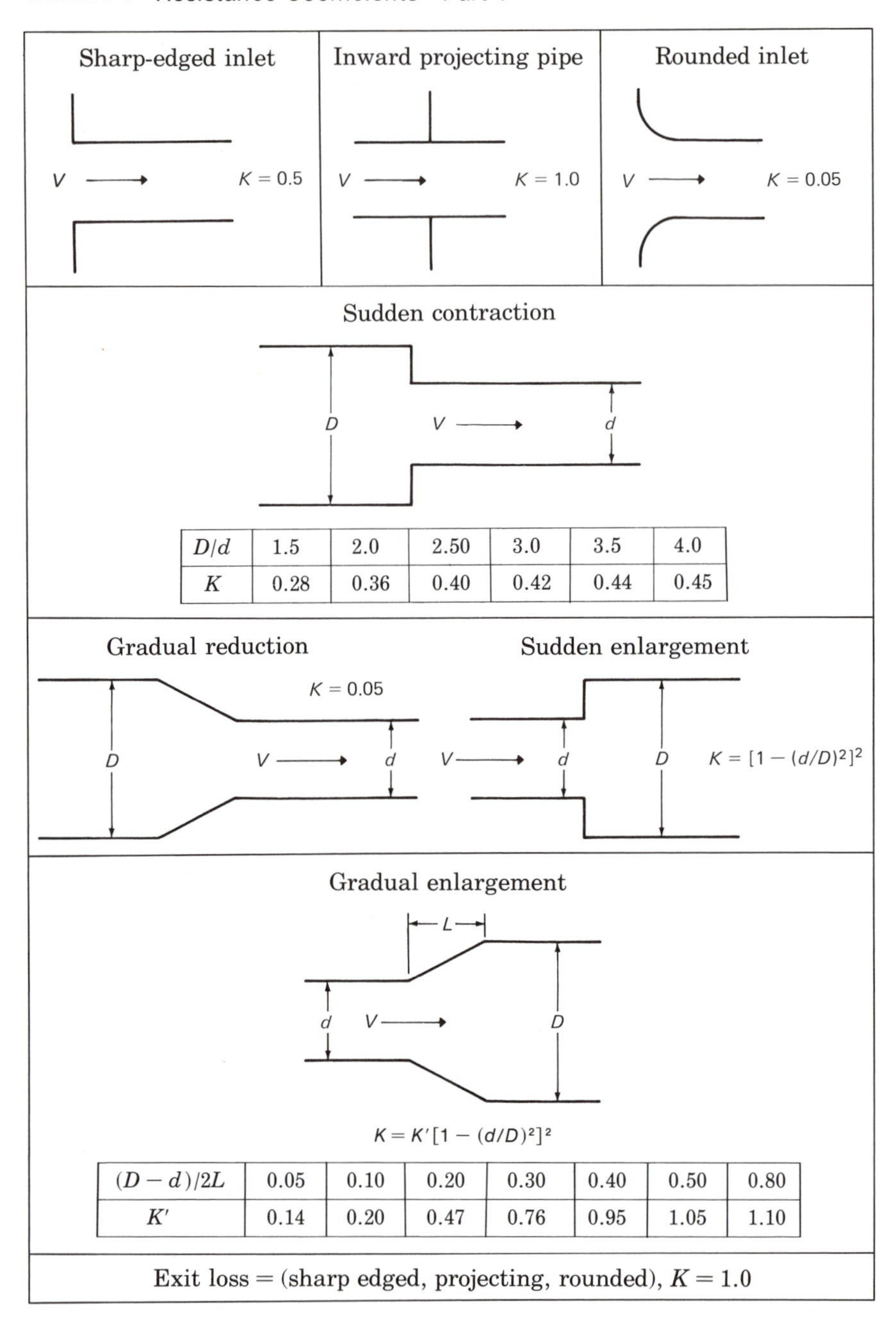

Sudden contraction

D/d	1.5	2.0	2.50	3.0	3.5	4.0
K	0.28	0.36	0.40	0.42	0.44	0.45

Gradual enlargement: $K = K'[1 - (d/D)^2]^2$

$(D-d)/2L$	0.05	0.10	0.20	0.30	0.40	0.50	0.80
K'	0.14	0.20	0.47	0.76	0.95	1.05	1.10

Exit loss = (sharp edged, projecting, rounded), $K = 1.0$

TABLE 8-3 *Resistance Coefficients—Part 2*

Type	Screwed				Flanged or welded					
NOMINAL DIAMETER (in.)	½	1	2	4	1	2	4	8	16	20
Valves										
Globe	14	8.2	6.9	5.7	13	8.5	6.0	5.8	5.5	5.5
Gate	0.30	0.24	0.16	0.11	0.80	0.35	0.16	0.07	0.04	0.03
Swing check	5.1	2.9	2.1	2.0	2.0	2.0	2.0	2.0	2.0	2.0
Angle	9.0	4.7	2.0	1.0	4.5	2.4	2.0	2.0	2.0	2.0
Elbows										
45° Regular	0.39	0.32	0.30	0.29						
45° Long radius					0.21	0.20	0.19	0.16	0.15	0.14
90° Regular	2.0	1.5	0.95	0.64	0.50	0.39	0.30	0.26	0.23	0.21
90° Long radius	1.0	0.72	0.41	0.23	0.40	0.30	0.19	0.15	0.11	0.10
180° Regular	2.0	1.5	0.95	0.64	0.41	0.35	0.30	0.25	0.21	0.20
180° Long radius					0.40	0.30	0.21	0.15	0.11	0.10
Tees										
Line flow	0.90	0.90	0.90	0.90	0.24	0.19	0.14	0.10	0.08	0.07
Branch flow	2.4	1.8	1.4	1.1	1.0	0.80	0.64	0.58	0.45	0.41

In pipe bends, due to the free vortex action (Section 4.21), a swirl is created. It takes about 50 D to restore the normal velocity profile. Loss takes place in the 50 D downstream.

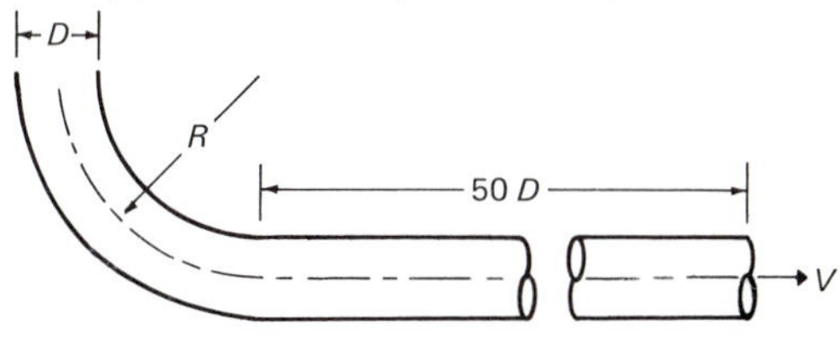

Resistance coefficients for 90° bends

	R/D	1	2	3	4	5	6	7	8	9	10
ϵ/D	0	0.24	0.15	0.12	0.10	0.09	0.08	0.08	0.08	0.09	0.09
	0.0005	0.30	0.18	0.16	0.14	0.12	0.11	0.10	0.11	0.12	0.13
	0.001	0.36	0.24	0.19	0.17	0.15	0.14	0.13	0.13	0.14	0.14
	0.002	0.50	0.30	0.26	0.24	0.20	0.18	0.17	0.18	0.18	0.19

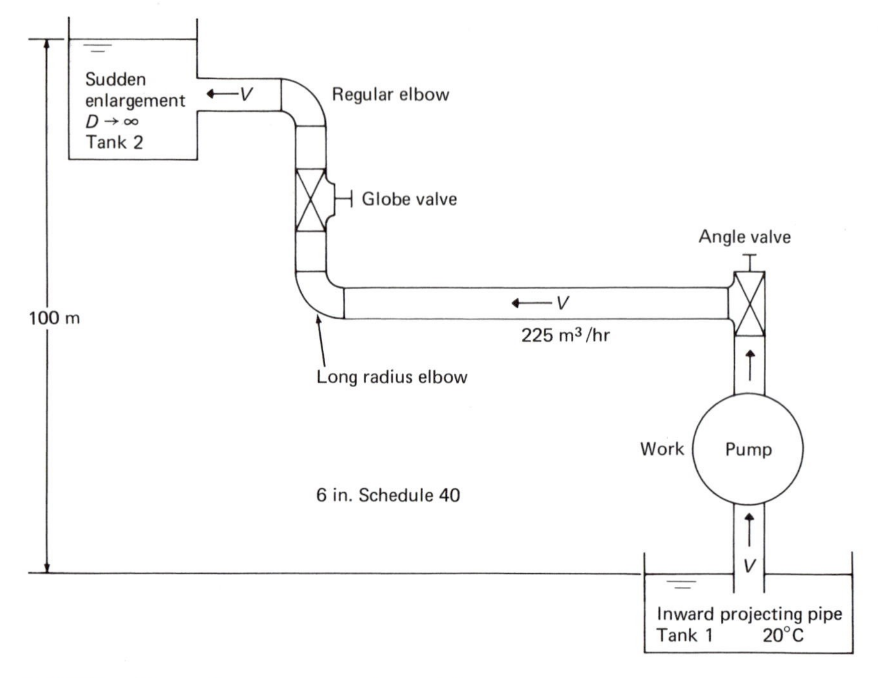

FIGURE 8-6 Notation for example problem

(3) *proportionality constant*

(1-9) $g_c = 1 \text{ kg} \cdot \text{m/N} \cdot \text{s}^2$

(4) *fluid properties,* water at 20°C

Table A-1 $\rho = 998.3 \text{ kg/m}^3$

Table A-4 $\mu = 10.02 \times 10^{-4} \text{ Pa} \cdot \text{s}$

(5) *pipe properties,* 6-in. schedule 40 wrought iron

Table A-11 $D = 154.1 \text{ mm}, A = 18{,}650 \text{ mm}^2$

Table A-11 $\frac{\epsilon}{D} = 296.8 \times 10^{-6}$

Table 8-3 (6) *resistance coefficients,* 6-in. flanged or welded

Projecting inlet pipe,	$K = 1.0$
Angle value,	$K = 2.0$
Long radius 90° elbow,	$K = 0.17$
Globe valve,	$K = 5.9$
Regular 90° elbow,	$K = 0.28$
Sudden enlargement $D \to \infty$,	
$K = \left[1 - \left(\frac{d}{\infty}\right)^2\right]^2$	$K = 1.0$
System total,	$\Sigma K = 10.35$

(7) *kinematic*

(3-4) $$V = \frac{Q}{A} = \frac{0.062\ 5}{18\ 650 \times 10^{-6}} = 3.351 \text{ m/s}$$

Step B *Compute System Resistance*

(6-18) $$\mathbf{R} = \frac{DV\rho}{\mu} = \frac{0.154\ 1 \times 3.351 \times 998.3}{10.02 \times 10^{-4}}$$

$$\mathbf{R} = 514\ 482 > 3\ 500 \therefore \text{flow is turbulent}$$

(8-15) $$f_a = 0.0055\left[1 + \left(20{,}000\left(\frac{\epsilon}{D}\right) + \frac{10^6}{\mathbf{R}}\right)^{1/3}\right]$$

$$f_a = 0.0055\left[1 + \left(20{,}000 \times 296.8 \times 10^{-6} + \frac{10^6}{514\ 482}\right)^{1/3}\right]$$

$$f_a = 0.0164\ 4$$

(8-16) $$\frac{1}{\sqrt{f}} = -2 \log_{10}\left(\frac{\epsilon/D}{3.7} + \frac{2.51}{\mathbf{R}\sqrt{f_a}}\right)$$

$$\frac{1}{\sqrt{f}} = -2 \log_{10}\left(\frac{296.8 \times 10^{-6}}{3.7} + \frac{2.51}{514\ 482\sqrt{0.016\ 44}}\right)$$

$$\sqrt{f} = 0.127\ 3, \qquad f = 0.016\ 21$$

$$Pipe = K_p = \frac{fL}{D} = \frac{0.016\ 21 \times 180}{0.154\ 1}$$

$$K_p = 18.93$$

Total Resistance Coefficient, $K_t = K_p + \Sigma K$

$$K_t = 18.93 + 10.35 = 29.28$$

(8-2) $$\overline{H}_f = \frac{K_t V^2}{2g} = \frac{29.28(3.351)^2}{2 \times 9.807}$$

$$\overline{H}_f = 16.76 \text{ m}$$

Step C *Evaluate Each Term of Eq. (8-39)*

(8-39) $$\overline{W} + (z_2 - z_1) + \frac{V_2^2 - V_1^2}{2g} + \frac{p_2 - p_1}{\rho g} + \overline{H}_f = 0$$

(1) *potential energy change,* $z_2 - z_1$

Fig. 8-6 $$(z_2 - z_1) = 100 \text{ m}$$

(2) *kinetic energy change,*

$$\frac{V_2^2 - V_1^2}{2g}$$

(3-8) $$\dot{m} = \rho_1 A_1 V_1 = \rho_2 A_2 V_2$$

For an incompressible fluid $\rho_1 = \rho_2$ for a constant area pipe $A_1 = A_2 \therefore V_1 = V_2$

$$\frac{V_2^2 - V_1^2}{2g} = 0$$

(3) *flow work change,*

$$\frac{p_2 - p_1}{\rho g}$$

Fig. 8-6 Tanks 1 and 2 are open to atmosphere and

$$p_1 = p_2 = p_b$$

$$\frac{p_2 - p_1}{\rho g} = 0$$

(4) *friction energy,* $\overline{H}_f$

Step B $\overline{H}_f = 16.76$ m

(5) *steady-flow shaft work*

(8-39)
$$\overline{W} + (z_2 - z_1) + \frac{V_2^2 - V_1^2}{2g} + \frac{p_2 - p_1}{\rho g} + \overline{H}_f = 0$$

$$\overline{W} + 100 + 0 + 0 + 16.76 = 0$$

$$\overline{W} = -116.76 \text{ m (work on fluid)}$$

Step D *Compute Power Required*

(3-8)
$$\dot{m} = \rho A V = 998.3 \times 0.018\,650 \times 3.351$$

$$\dot{m} = 62.39 \text{ kg/s}$$

(8-37)
$$W_{sf} = \frac{g}{g_c}\,\overline{W} = \left(\frac{9.807}{1}\right)116.76 = 1\,145 \text{ N} \cdot \text{m/kg}$$

$$p = \dot{m}\, W_{sf} = 62.39 \times 1\,145 = 71\,440\ W = 71.44 \text{ kW}$$

8-9 STANDARD PIPELINE METERS

This section covers the pipeline differential flow meters that have been standardized by the American Society of Mechanical Engineers Research Committee on Fluid Meters. These are the Venturi, the flow nozzle, and the square-edge thin-plate concentric orifice.

Dimensional Analysis

Consider an incompressible fluid of density ρ, viscosity μ, flowing with a velocity of V through a primary element (Venturi, nozzle, or orifice) whose diameter is d. The primary element is located in a horizontal meter tube whose roughness is ϵ and diameter is D. The flow through the primary element produces a pressure differential of Δp sensed by pressure taps located a distance L apart.

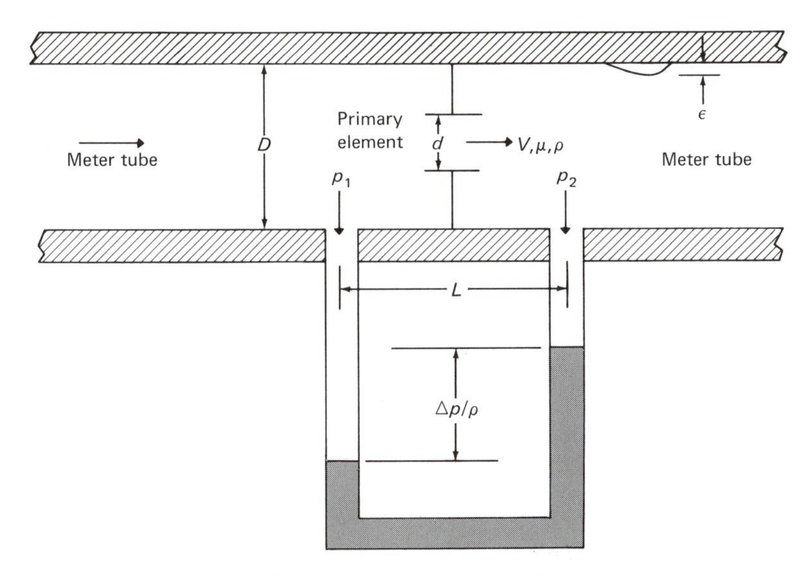

FIGURE 8-7 Notation for dimensional analysis of flow meters

Derive the dimensionless parameters involved in fluid metering, as shown in Fig. 8-7. Results of the application of the Buckingham Π theorem are shown in Table 8-4.

Physical Analysis

From dimensional analysis

$$\pi_5 = f(\pi_1, \pi_2, \pi_3, \pi_4)$$

or

$$\overline{K} = V/\sqrt{2\Delta p/\rho} \qquad \text{and} \qquad V = \overline{K}\sqrt{2\Delta p/\rho} \tag{8-40}$$

where

$$\overline{K} = f\left(\mathbf{R}_d, \frac{L}{D}, \mathrm{B}, \frac{\epsilon}{D}\right)$$

Substituting $V = Q/A_d$ from the continuity equation (3-4) in Eq. (8-40),

$$Q = \overline{K}A_d\sqrt{2\Delta p/\rho} \tag{8-41}$$

where A_d is the flow area of the primary element. Conventional practice is to base flow meter computations on the one-dimensional

TABLE 8-4 *Dimensional Analysis of Flow Meters*

Item		Variables considered							
		PRIMARY ELEMENT DIAMETER	FLUID			METER TUBE			PRESSURE DIFFERENTIAL
			Primary element velocity	*Density*	*Viscosity*	*Length*	*Diameter*	*Roughness*	
Symbol		d	V	ρ	μ	L	D	ϵ	Δp
Selection		B_G^x	B_K^y	B_D^z	A_1	A_2	A_3	A_4	A_5
π Ratios		Basic group			π_1	π_2	π_3	π_4	π_5
Dimensions		L	LT^{-1}	$FL^{-4}T^2$	$FL^{-2}T$	L	L	L	FL^{-2}
Exponents	F	0	0	z	1	0	0	0	1
for	L	x	y	$-4z$	-2	1	1	1	-2
equations	T	0	$-y$	$2z$	1	0	0	0	0

TABLE 8-4 *Solution*

π Ratio	Solved exponents d^x	V^y	ρ^z	Derived ratio	Conventional practice FORM	SYMBOL	NAME
π_1	-1	-1	-1	$\mu/dV\rho$	$dV\rho/\mu$	$\mathbf{R}_d$	Primary element Reynolds number
π_2	-1	0	0	L/d	L/D	—	Tap location ratio
π_3	-1	0	0	D/d	d/D	β	Beta ratio
π_4	-1	0	0	ϵ/d	ϵ/D	—	Relative roughness
π_5	-1	-2	0	$\Delta p/\rho V^2$	$V/\sqrt{2\Delta p/\rho}$	$\overline{K}$	Flow coefficient (pressure coefficient)$^{-1/2}$

frictionless flow of an incompressible fluid in a horizontal meter tube and to correct for actual conditions by use of coefficients and factors. In Sec. 5-5, Eq. (5-21) was developed for ideal one-dimensional flow of an incompressible fluid as

$$M_i = A_2\sqrt{\frac{2g_c(p_1 - p_2)/v_1}{1 - (A_2/A_1)^2}} \qquad \text{(5-21)}$$

Noting from Eq. (5-25) that

$$\beta = \left(\frac{A_2}{A_1}\right)^{1/2} \qquad \text{(5-25)}$$

and that $A_2 = A_d$, $p_1 - p_2 = \Delta p$, from Eq. (5-12)

$$\dot{M}_i = \frac{A_d V}{v_1} = \frac{Q_i}{v_1} \qquad \text{(5-12)}$$

and from Eq. (1-36)

$$v = \frac{1}{\rho g_c} \tag{1-36}$$

Substituting in Eq. (5-21),

$$\frac{Q_i}{v_1} = Q_i \rho g_c = A_d \sqrt{\frac{2g_c \Delta p(\rho g_c)}{1-\beta^4}}$$

or

$$Q_i = A_d \sqrt{\frac{\Delta p/\rho}{1-\beta^4}} \tag{8-42}$$

The coefficient of discharge C is defined as

$$C = \frac{\text{actual flow rate}}{\text{ideal flow rate}} = \frac{Q}{Q_i} \tag{8-43}$$

or

$$Q = CQ_i = CA_d \sqrt{\frac{2\Delta p/\rho}{1-\beta^4}} \tag{8-44}$$

It is also conventional practice to let

$$E = \frac{1}{\sqrt{1-\beta^4}} \tag{8-45}$$

where E is called the velocity of approach factor because it accounts for the one-dimensional kinetic energy at the inlet tap. Substituting Eq. (8-45) in Eq. (8-44),

$$Q = CEA_d \sqrt{2\Delta p/\rho} \tag{8-46}$$

Equating Eq. (8-42) to Eq. (8-46),

$$Q = \overline{K} A_d \sqrt{2\Delta p/\rho} = CEA_d \sqrt{2\Delta p/\rho} \qquad \text{or} \qquad \overline{K} = CE \tag{8-47}$$

For compressible fluids, Eq. (8-40) is modified by the expansion factor Y where Y is the ratio of the flow of a compressible fluid to that of an incompressible fluid at the same Reynolds number (Sec. 5-5). Calculations are then based on the inlet density ρ_1. The value of the expansion factor Y for liquids is by this definition unity. Equation (8-40) may be thus modified for the flow of any fluid:

$$Q = \overline{K} Y A_d \sqrt{2\Delta p/\rho_1} \tag{8-48}$$

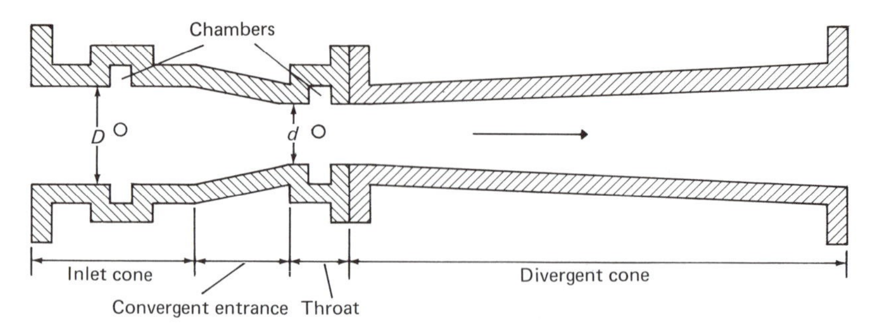

FIGURE 8-8 Venturi tube

Venturi Tubes

Figure 8-8 shows a typical Venturi tube consisting of a cylindrical inlet, convergent entrance, throat, and divergent outlet. The convergent entrance has an included angle of about 21° and the divergent cone 7 to 8°. The effect of the divergent cone is to reduce the overall pressure loss of the meter, its removal will have no effect on the coefficient of discharge. Pressure is sensed through a series of holes in the inlet cone and throat. These holes lead to an annular chamber, and the two chambers are connected to a pressure differential sensor.

Discharge coefficients for Venturi tubes as established by the American Society of Mechanical Engineers are given in Table 8-5. For values of Reynolds numbers outside the tabulated limits, the Venturi tube must be calibrated to obtain the numerical value of the discharge coefficient.

EXAMPLE 1 Venturi Tube Calculation Benzene at 86°F flows through a machined inlet Venturi tube whose inlet diameter is 8 in. and whose throat diameter is 3.5 in. The differential pressure is sensed by a U-tube manometer. The manometer contains benzene and mercury under the benzene. The mercury level in the throat leg is 4 in. (see Fig. 8-9). What is the volumetric flow rate?

Step A *Data Reduction*

(1) *unit conversion*

(B-55) $D = 8/12 = 2/3$ ft $\quad d = \dfrac{3.5}{12}$ ft $\quad h_m = 4/12 = 1/3$ ft

(2) *gravity*

Assume standard

(B-50) $g = 32.17$ ft/sec^2

TABLE 8-5 *ASME Coefficients for Venturi Tubes*

Type of Inlet Cone	R_d		Value of C	Tolerance %
	MINIMUM	MAXIMUM		
Machined	500,000	1,000,000	0.995	±1.00
Rough welded sheet metal	500,000	2,000,000	0.985	±1.50
Rough cast	500,000	2,000,000	0.984	±0.70

(3) *fluid properties*

benzene at 86°F

Table A-1 $\rho = 1.684$ slugs/ft^3

Table A-4 $\mu = 11.72 \times 10^{-6}$ lbf-sec/ft^2

mercury at 86°F

Table A-1 $\rho_m = 26.24$ slugs/ft^3

(4) *geometric*

Table A-10 $$A_d = \frac{\pi d^2}{4} = \frac{\pi(3.5/12)^2}{4} = 0.06681 \text{ ft}^2$$

(5-25) $$\beta = \frac{d}{D} = \frac{3.5/12}{2/3} = 0.4375$$

Step B *Determine $\Delta p/\rho$ by Manometer Balance*

(2-8) $$p_1 - p_2 = (\gamma_m - \gamma_f)(h)$$

From Fig. 8-9, noting from Eq. (1-35),

$\gamma = \rho g$, Eq. (2-8) becomes

$$\Delta p = (\rho_m g - \rho g)(h_m) \tag{a}$$

FIGURE 8-9 Notation for Example 1

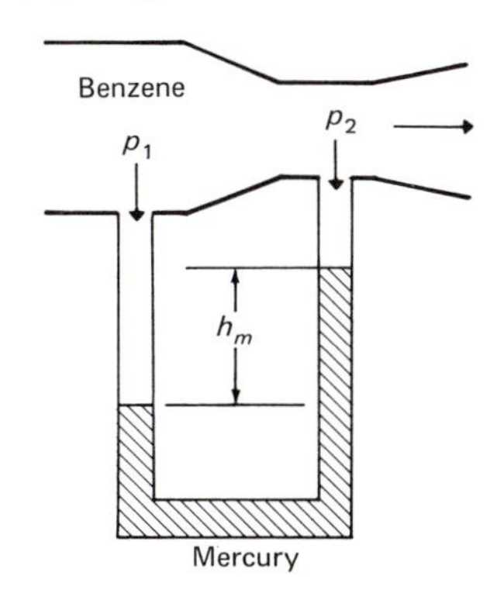

and

$$\frac{\Delta p}{\rho} = \left(\frac{\rho_m}{\rho} - 1\right) g h_m \quad \text{(b)}$$

$$= \left(\frac{26.24}{1.684} - 1\right) 32.17(1/3) = 156.37$$

Step C *Compute Flow Rate*

$$E = \frac{1}{\sqrt{1-\beta^4}} = \frac{1}{\sqrt{1-(0.4375)^4}} = 1.0188 \quad \text{(8-45)}$$

$Y = 1$ (by definition for liquids)

$C = 0.995$ (assume that Table 8-5 applies, then check $\mathbf{R}_d$)

$$\overline{K} = CE = 0.995 \times 1.0188 = 1.0137 \quad \text{(8-47)}$$

$$Q = \overline{K} Y A_d \sqrt{2\Delta p/\rho} \quad \text{(8-48)}$$

$$= (1.0137)(1)(0.06681)\sqrt{2 \times 156.37}$$

$$Q = 1.197 \text{ ft}^3/\text{sec}$$

$$V_d = \frac{Q}{A_d} = \frac{1.197}{0.06681} = 17.93 \text{ ft/sec} \quad \text{(3-4)}$$

$$\mathbf{R}_d = \frac{V_d d \rho}{\mu} = \frac{17.93 \times (3.5/12)(1.684)}{(11.72 \times 10^{-6})} \quad \text{(6-18)}$$

$$\mathbf{R}_d = 751{,}418$$

$\mathbf{R}_d$ falls within the limits of 500,000 and 1,000,000 for machined inlet cone of Table 8-5 ∴ solution is valid.

ASME Flow Nozzle

Figure 8-10 shows an ASME flow nozzle. This nozzle is built to rigid specifications, and pressure differential may be sensed by either throat taps or pipe wall taps. Each is located one pipe diameter upstream and one-half diameter downstream from the nozzle inlet.

Coefficient of discharge for ASME flow nozzles may be computed from Eq. (8-49) to within ±1 to 2% depending upon β and $\mathbf{R}_d$.

$$C = 0.9975 - 0.00653\left(\frac{10^6}{\mathbf{R}_d}\right)^a \quad \text{(8-49)}$$

where $a = 1/2$ for $\mathbf{R}_d < 10^6$ and $1/5$ for $\mathbf{R}_d > 10^6$

EXAMPLE 2 Flow Nozzle Calculation An ASME flow nozzle is to be designed to measure the flow of 100 m³/hr of water at 20°C in a 6-in. 250 psig cast iron pipe. The pressure differential across the

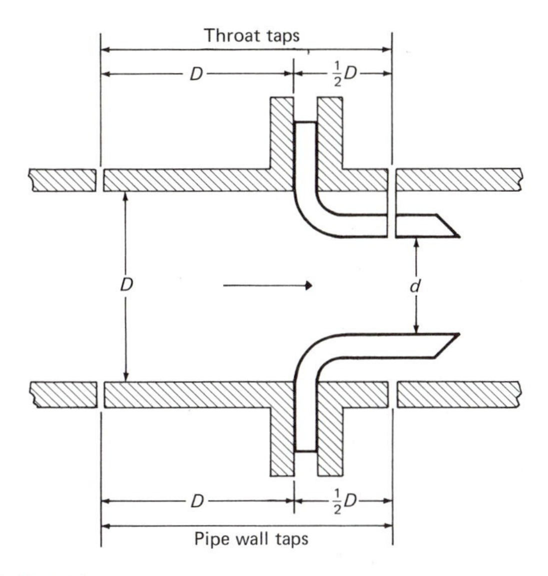

FIGURE 8-10 ASME flow nozzle

nozzle is not to exceed 20 kPa. What should be the throat diameter of the nozzle?

Step A *Data Reduction*

(1) *units*

(B-54) $$Q = \frac{100}{3\,600} = 0.027\,778 \text{ m}^3/\text{s}$$

(2) *fluid properties,* water at 20°C

Table A-1 $\rho = 998.3$ kg/m^3

Table A-4 $\mu = 10.02 \times 10^{-4}$ Pa · s

(3) *pipe properties,* 6-in. 250 psi cast iron

Table A-12 $D = 155.9$ mm

Step B *Compute a Trial Value of d*

A trial and error solution is necessary to establish values of C and E because they are dependent upon β and $\mathbf{R}_d$, both of which require that the value of d be known. Since $\overline{K} = CE$ is close to unity, assume for a trial value that $\overline{K} = 1$. Since the fluid is a liquid, Y also equals unity.

(8-48) $$A_d = \frac{Q}{\overline{K} Y \sqrt{2\Delta p/\rho}} = \frac{0.027\,778}{(1)(1)\sqrt{2 \times 20\,000/998.3}}$$

$$A_d = 4.388\,3 \times 10^{-3} \text{ m}^2$$

Table A-10 $d = \sqrt{4A_d/\pi} = \sqrt{4 \times 4.388\ 3 \times 10^{-3}/\pi} = 0.074\ 75$ m

$= 74.75$ mm

Step C *Check Trial Value*

(5-25) $$\beta = \frac{d}{D} = \frac{0.074\ 75}{0.155\ 9} = 0.479\ 5$$

(8-45) $$E = \frac{1}{\sqrt{1-\beta^4}} = \frac{1}{\sqrt{1-(0.479\ 5)^4}} = 1.027\ 5$$

(3-4) $$V_d = \frac{Q}{A_d} = \frac{0.027\ 778}{4.388\ 3 \times 10^{-3}}$$

$$V_d = 6.330 \text{ m/s}$$

(6-18) $$\mathbf{R}_d = \frac{V_d\, d\rho}{\mu} = \frac{6.330 \times 0.074\ 75 \times 998.3}{10.02 \times 10^{-4}}$$

$$\mathbf{R}_d = 471\ 420$$

(8-49) $$C = 0.9975 - 0.00653(10^6/\mathbf{R}_d)^a$$

$$\mathbf{R}_d < 10^6 \therefore a = 1/2$$

$$C = 0.9975 - 0.00653\left(\frac{10^6}{471\ 420}\right)^{1/2} = 0.988\ 0$$

(8-47) $$\overline{K} = CE = 0.988\ 0 \times 1.027\ 5 = 1.015\ 2$$

(8-48) $$A_d = \frac{Q}{\overline{K}Y\sqrt{2\Delta p/\rho}} = \frac{0.027\ 778}{(1.0152)(1)\sqrt{2 \times 20\ 000/998.3}}$$

$$A_d = 4.323 \times 10^{-3}$$

Table A-1 $$d = \sqrt{4A/\pi} = \sqrt{4 \times 4.323 \times 10^{-3}/\pi}$$

$$d = 0.074\ 19 = 74.19 \text{ mm}$$

Further trials not necessary.

ASME Orifice Meters

When a fluid flows through a square-edged thin-plate orifice, the minimum area is called the vena contracta and its location is a function of the beta ratio.

Figure 8-11 shows the relative pressure difference due to the presence of the orifice plate, and Figure 8-12 shows the location of the vena contracta with respect to β.

Because the location of the pressure taps is critical, it is necessary to specify the exact position of the downstream pressure tap. The jet contraction amounts to about 60% of the orifice area, so orifice coefficients are in the order of 0.6 compared with nearly unity obtained with Venturi tubes and flow nozzles.

Three tap locations are specified by the ASME for measuring pressure differential. These are the flange, vena contracta, and

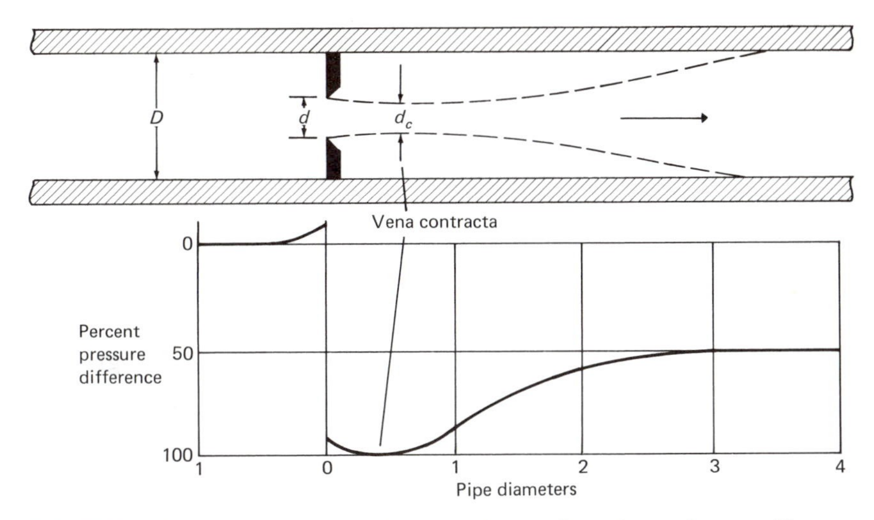

FIGURE 8-11 Relative pressure changes due to flow through an orifice

the $1D$ and $1/2\,D$. In the flange tap, the location is always one inch from either face of the orifice plate regardless of the size of the pipe. In the vena contracta tap, the upstream tap is located one pipe diameter from the inlet face of the orifice plate and the downstream tap, as shown in Fig. 8-12. In the $1D$ and $1/2\,D$ tap, the upstream tap is located one pipe diameter from the inlet face of the orifice plate and the downstream tap one-half pipe diameter from the inlet face of the orifice plate.

Flange taps are used because they can be prefabricated, and flanges with holes drilled at the correct locations may be purchased

FIGURE 8-12 Location of vena contracta

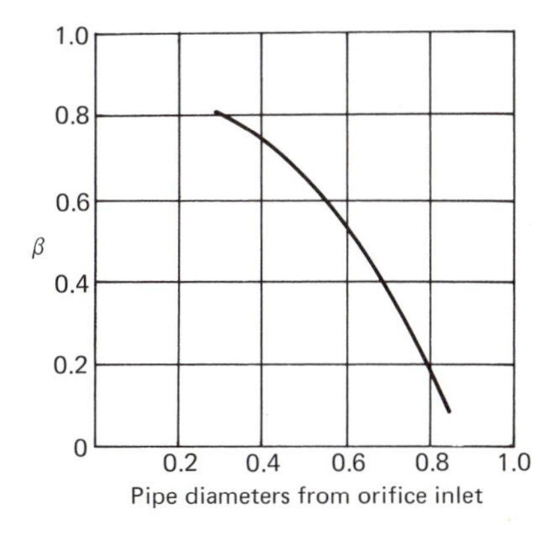

as off-the-shelf items, thus saving the cost of field fabrication. The disadvantage of flange taps is that they are not symmetrical with respect to pipe size. In a 1-in. pipe, the downstream location is one pipe diameter, but in a 30-in. pipe, they are 1/30 pipe diameter. Because of this, coefficients of discharge for flange taps vary greatly with pipe size.

Vena contracta taps are used because they give the maximum differential for any given flow. The disadvantage of the vena contracta tap is that, if the orifice size is changed, new holes must be drilled. The $1D$ and $\frac{1}{2}D$ taps incorporate the best features of the vena contracta taps and are symmetrical with respect to pipe size.

Discharge coefficients for orifices may be calculated from Eq. (8-50):

$$C = C_o + \frac{\Delta C}{\mathbf{R}_d^a} \qquad (\mathbf{R}_d > 10^4) \tag{8-50}$$

where C_o and a are obtained from Table 8-6.

TABLE 8-6 *Values of* C_o, ΔC *and a for use in Equation (8-50)*

β	D = 2 in. = 50 mm		D = 4 in. = 100 mm		D = 8 in. = 200 mm		D = 16 in. = 400 mm	
	C_o	ΔC	C_o	ΔC	C_o	ΔC	C_o	ΔC
				Flange Taps $a = 1$				
0.20	0.5972	127	0.5946	200	0.5951	327	0.5955	551
0.30	0.5978	144	0.5977	209	0.5978	307	0.5980	457
0.40	0.6014	181	0.6005	256	0.6002	362	0.6001	514
0.50	0.6050	260	0.6034	386	0.6026	584	0.6022	903
0.60	0.6078	392	0.6055	622	0.6040	1015	0.6032	1710
0.70	0.6068	573	0.6030	953	0.6006	1637	0.5991	2898
				Vena Contracta Taps $a = \frac{1}{2}$				
0.20	0.5938	1.61	0.5928	1.61	0.5925	1.61	0.5924	1.61
0.30	0.5939	1.78	0.5934	1.78	0.5933	1.78	0.5932	1.78
0.40	0.5970	2.01	0.5954	2.01	0.5953	2.01	0.5953	2.01
0.50	0.5994	2.29	0.5992	2.29	0.5992	2.29	0.5991	2.29
0.60	0.6042	2.68	0.6041	2.68	0.6041	2.69	0.6041	2.70
0.70	0.6069	3.34	0.6068	3.37	0.6067	3.44	0.6068	3.57
				$1D$ and $\frac{1}{2}D$ Taps $a = \frac{1}{2}$				
0.20	0.5909	2.03	0.5922	1.41	0.5936	1.10	0.5948	0.94
0.30	0.5915	2.02	0.5930	1.50	0.5944	1.24	0.5956	1.12
0.40	0.5936	2.17	0.5951	1.72	0.5963	1.49	0.5974	1.38
0.50	0.5979	2.40	0.5978	1.99	0.5999	1.79	0.6007	1.69
0.60	0.6036	2.67	0.6040	2.31	0.6044	2.12	0.6048	2.11
0.70	0.6078	3.19	0.6072	2.98	0.6068	3.07	0.6064	3.51

Tolerances for uncalibrated orifice meters are about the same as those for uncalibrated flow nozzles, being in the order of ±1 to ±2% depending upon β, D, and $\mathbf{R}_d$.

EXAMPLE 3 Orifice Meter Calculation Turpentine at 68°F flows at a rate of 12,000 lbm/hr in a 4-in. schedule 40 steel pipe. It is proposed to measure this flow with a 1.6-in. diameter ASME orifice equipped with vena contracta taps. What differential in inches of water at 68°F may be expected?

Step A — *Data Reduction*

(1) *unit conversions*

(B-55) $$d = \frac{1.6}{12} \text{ ft}$$

(B-93/54) $$\dot{m} = \frac{12{,}000}{(3{,}600 \times 32.17)} = 0.1036 \text{ slugs/sec}$$

(2) *gravity*

Assume standard

(B-50) $$g = 32.17 \text{ ft/sec}^2$$

(3) *fluid properties*

Turpentine at 68°F

Table A-1 $$\rho = 1.673 \text{ lbm/ft}^3$$

Table A-4 $$\mu = 31.06 \times 10^{-6} \text{ lbf-sec/ft}^2$$

Water at 68°F

Table A-1 $$\rho_w = 1.937 \text{ lbm/ft}^3$$

(1-35) $$\gamma_w = \rho g = 1.937 \times 32.17 = 62.31 \text{ lbf/ft}^3$$

(4) *pipe properties,* 4-in. schedule 40 steel

Table A-11 $$D = 0.3355 \text{ ft}$$

(5) *geometric*

Table A-10 $$A_d = \frac{\pi d^2}{4} = \frac{\pi(1.6/12)^2}{4} = 0.01396 \text{ ft}^2$$

(5-25) $$\beta = \frac{d}{D} = \frac{1.6/12}{0.3355} = 0.3974$$

(6) *kinematic*

(3-8) $$V_d = \frac{\dot{m}}{\rho A_d} = \frac{0.1036}{1.673 \times 0.01396} = 4.436 \text{ ft/sec}$$

(3-4) $$Q = A_d V_d = 0.01396 \times 4.436 = 0.06193 \text{ ft}^3\text{/sec}$$

Step B *Compute Δp*

(6-18)
$$\mathbf{R}_d = \frac{V_d\, d\rho}{\mu} = \frac{4.436 \times (1.6/12) \times 1.673}{31.06 \times 10^{-6}}$$

$$\mathbf{R}_d = 31{,}858$$

Table 8-6 Interpolated for $\beta = 0.3974$ and $D = 0.3355 = 4.026$ in.

$a = 1/2$, $C_o = 0.5952$ $\Delta C = 2.00$

(8-50)
$$C = C_o + \frac{\Delta C}{\mathbf{R}_d^a} = 0.5952 + \frac{2.00}{(31{,}858)^{1/2}}$$

$$C = 0.6065$$

(8-45)
$$E = \frac{1}{\sqrt{1-\beta^4}} = \frac{1}{\sqrt{1-(0.3974)^4}}$$

$$E = 1.0127$$

(8-47)
$$\overline{K} = CE = 0.6065 \times 1.0127 = 0.6142$$

and $Y = 1$ for liquids

(8-48)
$$\Delta p = \frac{\rho Q^2}{2\overline{K}^2 Y^2 A_d^2}$$

$$\Delta p = \frac{1.673(0.06193)^2}{2(0.6142)^2(1)^2(0.01396)^2}$$

$$\Delta p = 43.65 \text{ lbf/ft}^2$$

Step C *Compute Inches of Water Differential*

(2-6)
$$h_w = \frac{\Delta p}{\gamma_w} = \frac{43.65}{62.31} = 0.7005 \text{ ft}$$

(B-55)
$$h_w = 0.7005 \times 12 = 8.406 \text{ in.}$$

8-10 COMPRESSIBLE FLOW THROUGH METERS

Compressible Flow Through Venturi Tubes and Flow Nozzles

For Venturi tubes and flow nozzles, the relation for the expansion factor Y may be established for ideal gases from theoretical considerations only. The expansion factor may be calculated from Eq. (5-23) or may be taken from Table A-15.

Maximum flow is obtained when the critical pressure ratio is reached. The critical pressure ratio may be calculated from Eq. (5-19). Conventional practice is to use tables such as Table A-15 to avoid tedious computations.

EXAMPLE 1 Compressible Flow Through an ASME Flow Nozzle A piping system consists of a compressor, a horizontal straight length of 2-in. schedule 40 steel pipe, and a 25-mm throat diameter ASME flow nozzle attached to the end of the pipe and discharging into the atmosphere. The compressor is operated to maintain a flow of air with 800 kPa and 40°C conditions in the pipe just one pipe diameter before the nozzle inlet. Barometric pressure is 101.3 kPa. Estimate the flow rate of the air.

Step A *Data Reduction*

(1) *units*

(B-27) $T_1 = 40 + 273.15 = 313.15 \text{ K}$

(2) *proportionality constant*

(1-9) $g_c = 1 \text{ kg} \cdot \text{m/N} \cdot \text{s}^2$

(3) *fluid properties,* air at 40°C

Table A-7 $R = 287.1 \text{ J/kg} \cdot \text{K}, \text{k} = 1.400$

(1-44) $\rho_1 = \dfrac{p_1}{g_c R T_1} = \dfrac{800\,000}{1 \times 287.1 \times 313.15}$

$\rho_1 = 8.898 \text{ kg/m}^3$

Table A-8 $\mu_1 = 19.04 \times 10^{-6} \text{ Pa} \cdot \text{s}$

(4) *pipe properties,* schedule 40 steel pipe

Table A-11 $D = 52.52 \text{ mm}$

(5) *geometric*

Table A-10 $A_d = \dfrac{\pi d^2}{4} = \dfrac{\pi (0.025)^2}{4} = 4.909 \times 10^{-4} \text{ m}^2$

(5-25) $\beta = \dfrac{d}{D} = \dfrac{25}{52.52} = 0.4760$

(8-45) $E = \dfrac{1}{\sqrt{1 - \beta^4}} = \dfrac{1}{\sqrt{1 - (0.4760)^4}} = 1.0267$

Step B *Determine Type of Flow*

(5-24) $r = \dfrac{p_2}{p_1} = \dfrac{101\,300}{800\,000} = 0.1263$

Interpolation of Table A-15 indicates that the critical pressure ratio for $k = 1.400$ and $\beta = 0.4760$ is $r_c = 0.5356 > 0.1263$ $\therefore$ flow is sonic

(5-24) $p_c = 800\,000 \times 0.5356 = 428\,480$

Table A-15 $Y = Y_c = 0.6979$

Step C *First Trial*

Assume $C = 1$

(8-47) $\bar{K} = CE = 1 \times 1.0267 = 1.0267$

(8-48) $Q_1 = \bar{K}YA_d\sqrt{2\Delta p/\rho_1} = \bar{K}Y_cA_d\sqrt{2(p_1 - p_c)/\rho_1}$

$$Q_1 = 1.0267 \times 0.6979 \times 4.909 \times 10^{-4} \times \sqrt{2(800\ 000 - 428\ 400)/8.898}$$

$Q_1 = 0.101\ 7\ \text{m}^3/\text{sec}$ (at inlet conditions)

(3-8) $\dot{m} = \rho_1 A_d V_d = \rho_1 Q_1 = 8.898 \times 0.101\ 7 = 0.904\ 9\ \text{kg/s}$

(8-26) $\mathbf{R}_d = \dfrac{\dot{m}d}{A_d \mu_d}$

$$\mathbf{R}_d = \frac{(0.904\ 9)(0.025)}{(4.909 \times 10^{-4})(19.04 \times 10^{-6})}$$

$\mathbf{R}_d = 2\ 420\ 364$ (assumes $\mu_d = \mu_1$)

(8-4) $C = 0.9975 - 0.00653\left(\dfrac{10^6}{\mathbf{R}_d}\right)^a$ (for $\mathbf{R}_d > 10^6$, $a = 1/5$)

$$C = 0.9975 - 0.00653\left(\frac{10^6}{2\ 420\ 364}\right)^{1/5}$$

$C = 0.992\ 0$

Step D *Second Trial*

(3-8) $\dot{m} = 0.992\ 0 \times 0.904\ 9 = 0.897\ 7\ \text{kg/s}$

Further trial unnecessary.

Compressible Flow through ASME Orifices

As shown in Fig. 8-11, the minimum flow area for an orifice is at the vena contracta located downstream of the orifice. The stream of compressible fluid is not restrained as it leaves the orifice throat and is free to expand transversely and longitudinally to the point of minimum flow area. Thus, the maximum contraction of the jet will be less for a compressible fluid than for a liquid. Because of this, the theoretical expansion factor, Eq. (5-23), may not be used with orifices. Neither may the critical pressure ratio equation, Eq. (5-19), be used, as the phenomenon of critical flow has not been observed during testing of orifice meters.

For orifice meters, Eq. (8-51), which is based on experimental data, is used:

$$Y = 1 - \frac{(0.41 + 0.35\beta^4)(1 - r)}{k} \qquad \text{(8-51)}$$

EXAMPLE 1 Compressible Flow Through an ASME Orifice Air at 68°F and 150 psia flows in a 2-in. schedule 40 steel pipe at a volumetric rate of 15 ft³/min. An ASME orific equipped with flange taps is to be used to meter this flow with a readout device that has a maximum indication of 150 in. of water at 68°F. What diameter of orifice should be used if the differential is not to exceed the 150 in. of water?

Step A *Data Reduction*

(1) *unit conversions*

(B-28) $T_1 = 68 + 459.67 = 527.67°\text{R}$

(B-89) $p_1 = 150 \times 144 = 21{,}600 \text{ lbf/ft}^2$

(B-68) $Q_1 = {}^{15}\!/_{60} = 0.25 \text{ ft}^3\text{/sec}$

(B-55) $h_w = {}^{150}\!/_{12} = 12.5 \text{ ft}$

(2) *gravity*

Assume standard

(B-50) $g = 32.17 \text{ ft/sec}^2$

(3) *proportionality constant*

(1-11) $g_c = 32.17 \text{ lbm-ft/lbf-sec}^2$

(4) *fluid properties*

Air at 68°F

Table A-7 $R = 53.36 \text{ ft-lbf/lbm-°R}, \ k = 1.400$

(1-44) $$\rho_1 = \frac{p_1}{g_c R T_1} = \frac{21{,}600}{32.17 \times 53.36 \times 527.67}$$

$\rho_1 = 0.02385 \text{ slugs/ft}^3$

Table A-8 $\mu_1 = 38.31 \times 10^{-8} \text{ lbf-sec/ft}^2$

Water at 68°F

Table A-1 $\rho_w = 1.937 \text{ slugs/ft}^3$

(1-35) $\gamma_w = \rho_w g = 1.937 \times 32.17 = 62.31 \text{ lbf/ft}^3$

(5) *pipe properties* 2-in. schedule 40

Table A-11 $D = 0.1723 \text{ ft}$

(6) *dynamic*

(3-8) $\dot{m} = \rho_1 A_1 V_1 = \rho_1 Q_1 = 0.02385 \times 0.25$

$\dot{m} = 5.9625 \times 10^{-3} \text{ slugs/sec}$

(2-6) $\Delta p = h_w \gamma_w = 12.5 \times 62.31 = 778.88 \text{ lbf/ft}^2$

(5-24) $$(1 - r) = \frac{\Delta p}{p_1} = \frac{778.88}{21{,}600} = 0.03606$$

Step B *First Trial*

Assume $Y = 1$, $E = 1$, $C = 0.6$

(8-47) $\bar{K} = CE = 0.6 \times 1 = 0.6$

(8-48)
$$A_d = \frac{Q_1}{\bar{K}Y\sqrt{2\Delta p/\rho_1}}$$

$$A_d = \frac{0.25}{0.6 \times 1 \times \sqrt{2 \times 778.88/0.02385}}$$

$$A_d = 1.6304 \times 10^{-3}\ \text{ft}^2$$

Table A-10
$$d = \sqrt{\frac{4A_d}{\pi}} = \sqrt{\frac{4 \times 1.6304 \times 10^{-3}}{\pi}} = 0.04556\ \text{ft}$$

Step C *Second Trial*

Use first trial values

(5-25)
$$\beta = \frac{d}{D} = 0.04556/0.1723 = 0.2644$$

(8-45)
$$E = \frac{1}{\sqrt{1-\beta^4}} = \frac{1}{\sqrt{1-(0.2644)^4}}$$

$$E = 1.0025$$

(8-51)
$$Y = 1 - \frac{(0.41 + 0.35\beta^4)(1-r)}{k}$$

$$Y = 1 - \frac{[0.41 + 0.35(0.2644)^4](0.03606)}{1.400}$$

$$Y = 0.9894$$

(8-20)
$$\mathbf{R}_d = \frac{\dot{m}d}{A_d\mu_1}\ (\text{assumes } \mu_1 = \mu_d)$$

$$\mathbf{R}_d = \frac{(5.9625 \times 10^{-3})(0.04556)}{(1.6304 \times 10^{-3})(38.31 \times 10^{-8})}$$

$$\mathbf{R}_d = 434{,}916$$

Table 8-6 Flange Taps $a = 1$

$C_o = 0.5976$, $\Delta C = 140$ (interpolated)

(8-50)
$$C = C_o + \frac{\Delta C}{\mathbf{R}_d^a}$$

$$C = 0.5976 + \frac{140}{434{,}916} = 0.5979$$

(8-47) $\bar{K} = CE = 0.5979 \times 1.0025 = 0.5994$

(8-48)
$$A_d = \frac{Q_1}{\bar{K}Y\sqrt{2\Delta p/\rho_1}}$$

$$A_d = \frac{0.25}{(0.5994 \times 0.9894)\sqrt{2 \times 778.88/0.02385}}$$

$$A_d = 1.6495 \times 10^{-3}\ \text{ft}^2$$

Table A-10

$$d = \sqrt{4A_d/\pi} = \sqrt{4 \times 1.6495 \times 10^{-3}/\pi} = 0.04583\ \text{ft}$$

Step D *Third Trial; Use Second Trial Values*

(5-25)

$$\beta = \frac{d}{D} = \frac{0.04583}{0.1723} = 0.2660$$

(8-45)

$$E = \frac{1}{\sqrt{1-\beta^4}} = \frac{1}{\sqrt{1-(0.2660)^4}} = 1.0025\ \text{(no change)}$$

(8-51)

$$Y = 1 - \frac{(0.41 + 0.35\beta^4)(1-r)}{k}$$

$$Y = 1 - \frac{[0.41 + 0.35(0.2660)^4](0.03606)}{1.400}$$

$$Y = 0.9894$$

(8-26)

$$\mathbf{R}_d = \frac{\dot{m}d}{A_d\mu_1}\ \text{(assumes } \mu_1 = \mu_d\text{)}$$

$$\mathbf{R}_d = \frac{(5.9625 \times 10^{-3})(0.04583)}{(1.6496 \times 10^{-3})(38.31 \times 10^{-8})}$$

$$\mathbf{R}_d = 432{,}401$$

Table 8-6

Flange Taps $a = 1$

$C_o = 0.5776$, $\Delta C = 140$ (interpolated)

(8-50)

$$C = C_o + \frac{\Delta C}{\mathbf{R}_d^a}$$

$$C = 0.5976 + \frac{140}{432{,}401} = 0.5979\ \text{(no change)}$$

(8-47)

$$\overline{K} = CE = 0.5979 \times 1.0025 = 0.5994$$

(8-48)

$$A_d = \frac{Q_1}{\overline{K}Y\sqrt{2\Delta p/\rho_1}}$$

$$A_d = \frac{0.25}{(0.5994 \times 0.9894)\sqrt{2 \times 778.88/0.02385}}$$

$$A_d = 1.6495 \times 10^{-3}$$

Table A-10

$$d = \sqrt{4A_d/\pi} = \sqrt{4 \times 1.6495 \times 10^{-3}/\pi} = 0.04583\ \text{ft}$$

(no change from second trial)

(B-55)

$$d = 0.04583 \times 12 = 0.5499\ \text{in.}$$

8-11 PITOT TUBE

A Pitot tube is a device that is shaped in such a manner that it senses stagnation pressure. The first description of a tube used for measurement of stagnation pressure for the determination of velocity was given by Henri Pitot in 1732. The name "Pitot tube" has been applied to two general classifications of instruments, the first being a tube that measures the impact or stagnation pressures only, and the second a combined tube that measures both impact and static pressures with a single primary instrument. The combined sensor is also called a Pitot-static tube.

Incompressible Flow

From Fig. 8-13, it is evident that the Pitot tube can sense only the local streamtube velocity, U. Conventional practice is to base computations on the assumption that the conversion of kinetic energy to flow work is frictionless. Application of the Bernoulli equation (4-17) for constant elevation leads to

$$\frac{U_s^2 - U_i^2}{2g} + \frac{p_s - p_o}{\rho_o g} = 0 \tag{8-52}$$

where

U_s = velocity at stagnation point
U_i = ideal streamtube velocity
p_s = stagnation pressure
p_o = static pressure

Solving Eq. (8-52) for U_i, noting that by definition $U_s = 0$, results in

FIGURE 8-13 Notation for Pitot tube study

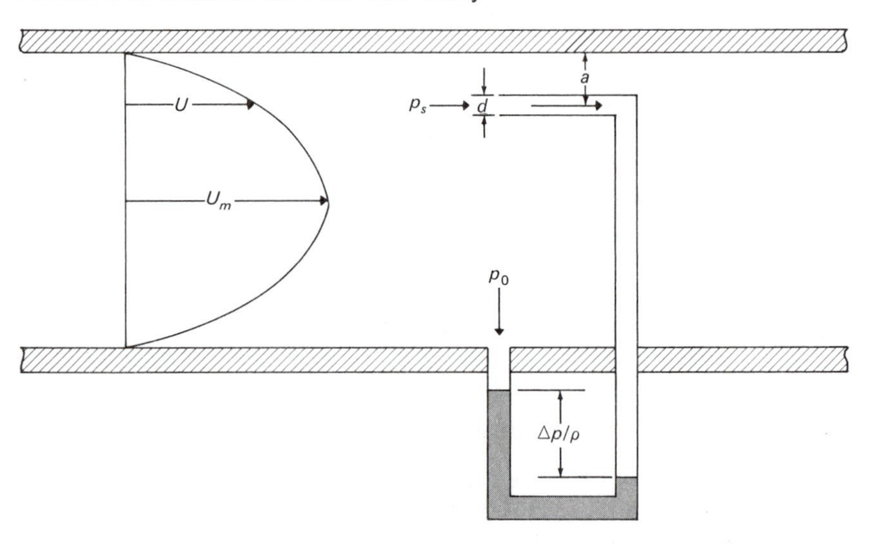

$$U = \sqrt{2(p_s - p_o)/\rho_o} = \sqrt{2\Delta p/\rho_o} \tag{8-53}$$

Conventional practice is to define the tube coefficient C_t as follows:

$$C_t = \frac{\text{actual streamtube velocity}}{\text{ideal streamtube velocity}} = \frac{U}{U_i}$$

or

$$U = C_t U_i = C_t \sqrt{2\Delta p/\rho_o} \tag{8-54}$$

From the equation of continuity (3-4), $Q = AV$, so again conventional practice is to define a "pipe" coefficient C_p as follows:

$$C_p = \frac{\text{average velocity}}{\text{streamtube velocity}} = \frac{V}{U} \tag{8-55}$$

If Eqs. (8-54) and (8-55) are substituted in Eq. (3-4),

$$Q = \int U\,dA = AV = AC_pC_t\sqrt{2\Delta p/\rho_o} \tag{8-56}$$

Numerical values of C_t depend upon the geometry of the tube and values of C_p on the velocity profile. If the value of C_t is not known, it is conventional practice to assume it to be unity. Values of C_p may be obtained from equations given in Sec. 8-5.

EXAMPLE 1 Incompressible Flow The Pitot tube shown in Fig. 8-14 is located in the center of a 6-in. schedule 40 galvanized iron pipe in which carbon tetrachloride flows at 30°C. The attached manometer containing mercury and carbon tetrachloride shows a deflection of 150 mm. Estimate the volumetric flow rate.

FIGURE 8-14 Notation for Example 1

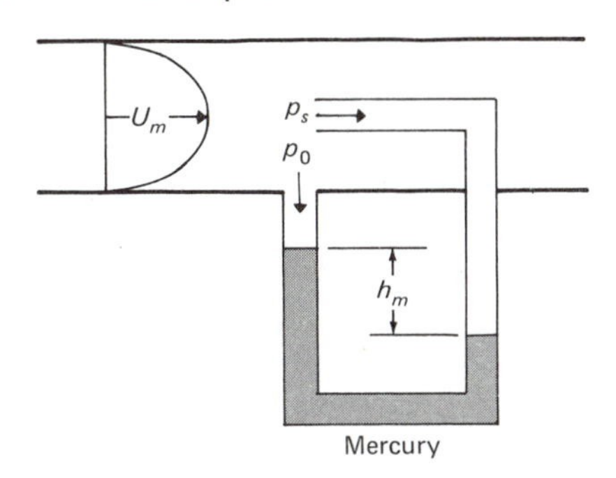

Step A *Data Reduction*

(1) *gravity*

Assume standard

(B-51) $g = 9.807 \text{ m/s}^2$

(2) *fluid properties*

Carbon Tetrachloride at 30°C

Table A-1 $\rho_o = 1\ 575 \text{ kg/m}^3$

Table A-4 $\mu = 8.418 \times 10^{-4} \text{ Pa} \cdot \text{s}$

(1-35) $\gamma_f = \rho_o g = 1\ 575 \times 9.807 = 15\ 446 \text{ N/m}^3$

Mercury at 30°C

Table A-1 $\mu_m = 13\ 520 \text{ kg/m}^3$

(1-35) $\gamma_m = \rho_m g = 13\ 520 \times 9.807 = 132\ 590 \text{ N/m}^3$

(3) *pipe properties*, 6-in. schedule 40 galvanized iron

Table A-11 $D = 154.1 \text{ mm},\ A = 18\ 650 \text{ mm}^2$

Table 8-2 $\epsilon = 500 \times 10^{-6} \times 0.304\ 8 = 152 \times 10^{-6} \text{ m}$

$$\frac{\epsilon}{D} = \frac{152 \times 10^{-6}}{0.154\ 1} = 989 \times 10^{-6}$$

Step B *Compute Centerline Velocity, U_m*

(2-8) $p_1 - p_2 = (\gamma_m - \gamma_f)h$

Fig. 8-14 $\Delta p = (\gamma_m - \gamma_f)h_m = (132\ 590 - 15\ 446)(0.150)$

$\Delta p = 17\ 572 \text{ N/m}^2$

(8-53) $U_i = \sqrt{2\Delta p/\rho_o} = \sqrt{2(17\ 572)/1\ 575}$

$U_i = 4.724 \text{ m/s}$

Since no tube coefficient was given, assume $C_t = 1$

(8-55) $U_m = C_t U_i = 1 \times 4.724 = 4.724 \text{ m/s}$

Step C *First Trial*

Assume $U_m = V$

(6-18) $$\mathbf{R} = \frac{DV\rho}{\mu} = \frac{0.154\ 1 \times 4.724 \times 1\ 575}{8.418 \times 10^{-4}}$$

$\mathbf{R} = 1\ 362\ 022 > 3\ 500\ \therefore$ flow is turbulent

(8-15) $$f_a = 0.0055\left\{1 + \left[20{,}000\left(\frac{\epsilon}{D}\right) + \frac{10^6}{\mathbf{R}}\right]^{1/3}\right\}$$

$$f_a = 0.0055\left\{1 + \left(20{,}000 \times 989 \times 10^{-6} + \frac{10^6}{1\ 362\ 022}\right)^{1/3}\right\}$$

$f_a = 0.0153\ 0$

(8-16) $$\frac{1}{\sqrt{f}} = -2 \log_{10}\left(\frac{\epsilon/D}{3.7} + \frac{2.51}{\mathbf{R}\sqrt{f_a}}\right)$$

$$\frac{1}{\sqrt{f}} = -2 \log_{10}\left(\frac{989 \times 10^{-6}}{3.7} + \frac{2.51}{1\ 362\ 022\sqrt{0.0153\ 0}}\right)$$

$$\sqrt{f} = 0.140\ 9, \qquad f = 0.0198\ 4$$

(8-23) $$C_p = \frac{V}{U_m} = \frac{1}{(1 + 1.326\sqrt{f}\,)}$$

$$C_p = \frac{1}{(1 + 1.326 \times 0.140\ 9)} = 0.842\ 6$$

(8-55) $$V = C_p U_m = 0.842\ 6 \times 4.724 = 3.980 \text{ m/s}$$

Step D *Second Trial*

Use first trial values

(6-18) $$\mathbf{R} = \frac{DV\rho}{\mu} = \frac{0.154\ 1 \times 3.980 \times 1\ 575}{8.418 \times 10^{-4}}$$

$\mathbf{R} = 1\ 147\ 512 > 3\ 500$ $\therefore$ flow is turbulent

Assume $f_a = 0.0198\ 4$ from first trial f

(8-16) $$\frac{1}{\sqrt{f}} = -2 \log_{10}\left(\frac{\epsilon/D}{3.7} + \frac{2.51}{\mathbf{R}\sqrt{f_a}}\right)$$

$$\frac{1}{\sqrt{f}} = -2 \log_{10}\left(\frac{989 \times 10^{-6}}{3.7} + \frac{2.51}{1147\ 512\sqrt{0.0198\ 4}}\right)$$

$$\sqrt{f} = 0.140\ 9, \qquad f = 0.0198\ 5$$

(8-23) $$C_p = \frac{1}{(1 + 1.326\sqrt{f}\,)}$$

$$= \frac{1}{(1 + 1.326 \times 0.01985)} = 0.842\ 6 \qquad \text{(checks)}$$

Step E *Compute Flow Rate*

(3-4) $$Q = AV = 0.018\ 65 \times 3.980 = 0.074\ 23 \text{ m}^3\text{/s}$$

Compressible Flow

For compressible flow, the ideal streamtube velocity U_{ic} is computed on the assumption that isentropic compression takes place between the moving fluid and the tube. Writing the general energy equation for a streamtube,

(4-62) $$q = W_{sf} + \frac{g}{g_c}(z_s - z_0) + \frac{U_s^2 - U_{ic}^2}{2g_c} + c_p(T_s - T_o)$$

Application of Eq. (4-62) to the streamtube of Fig. 8-13 indicates:

(a) *heat transfer* — isentropic compression $\therefore q = O$

(b) *steady flow work* — none in system $\therefore W_{sf} = 0$

(c) *potential energy change* — $z_s = z_o \therefore \dfrac{g}{g_c}(z_s - z_o) = 0$

(d) *kinetic energy change* — $U_s = 0 \therefore \dfrac{U_s^2 - U_{ic}^2}{2g_c} = \dfrac{U_{ic}^2}{2g_c}$

or

$$-\frac{U_{ic}^2}{2g_c} + c_p(T_s - T_o) = 0$$

and

$$U_{ic} = \sqrt{2g_c c_p(T_s - T_o)} = \sqrt{2g_c T_o c_p(T_s/T_o - 1)} \tag{8-57}$$

Substituting from Eq. (1-47),

$$\left(\frac{p_s}{p_o}\right)^{(k-1)/k} = \frac{T_s}{T_o} \tag{1-47}$$

from Eq. (1-44),

$$T_o = \frac{p_o}{g_c R \rho_o} \tag{1-44}$$

and from Eq. (4-56)

$$c_p = \frac{kR}{k-1} \tag{4-56}$$

in equation (8-57):

$$U_{ic} = \sqrt{2g_c\left(\frac{kR}{k-1}\right)\left(\frac{p_o}{g_c R \rho_o}\right)\left[\left(\frac{p_s}{p_o}\right)^{(k-1)/k} - 1\right]}$$

$$= \sqrt{\frac{2k}{(k-1)}\left(\frac{p_o}{\rho_o}\right)\left[\left(\frac{p_s}{p_o}\right)^{(k-1)/k} - 1\right]} \tag{8-58}$$

Let

$$Z = \frac{\text{ideal compressible velocity}}{\text{ideal incompressible velocity}} = \frac{U_{ic}}{U_i} \tag{8-59}$$

where Z is the compression factor.

Substituting from Eqs. (8-58) and (8-53) in Eq. (8-59),

$$Z = \frac{\sqrt{\dfrac{2k}{k-1}\left(\dfrac{p_o}{\rho_1}\right)\left[\left(\dfrac{p_s}{p_o}\right)^{(k-1)/k} - 1\right]}}{\sqrt{2(p_s - p_o)/\rho_0}}$$

$$Z = \sqrt{\left(\frac{k}{k-1}\right)\left[\frac{(p_s/p_o)^{(k-1)/k} - 1}{(p_s/p_o) - 1}\right]} \qquad \text{(8-60)}$$

The volume flow rate then becomes, from Eq. (8-56)

$$Q = AC_pC_tZ\sqrt{2\Delta p/\rho_o} \qquad \text{(8-61)}$$

EXAMPLE 2 Compressible Flow Carbon dioxide flows at 68°F and 20 psia at an average velocity of 500 ft/sec through an 8-in. schedule 40 wrought iron pipe. What pressure differential should be indicated by a pitot tube located on the pipe center line? Assume $C_t = 0.98$.

Step A *Data Reduction*

(1) *unit conversions*

(B-28) $T_o = 68 + 459.67 = 527.67°R$

(B-89) $p_o = 20 \times 144 = 2{,}880$ lbf/ft²

(2) *proportionality constant*

(1-11) $g_c = 32.17$ lbm-ft/lbf-sec²

(3) *fluid properties,* carbon dioxide at 68°F

Table A-7 $R = 35.11$ ft-lbf/lbm-°R, $k = 1.304$

(1-44) $\rho_o = \dfrac{p_o}{g_cRT_o} = \dfrac{2{,}880}{32.17 \times 35.11 \times 527.67}$

$\rho_o = 0.004832$ slugs/ft³

Table A-8 $\mu_o = 30.91 \times 10^{-8}$ lbf-sec/ft²

(4) *pipe properties,* 8-in. schedule 40 wrought iron

Table A-11 $D = 0.6651$ ft, $\dfrac{\epsilon}{D} = 225.5 \times 10^{-6}$

Step B *Compute Pipe Coefficient*

(6-18) $\mathrm{R} = \dfrac{VD\,\rho_o}{\mu_o} = \dfrac{500 \times 0.6651 \times 0.004832}{30.91 \times 10^{-8}}$

$\mathrm{R} = 5{,}198{,}582$

(8-15) $$f_a = 0.0055\left[1 + \left(20{,}000\,\frac{\epsilon}{D} + \frac{10^6}{\mathbf{R}}\right)^{1\ 3}\right]$$

$$f_a = 0.0055\left[1 + \left(20{,}000 \times 225.5 \times 10^{-6} + \frac{10^6}{5{,}198{,}582}\right)^{1/3}\right]$$

$$f_a = 0.01471$$

(8-16) $$\frac{1}{\sqrt{f}} = -2\log_{10}\left(\frac{\epsilon/D}{3.7} + \frac{2.51}{\mathbf{R}\sqrt{f_a}}\right)$$

$$\frac{1}{\sqrt{f}} = -2\log_{10}\left(\frac{225.5 \times 10^{-6}}{3.7} + \frac{2.51}{5{,}198{,}581\sqrt{0.01471}}\right)$$

$$\sqrt{f} = 0.1194, \qquad f = 0.01426$$

(8-23) $$C_p = \frac{V}{U_m} = \frac{1}{1 + 1.326\sqrt{f}} = \frac{1}{1 + 1.326 \times 0.1194}$$

$$C_p = 0.8633$$

Step C *First Trial*

Assume $Z = 1$

(8-61) $$\Delta p = \left(\frac{Q}{AC_pC_tZ}\right)^2 \frac{\rho_o}{2} = \left(\frac{V}{C_pC_tZ}\right)^2 \frac{\rho_o}{2}$$

$$\Delta p = \left(\frac{500}{0.8633 \times 0.98 \times 1}\right)^2 \frac{0.004832}{2} = 843.8 \text{ lbf/ft}^2$$

$$\Delta p = p_s - p_o = p_s - 2{,}880 = 843.8$$

$$p_s = 3{,}724 \text{ lbf/ft}^2$$

Step D *Second Trial*

Use first trial values

(8-60) $$Z = \sqrt{\left(\frac{k}{k-1}\right)\left(\frac{(p_s/p_o)^{(k-1)/k} - 1}{p_s/p_o - 1}\right)}$$

$$Z = \sqrt{\frac{1.304}{1.304 - 1}\left[\frac{(3{,}724/2{,}880)^{(1.304-1)/1.304} - 1}{3{,}724/2{,}880 - 1}\right]}$$

$$Z = 0.9507$$

(8-61) $$\Delta p = \left(\frac{V}{C_pC_tZ}\right)^2 \frac{\rho_o}{2} = \left(\frac{500}{0.8633 \times 0.98 \times 0.9507}\right)^2 \times \frac{0.004832}{2}$$

$$\Delta p = 933.6 \text{ lbf/ft}^2$$

$$\Delta p = p_s - p_o = p_s - 2{,}880 = 933.6$$

$$p_s = 3{,}814 \text{ lbf/ft}^2$$

Step E *Third Trial*

Use second trial values

(8-60)

$$Z = \sqrt{\left(\frac{k}{k-1}\right)\left[\frac{(p_s/p_o)^{(k-1)/k} - 1}{p_s/p_o - 1}\right]}$$

$$Z = \sqrt{\left(\frac{1.304}{1.304-1}\right)\left[\frac{(3{,}814/2{,}880)^{(1.304-1)/1.304} - 1}{3{,}814/2{,}880 - 1}\right]}$$

$$Z = 0.9461$$

(8-61)

$$\Delta p = \left(\frac{V}{C_p C_t Z}\right)^2 \frac{\rho_o}{2}$$

$$\Delta p = \left(\frac{500}{0.8633 \times 0.98 \times 0.9461}\right)^2 \frac{0.004832}{2}$$

$$\Delta p = 942.7 \text{ lbf/ft}^2$$

$$\Delta p = p_s - p_o = p_s - 2.880 = 942.7$$

$$p_s = 3{,}823 \text{ lbf/ft}^2$$

Step F *Further Trials*

No.	Z	Δp	p_s
4	0.9456	943.9	3,824
5	0.9456	(no change)	

$$\Delta p = \frac{943.9}{144} = 6.55 \text{ lbf/in}^2$$

PROBLEMS

1. Water at 20°C flows through a horizontal 12-in. schedule 40 pipe 300 m long at a velocity of 1.5 m/s. The inlet pressure is 700 kPa and the outlet 677 kPa. Compute (a) the resistance coefficient K and (b) the friction factor f.
2. Water at 86°F flows in a 12-in. internal diameter pipe 2000 ft long. The head "lost" due to friction in this pipe is 150 ft. Compute the wall shear stress.
3. Glycerine at 30°C flows at a rate of 1 m/s in a 3-in. diameter 250 psi cast iron horizontal pipe. What is the maximum length of pipe for this flow if the pressure differential may not exceed 240 kPa?
4. A liquid whose density is 1.86 slugs/ft³ flows through a ¼-in. diameter tube at a rate of 792 in³/hr. The measured pressure drop in a length of 20 ft is 48.70 lbf/ft². What is the dynamic viscosity of this liquid?
5. Water at 30°C flows at a rate of 1 200 m³/hr in a 300-mm internal diameter riveted steel pipe. Estimate the pressure loss per 100 m of pipe.

6. Air at 175 psia and 86°F flows in a horizontal 3-in. schedule 40 galvanized iron pipe. The "lost" head due to friction is 225 ft per 100 ft of straight pipe. Assume that the air is incompressible and estimate the mass flow of air in the pipe.
7. Water at 20°C flows in a ½-in. schedule 40 galvanized iron pipe. The pressure at a section of this pipe is 1 400 kPa. Three metres downstream from this section, the pressure is 1 000 kPa. Estimate the volumetric flow rate if the pipe is horizontal.
8. Two cubic feet per second of 68°F water are to flow through a 1,000 ft long straight horizontal 250 psi cast iron pipe. If the head "lost" to friction may not exceed 74.5 ft, what size pipe should be used?
9. Turpentine at 20°C flows at a rate of 3.90 m^3/min through an 18-in. standard size pipe. The measured velocity at the pipe center line is 0.51 m/s. Estimate (a) the friction factor and (b) the pipe surface roughness.
10. At a certain section in a 3-in. schedule 40 wrought iron pipe, carbon dioxide flows at 50 psia and 68°F. The center line velocity of the carbon dioxide is 2 ft/sec. Estimate the volume rate of flow at this section.
11. Air enters a 6-in. horizontal schedule 80 pipe at 550 kPa and 20°C with a velocity of 50 m/s. The pipe length is 215 m. At the pipe outlet the pressure is 412 kPa and the temperature is 20°C. Compute the friction factor.
12. Carbon monoxide flows in a horizontal 3-in. schedule 40 pipe at a constant temperature of 104°F. At a section of this pipe, the pressure is 50 psia. The pressure at a section 2,000 feet downstream is 39 psia. If the friction factor of the pipe is 0.015, estimate the mass flow rate of carbon monoxide.
13. A square steel duct 225 mm on a side is to be used to carry 20°C water. If the maximum head "lost" due to friction is 25 m per 300 m of straight duct, what is the maximum flow of water that can be transported?
14. An incompressible fluid flows in a 1 ft by 3 ft rectangular duct. It is planned to replace this duct with a circular duct made of the same amount of material. The pressure loss with the rectangular duct is 40 lbf/in^2. Estimate the pressure loss in the circular duct assuming that the ducts have identical friction factors.
15. A 3-in. schedule 40 horizontal steel-screwed piping system consists of a pump, a globe valve, a gate valve, four 90° regular elbows, and 150 m of straight pipe, and it discharges into the atmosphere. Estimate the pump discharge pressure required to maintain a flow of 0.379 m^3/min of water at 20°C in this system.
16. Water at 68°F flows through a schedule 40 steel piping system at the rate of 400 gal/min. The system consists of 110 ft of hori-

zontal straight 4-in. pipe, a 4-in. to 5-in. sudden enlargement, a 90° flanged regular 5-in. elbow, 75 ft of vertical straight 5-in. piping, a 90° flanged long radius 5-in. elbow and 150 ft of horizontal straight 5-in. piping. A pressure gage at system inlet in the 4-in. pipe indicates a pressure of 100 psig. Estimate the indication of a pressure gage located in the 5-in. pipe at the system outlet.

17. A 75 mm × 150 mm diameter Venturi tube located in a 6-in. schedule 40 diameter pipeline measures the flow of water at 20°C. The pressure differential is sensed by a U-tube manometer. The manometer contains water and mercury under water. It is observed that when the flow of water is 1.982 m^3/min, the unbalance of the manometer is 203.2 mm. What is the coefficient of discharge of the Venturi tube?
18. Water at 122°F flows in a 4-in. schedule 40 pipe through a 2.25 in. ASME orifice equipped with flange taps. The differential across the meter is equivalent to 125 in. of water at 68°F. Estimate the flow rate.
19. An ASME flow nozzle 25 mm in diameter measures the flow of water at 20°C in a 3-in. schedule 40 pipe. Estimate the flow rate when the differential across the nozzle tap is 9.377 kPa.
20. Carbon monoxide enters a Venturi Tube at 64.7 psia and 68°F. The inlet cone is 6 in., and the throat diameter is 3.6 in. The throat pressure is 19.4 psi lower than that at the inlet. Calibration of this meter showed that its coefficient of discharge varied with Reynolds number as follows:

$$C = 1 - 0.015\left(\frac{10^6}{\mathbf{R}_d}\right)^{1/2}$$

Determine the mass flow rate through this meter.

21. Air at 800 kPa and 40°C enters an ASME nozzle 25 mm in diameter installed in a 2-in. standard diameter steel pipe. The pressure differential across the nozzle taps is 138 kPa. Estimate the mass flow rate through the nozzle.
22. Air at 140 psia and 104°F flows in a 10-in. standard steel pipe through a 6.250-in. ASME orifice equipped with $1D$ and $\frac{1}{2}D$ taps. The differential head across the orifice is 30 in. of water at 68°F. Estimate the flow rate.
23. A pitot-static tube is located on the center line of a 10-in. 250 psi cast iron pipe in which glycerine flows at 20°C. The attached pressure differential sensing device indicates a difference in pressure equivalent to 38 mm of water at 20°C. Estimate the volumetric flow rate if the tube coefficient is 0.99.

24. Water at 68°F flows in a 10-in. standard pipe. A pitot-static tube traverse of this pipe yielded the following data:

Station	Location $-r/r_0$	Differential—in. of water at 68°F
1	0 (center line)	18.65
2	0.4472	15.44
3	0.6324	14.12
4	0.7746	11.94
5	0.8944	9.67
6	1.000 (wall)	0

Estimate (a) the flow rate, and (b) the friction factor of the pipe.

25. Air at 68°F and 95 psia flows in a duct. A pitot-static tube indicates a pressure difference of 20 lbf/in^2 at a certain location. Estimate the velocity of the air at the sensing point.

26. Air at 48°C and 55 kPa flows in a duct with a center line velocity of 250 m/s. A pitot-static tube is to be used to sense this velocity. Estimate the maximum range of sensing required in millimeters of water at 20°C.

REFERENCES

Moody, L. F., *Friction Factors for Pipe Flow.* Transactions of the American Society of Mechanical Engineers, New York, November 1974, pp. 671–684.

Murdock, J. W., "A Rational Equation for ASME Coefficients for Long-Radius Flow Nozzles Employing Pipe Wall Taps at 1*D* and ½*D*." American Society of Mechanical Engineers Paper No. 64-WA/FM-7.

Murdock, J. W., "Tables for the Interpolation and Extrapolation of ASME Coefficients for Square-Edged Concentric Orifices." American Society of Mechanical Engineers Paper No. 64-WA/FM-6.

Pigott, R. J. S., *Pressure Losses in Tubing, Pipe, and Fittings.* Transactions of the American Society of Mechanical Engineers, New York, July 1950, pp. 679–688.

Fluid Meters—Their Theory and Application, 6th ed. Edited by H. S. Bean. American Society of Mechanical Engineers, New York, 1971.

Pipe Friction Manual, 3d ed. Hydraulic Institute, Cleveland, Ohio, 1961.

9 Flow in Open Channels

9-1 INTRODUCTION

An open channel is a conduit in which a liquid flows with a free surface subjected to a constant pressure. Flow of water in natural streams, artificial canals, irrigation ditches, sewers, and flumes are examples where the water surface is subject to atmospheric pressure. The flow of any liquid in a pipe where there is a free liquid surface is an example of open channel flow where the liquid surface may be subjected to the pressure existing in the pipe.

The flow of fluids in open channels was first described in detail by Sextus Julius Frontinus (A.D. 40–103), who published a treatise on Roman methods of water distribution. Antoine Chezy (1718–1798) formulated similarity parameters for predicting flow characteristics of one channel from measurements carried out on another.

9-2 DEFINITIONS

Energy Relations for open channel flow may be determined by the application of Eq. (8-39), noting from Fig. 9-1 that no shaft work is performed in this open channel system. The modified equation becomes

$$(z_2 - z_1) + \frac{V_2^2 - V_1^2}{2g} + \frac{p_2 - p_1}{\rho g} + \overline{H}_f = 0 \qquad \textbf{(9-1)}$$

Figure 9-1 is a plot of Eq. (9-1), assuming that the channel is of uniform cross-section.

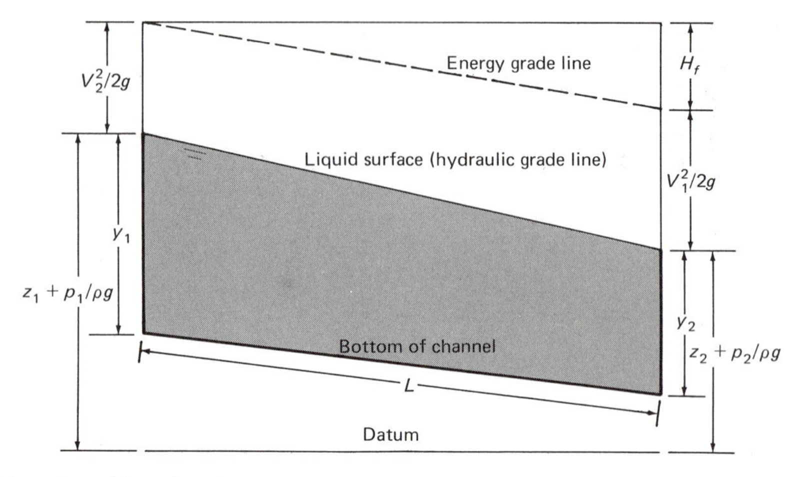

FIGURE 9-1 Notation for open-channel flow study

The *energy grade line* is the sum of all the available energy $(z + V^2/2g + p/\rho g)$.

The *hydraulic grade line* is the sum of the potential and pressure energies and is also the liquid surface $(z + p/\rho g)$.

The distance between the liquid surface and the bottom of the channel is sometimes called the *stage* and is denoted by the symbol y in Fig. 9-1.

When the stages between the sections are not uniform, that is $y_1 \neq y_2$ or the cross-section of the channel changes, or both, then the flow is said to be *varied.*

When a liquid flows in a channel of uniform cross-section and the slope of the surface is the same as the slope of the bottom of the channel $(y_1 = y = y_2)$, then the flow is said to be uniform.

The *slope* of a channel, S, is the change in elevation per unit of horizontal distance. For small slopes, this is equivalent to dividing the change in elevation by the distance, L, measured along the channel bottom between two sections. For steady, uniform flow, the velocity is uniform at all sections of the channel, so that the energy grade line has the same angle as the bottom of the channel, thus

$$S = \frac{z_1 - z_2}{L} \tag{9-2}$$

For uniform flow $V_1 = V_2$, and at any streamtube $p_1 = p_2$, so that Eq. (9-1) reduces to

$$(z_2 - z_2) + \overline{H}_f = 0 \tag{9-3}$$

Substituting in Eq. (9-1)

$$S = \frac{z_1 - z_2}{L} = \frac{\overline{H}_f}{L} \tag{9-4}$$

9-3 PARAMETERS FOR OPEN CHANNEL FLOW

Dimensional Analysis

Consider the steady uniform flow of a liquid in an open channel produced by a gravity of g. The liquid flows with a velocity of V and has a density of ρ, viscosity of μ, and a surface tension of σ. The channel has an absolute surface roughness of ϵ. The flow has a hydraulic radius of R_h. Derive the dimensionless parameters involved in this phenomenon. The results of the application of the Buckingham Π theorem are shown in Table 9-1.

TABLE 9-1 *Dimensional Analysis of Open Channel Flow*

Item		Variables considered						
		HYDRAULIC RADIUS	FLUID					
			Velocity	*Density*	*Viscosity*	*Surface tension*	*Rough-ness*	*Gravity*
Symbol		R_h	V	ρ	μ	σ	ϵ	g
Selection		B_G^x	B_K^y	B_D^z	A_1	A_2	A_3	A_4
π Ratios		Basic group			π_1	π_2	π_3	π_4
Dimensions		L	LT^{-1}	$FL^{-4}T^2$	$FL^{-2}T$	FL^{-1}	L	LT^{-2}
Exponents	F	0	0	z	1	1	0	0
for	L	x	y	$-4z$	-2	-1	1	1
equations	T	0	$-y$	$2z$	1	0	0	-2

TABLE 9-1 *Solution*

π Ratio	Solved exponents R_h^x	V^y	ρ^z	Derived ratio	Conventional practice FORM	SYMBOL	NAME
π_1	-1	-1	-1	$\mu/R_hV\rho$	$\frac{4R_hV\rho}{\mu}$	R	Reynolds number
π_2	-1	-2	-1	$\sigma/R_hV^2\rho$	$\frac{4R_hV^2\rho}{\sigma}$	W	Weber number
π_3	-1	0	0	ϵ/R_h	$\epsilon/4R_h$	—	Relative roughness
π_4	1	-2	0	R_hg/V^2	$\frac{V}{\sqrt{R_hg}}$	F	Froude number

Physical Analysis

From dimensional analysis:

$$\pi_4 = f(\pi_1, \pi_2, \pi_3)$$

or

$$V = \mathbf{F}\sqrt{R_h g} \tag{9-5}$$

where

$$\mathbf{F} = f\left(\mathbf{R}, \mathbf{W}, \frac{\epsilon}{4R_h}\right)$$

It is conventional practice to write Eq. (9-5) in the following form:

$$V = C\sqrt{R_h S} \tag{9-6}$$

where

$$C = \text{the Chezy coefficient}$$

and

$$C = f\left(\mathbf{R}, \mathbf{W}, \frac{\epsilon}{4R_h} \text{ and } g\right)$$

The relation of the Chezy coefficient C to the friction factor f may be derived by writing the friction factor equation (8-4) and substituting for D the hydraulic radius relation of Eq. (4-10) $D = 4R_h$ and the slope from Eq. (9-4) $S = \overline{H}_f/L$:

$$\frac{\overline{H}_f}{L} = S = \frac{(f/D)V^2}{2g} = \frac{(f/4R_h)V^2}{2g} = \left(\frac{f}{8R_h g}\right)V^2$$

or

$$V = \left(\frac{8g}{f}\right)^{1/2}\sqrt{R_h S} \tag{9-7}$$

Comparing Eq. (9-6) with Eq. (9-7),

$$V = C\sqrt{R_h S} = \left(\frac{8g}{f}\right)^{1/2}\sqrt{R_h S}$$

or

$$C = \left(\frac{8g}{f}\right)^{1/2} \tag{9-8}$$

For open channel flow, the Chezy coefficient is calculated by the Manning equation, which was developed from examination of experimental results of water tests. The Manning relation is stated as

$$C = \frac{0.2620 \sqrt{g} R_h^{1/6}}{n} \tag{9-9}$$

where n is a roughness factor.

From the dimensional analysis, n should be a function of Reynolds number, Weber number, and relative roughness. Since only water test data obtained at ordinary temperatures support this value, it must be assumed that n is the value for turbulent flow only. Since surface tension is a weak property, the effects of Weber number variation are negligible, leaving n to be some function of surface roughness. Design values of n are given in Table 9-2.

EXAMPLE Water flows in a rubble-lined trapezoidal channel at a depth of 4 m. The sides slope at 45°, and the bottom width is 20 m. If the channel drops 9 m per 10 km, estimate the flow rate.

Step A *Data Reduction*

(1) *gravity*

assume standard

(B-51) $g = 9.807$ m/s²

TABLE 9-2 *Values of Roughness Factor* n *for Use in Manning Equation* (9-9)

Type of Surface	Value of n	
	ft$^{1/6}$	m$^{1/6}$
Brick	0.016	0.013
Cast iron	0.015	0.012
Concrete, finished	0.012	0.010
Concrete, unfinished	0.014	0.012
Corrugated metal	0.022	0.018
Earth, good condition	0.025	0.021
Earth, with stones and weeds	0.035	0.029
Gravel	0.029	0.024
Riveted steel	0.018	0.015
Rubble	0.025	0.021
Wood, planed	0.012	0.010
Wood, unplaned	0.013	0.011

(2) *channel properties,* trapezoid $\alpha = \beta = 45°$

Table A-14 $A = (b + h)h = (20 + 4)4 = 96\ \text{m}^2$

Table A-14 $R_h = \dfrac{(b + h)h}{b + 2.828\ h}$

$$R_h = \frac{(20 + 4) \times 4}{20 + 2.828 \times 4} = 3.066\ \text{m}$$

(3) *Manning coefficient,* rubble

Table 9-2 $n = 0.021\ \text{m}^{1/6}$

(4) *Chezy coefficient*

(9-9) $C = \dfrac{0.2620\sqrt{g}\,R_h^{1/6}}{n} = \dfrac{0.2620\sqrt{9.807}\,(3.066)^{1/6}}{0.021}$

$C = 47.09$

(5) *geometric*

(9-4) $S = \dfrac{z_2 - z_1}{L} = \dfrac{9}{10\ 000} = 0.000\ 9$

Step B *Compute Flow Rate*

(9-6) $V = C\sqrt{R_h S} = 47.09\sqrt{3.066 \times 0.000\ 9}$

$= 2.474\ \text{m/s}$

(3-3) $Q = AV = 96 \times 2.474 = 237.5\ \text{m}^3/\text{sec}$

9-4 MAXIMUM HYDRAULIC RADIUS

If Eq. (9-6) is substituted for V in Eq. (3-4) and in Eq. (9-9) for C, then

$$Q = AV = AC\sqrt{R_h S} = A\left(\frac{0.2620}{n}\right)\sqrt{g}\,R_h^{1/6}\sqrt{R_h S}$$

or

$$Q = \frac{0.2620}{nA}\,R_h^{2/3}\sqrt{gS} \qquad \text{(9-10)}$$

Examination of Eq. (9-10) indicates that, for a given roughness, slope, and area, the volume rate of flow will be a maximum when the hydraulic radius is a maximum.

The maximum hydraulic radius may be determined as follows: From Table A-14 the hydraulic radius of a trapezoid is given by

$$R_h = \frac{A}{b + h(\text{cosec}\ \alpha + \text{cosec}\ \beta)} \qquad \text{(9-11)}$$

and the area by

$$A = [b + \tfrac{1}{2}\, h(\cot \alpha + \cot \beta)]h \tag{9-12}$$

Solving Eq. (9-12) for b and substituting in Eq. (9-11),

$$R_h = \frac{A}{A/h - \tfrac{1}{2}\, h(\cot \alpha + \cot \beta) + h(\operatorname{cosec} \alpha + \operatorname{cosec} \beta)}$$

$$R_h = \frac{Ah}{A + h^2\left(\operatorname{cosec} \alpha + \operatorname{cosec} \beta - \dfrac{\cot \alpha + \cot \beta}{2}\right)} \tag{9-13}$$

Let

$$\operatorname{cosec} \alpha + \operatorname{cosec} \beta - \frac{\cot \alpha + \cot \beta}{2} = \overline{C} \text{ (a constant)} \tag{9-14}$$

Substituting Eq. (9-14) in Eq. (9-13),

$$R_h = \frac{Ah}{A + \overline{C}h^2} \tag{9-15}$$

Differentiating Eq. (9-15) with respect to h,

$$dR_h = \frac{(A + \overline{C}h^2)A\, dh - Ah(2\overline{C}h\, dh)}{(A + \overline{C}h^2)^2} \tag{9-16}$$

Setting Eq. (9-16) equal to zero for maximum,

$$\frac{dR_h}{dh} = 0 = \frac{(A + \overline{C}h^2)A - 2A\overline{C}h^2}{(A + \overline{C}h^2)^2}$$

or

$$A = \overline{C}h^2 \tag{9-17}$$

Substituting Eq. (9-17) in Eq. (9-15),

$$R_{h_{\max}} = \frac{\overline{C}h^2 h}{\overline{C}h^2 + \overline{C}h^2} = \frac{h}{2} \tag{9-18}$$

Since rectangular, square, and triangular channels are a special case of the trapezoidal channel, $R_{h_{\max}} = h/2$ must also apply. Table A-14 gives values of $R_{h_{\max}}$ for various channel shapes.

EXAMPLE A brick-lined rectangular channel is to carry 200 ft³/sec of water. If the channel slope is 1 in 10,000, what should be the size of the channel?

Step A *Data Reduction*

(1) *gravity*

assume standard

(B-50) $g = 32.17\ \text{ft/sec}^2$

(2) *roughness factor,* brick

Table 9-2 $n = 0.016\ \text{ft}^{1/6}$

Step B *Compute Proportions*

(9-10) $$Q = \frac{0.2620}{n} A R_h^{2/3} \sqrt{gS}$$

$$200 = \frac{0.2620}{0.016} A R_h^{2/3} \sqrt{32.17 \times 1/10{,}000}$$

$$A R_h^{2/3} = 215.34 \qquad \text{(a)}$$

(9-18) $$R_{h_{\max}} = \frac{h}{2}$$

Table A-14 $$R_{h_{\max}} = \frac{bh}{2h + b} = \frac{h}{2} \qquad \text{(b)}$$

or

$$b = 2h \qquad \text{(c)}$$

Table A-14 $$A = bh = 2h^2$$

Substituting Eq. (9-18) and (c) in (a),

$$A R_h^{2/3} = 215.34 = 2h^2 \left(\frac{h}{2}\right)^{2/3} \qquad \text{(d)}$$

$$h = 6.875\ \text{ft}$$

$$b = 2h = 2 \times 6.875 = 13.750\ \text{ft} \qquad \text{(c)}$$

9-5 SPECIFIC ENERGY

Specific energy is defined as the energy of the fluid referred to the bottom of the channel as the datum, as shown in Fig. 9-2. Thus, the specific energy E at any section is given by

$$E = y + \frac{V^2}{2g} \qquad \text{(9-19)}$$

Equation (9-19) may be written in terms of volumetric flow rate by substituting from the continuity equation (3-3) $V = Q/A$ or

$$E = y + \frac{(Q/A)^2}{2g} \qquad \text{(9-20)}$$

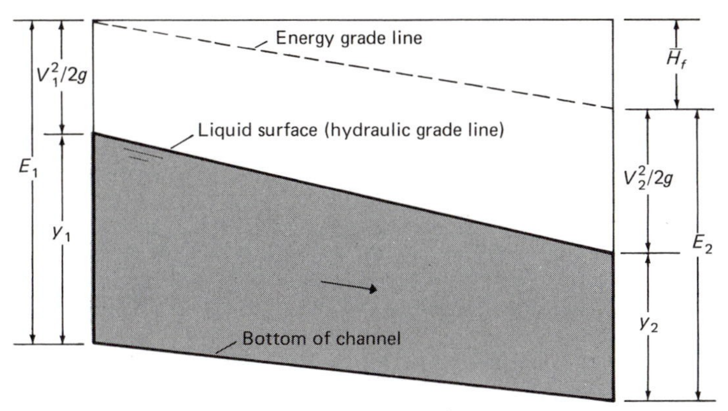

FIGURE 9-2 Notation for specific energy study

For a rectangular channel whose width is b, $A = by$, and if q is defined as the flow rate per unit width, $q = Q/b$, then Eq. (9-10) may be written as

$$E = y + \frac{(Q/by)^2}{2g} = y + \frac{(qb/by)^2}{2g}$$

$$E = y + \frac{(q/y)^2}{2g} \tag{9-21}$$

Critical Values for Rectangular Channels

Critical values of specific energy E_c, depth y_c, and unit flow rate q_c may be derived by differentiating Eq. (9-21) with respect to y and setting the first derivative equal to zero:

$$\frac{dE}{dy} = \frac{d}{dy}\left[\frac{y + (q/y)^2}{2g}\right] = 0 = 1 + \left(\frac{q^2}{2g}\right)(-2)y^{-3}$$

$$= 1 - \frac{q^2}{y^3 g}$$

or

$$q_c^2 = y_c^3 g \tag{9-22}$$

Substituting Eq. (9-22) in Eq. (9-21),

$$E_c = y_c + \frac{(q_c/y_c)^2}{2g} = y_c + \frac{(y_c^3 g)}{2gy_c^2}$$

$$E_c = y_c + \frac{y_c}{2} = \frac{3y_c}{2} \tag{9-23}$$

Critical velocity V_c may be obtained by substituting in Eq. (9-19) for V_c and from Eq. (9-23)

$$E_c = \frac{3y_c}{2}$$

$$E_c = y_c + \frac{V_c^2}{2g} = \frac{3y_c}{2},$$

or

$$\frac{V_c^2}{2g} = \frac{y_c}{2}$$

and

$$V_c = \sqrt{gy_c} \tag{9-24}$$

Figure 9-3 shows the relation of depth to specific energy for a constant flow rate, and Fig. 9-4 shows the relation between depth and flow rate for a constant specific energy. If the depth is greater than critical then the flow is *subcritical,* at critical depth it is *critical,* and at depths below critical the flow is *supercritical.* For a given specific energy there is a maximum unit flow rate that can exist. It should be noted from Eq. (9-24) that

$$\frac{V_c}{\sqrt{gy_c}} = 1 = \mathbf{F}_c \qquad \text{(critical Froude number)} \tag{6-27}$$

Substituting the general relation for Froude number from Eq. (6-27) in Eq. (9-19),

$$E = y + \frac{V^2}{2g} = y + \frac{(gy\mathbf{F}^2)}{2g} = y\left(1 + \frac{\mathbf{F}^2}{2}\right)$$

FIGURE 9-3 Specific energy diagram; q = constant

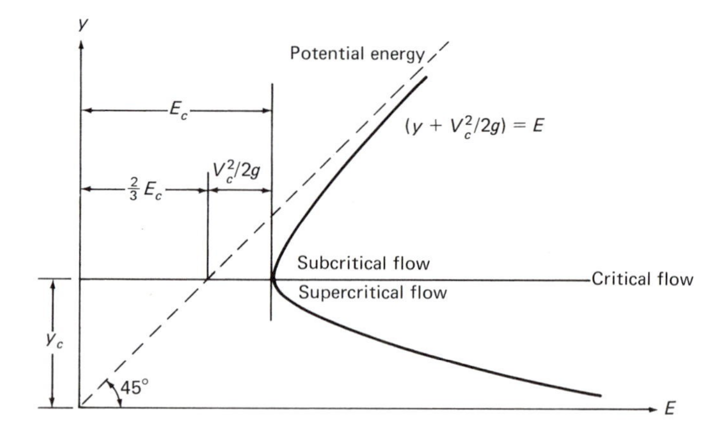

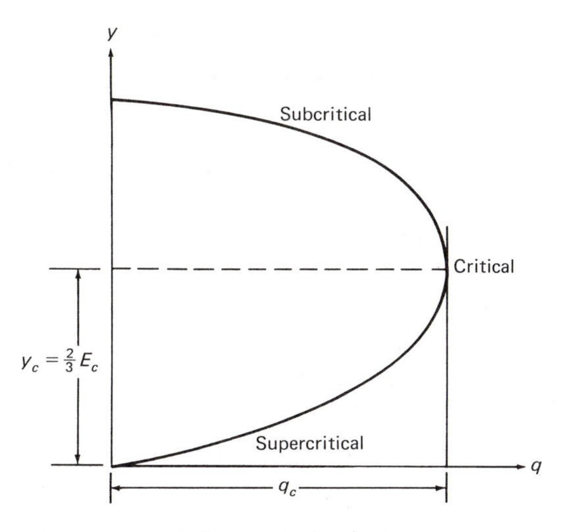

FIGURE 9-4 Depth versus unit flow rate for constant specific energy

and

$$\frac{y}{E} = \frac{1}{(1 + \mathbf{F}^2/2)} \tag{9-25}$$

Examination of Eq. (9-25) indicates the following:

$\mathbf{F} < 1$	$\frac{y}{E} < \frac{2}{3}$	flow is subcritical
$\mathbf{F} = 1$	$\frac{y}{E} = \frac{2}{3}$	flow is critical
$\mathbf{F} > 1$	$\frac{y}{E} > \frac{2}{3}$	flow is supercritical

It is seen that for open channel flow the Froude number determines the type of flow in the same manner as Mach number for compressible flow.

EXAMPLE 1 Water flows at a rate of 300 m^3/s in a channel 10 m wide at a depth of 4 m. Determine (a) specific energy and (b) type of flow.

Step A *Data Reduction*

(1) *gravity*

assume standard

(B-51) $g = 9.807\ m/s^2$

(2) *geometric*-rectangle

Table A-10 $A = by = 10 \times 4 = 40\ \text{m}^2$

(3) *kinematic*

(3-3) $V = Q/A = 300/40 = 7.5\ \text{m/s}$

Step B *Determine Specific Energy*

(9-19) $$E = y + \frac{V^2}{2g}$$

$$E = 4 + \frac{(7.5)^2}{2 \times 9.807} = 6.868\ \text{m}$$

Step C *Determine Type of Flow*

(6-27) $$\mathbf{F} = \frac{V}{\sqrt{gy}} = \frac{7.5}{\sqrt{9.807 \times 4}}$$

$$\mathbf{F} = 1.197 > 1 \therefore \text{flow is supercritical}$$

EXAMPLE 2 Compute the maximum flow that can exist in a rectangular channel 32 ft wide for a specific energy of 7 ft.

Step A *Data Reduction*

Gravity

assume standard

(B-50) $g = 32.17\ \text{ft/sec}^2$

Step B *Compute Maximum Flow Rate*

(9-23) $$y_c = \frac{2E_c}{3} = 2 \times 7/3 = 4.667\ \text{ft}$$

(9-24) $$V = \sqrt{gy} = \sqrt{32.17 \times 4.667} = 12.25\ \text{ft/sec}$$

(3-3) $$Q = A_c V_c = b y_c V_c$$

$$Q = 32 \times 4.667 \times 12.25 = 1{,}830\ \text{ft}^3/\text{sec}$$

9-6 HYDRAULIC JUMP

Under certain conditions in open channel flow a stream of water flowing at supercritical velocity may change abruptly to subcritical flow. This phenomenon is accompanied by a change in surface elevation and is known as a hydraulic jump.

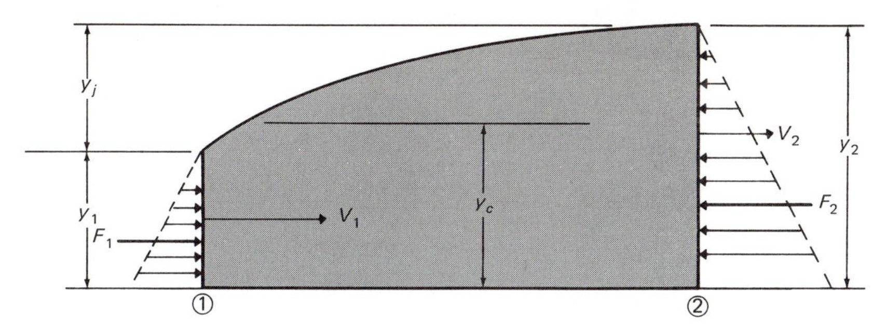

FIGURE 9-5 Notation for hydraulic jump study

Consider the hydraulic jump shown in Fig. 9-5 as a free body diagram. Application of the impulse-momentum equation (1-34) to Fig. 9-5 results in

$$\Sigma F = F_1 - F_2 = \dot{m}(V_2 - V_1) \quad \textbf{(9-26)}$$

From Eq. (2-21) $F = \gamma h_c A$, noting from Eq. (1-35) $\gamma = \rho g$, and by definition $h_c = y/2$, and for a channel of b width, $A = by$, Eq. (2-21) may be written as

$$F = \gamma h_c A = \rho g\left(\frac{y}{2}\right)by = \frac{\rho g b y^2}{2} \quad \textbf{(9-27)}$$

From Eq. (3-8) $\dot{m} = \rho AV$, substituting $A = by$, noting from Eq. (3-3) $Q = AV$ and from Sec. 9-5 $Q = bq$. With these considerations Eq. (3-8) may be written as

$$\dot{m} = \rho AV = \rho Q = \rho bq \quad \textbf{(9-28)}$$

Also from Eq. (3-3) $V = Q/\rho$ and from Sec. 9-5, $Q = bq$, noting that $A = by$, we can write Eq. (3-3) as

$$V = \frac{Q}{A} = \frac{bq}{by} = \frac{q}{y} \quad \textbf{(9-29)}$$

Substituting Eqs. (9-27), (9-28), and (9-29) in Eq. (9-26),

$$F_1 - F_2 = \dot{m}(V_2 - V_1) = \frac{\rho g b y_1^2}{2} - \frac{\rho g b y_2^2}{2}$$

$$= (\rho bq)\left(\frac{q}{y_2} - \frac{q}{y_1}\right)$$

which reduces to

$$\frac{q_1^2}{gy_1} + \frac{y_1^2}{2} = \frac{q_2^2}{gy_1} + \frac{y_2^2}{2} \tag{9-30}$$

Equation (9-30) may be solved for either y_1 or y_2

$$y_1 = \frac{y_2}{2}\,[-1 + \sqrt{1 + (8q^2/gy_2^3)}] \tag{9-31}$$

or

$$y_2 = \frac{y_1}{2}\,[-1 + \sqrt{1 + (8q^2/gy_1^3)}] \tag{9-32}$$

The lost energy due to a hydraulic jump may be computed from Eq. (9-21)

$$\Delta E - E_1 - E_2 = [y_1 + (q/y_1)^2/2g] - [y_2 + (q/y_2)^2/2g] \tag{9-33}$$

EXAMPLE Water flows at a rate of 10 m/s in a rectangular channel 5.5 m wide at a depth of 300 mm. What is the energy loss in a hydraulic jump that has occurred from this flow?

Step A *Data Reduction*

(1) *gravity*

Assume standard

(B-51) $g = 9.807\ \text{m/s}^2$

(2) *geometric,* rectangle

Table A-10 $A_1 = by_1 = 5.5 \times 0.300 = 1.65\ \text{m}^2$

(3) *kinematic*

Sec. 9-5 $q = \dfrac{Q}{b} = \dfrac{10}{5.5} = 1.818\ \text{m}^2/\text{s}$

Step B *Determine If Jump Can Take Place*

(9-22) $y_c = \left(\dfrac{q_c^2}{g}\right)^{1/3} = \left(\dfrac{1.818^2}{9.807}\right)^{1/3}$

$y_c = 0.696\,0\ \text{m} > y_1\ \therefore$ jump can take place

Step C *Compute Conditions at Section 2*

(9-32) $y_2 = \dfrac{y_1}{2}\,[-1 + \sqrt{1 + (8q^2/gy_1^3)}]$

$$y_2 = \left(\frac{0.300}{2}\right)[-1 + \sqrt{1 + 8(1.818)^2/9.807 \times (0.300)^3}]$$

$$y_2 = 1.356 \text{ m} > y_c \quad \text{(subcritical)}$$

Step D *Compute Energy Lost*

(9-33)
$$\Delta E = [y_1 + (q/y_1)^2/2g] - [y_2 + (q/y_2)^2/2g]$$

$$\Delta E = \left[0.300 + \frac{(1.818/0.300)^2}{2 \times 9.807}\right] - \left[1.356 + \frac{(1.818/1.356)^2}{2 \times 9.807}\right]$$

$$= 2.172 - 1.448 = 0.724 \text{ m}$$

9-7 WEIRS

Weirs are used to measure the flow of liquids in open channels or in conduits which do not flow full, i.e., in which there is a free liquid surface. A weir is in effect a dam over which the liquid is forced to flow. Weirs are almost exclusively used for measuring water flow, although small ones have been used for metering other liquids.

Weirs are classified according to the form of their notch or opening as follows: rectangular notch, the original form; the V or triangular notch; and special notches such as the trapezoidal, hyperbolic, and parabolic notches, which are designed to have a constant discharge coefficient or to have the head vary directly with the flow. This section will be limited to the rectangular and triangular weirs.

Velocity-Height Relations

Conventional practice is to base weir computations on the ideal flow of an incompressible fluid and to correct to actual flow conditions by use of a coefficient of discharge and adjustment lengths for the rectangular weir and head corrections for both types.

Consider the flow of a liquid over a weir as shown in Fig. 9-6. A is a fluid particle located on the surface of the liquid a distance upstream from the weir. B is a fluid particle in the free jet issuing from the weir and is a distance y below the surface point A. Both A and B are at atmospheric pressure. Writing the equation of motion (4-8) for constant pressure ($dp = 0$) and for frictionless flow ($\tau = 0$):

(4-8)
$$\frac{g}{g_c}\,dz + \frac{U\,dU}{g_c} + v\,dp + v\tau\,dL\left(\frac{dP}{dA}\right) = 0 = \frac{g}{g_c}\,dz + \frac{U\,dU}{g_c} + 0 + 0$$

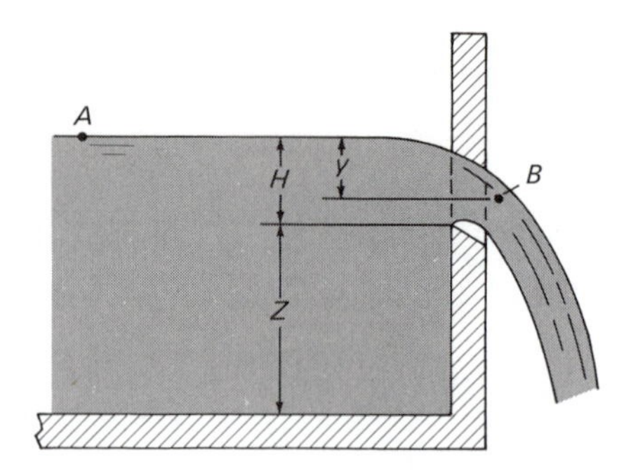

FIGURE 9-6 Flow over a weir

or

$$g\,dz + U\,dU = 0 \tag{9-34}$$

Integrating Eq. (9-34) between the limits of A and B:

$$g\int_A^B dz + \int_A^B U\,dU = 0 = g(z_B - z_A) + \tfrac{1}{2}(U_B^2 - U_A^2)$$

or

$$U_B^2 = 2g(z_A - z_B) + U_A^2 \tag{9-35}$$

Conventional practice is to assume that point A is located in a channel of infinite length, so that $U_A \to 0$, and to correct for the upstream kinetic energy in the coefficient of discharge. From Fig. 9-6, $z_A - z_B = y$ and let $U_B = U_i$ (ideal jet velocity); then Eq. (9-35) may be written as

$$U_B^2 = 2g(z_A - z_B) + U_A^2 = U_i^2 = 2gy + 0$$

or

$$U_i = \sqrt{2gy} \tag{9-36}$$

Rectangular Weirs

Consider the jet issuing from the rectangular weir shown in Fig. 9-7. The flow area of the fluid element shown at a distance y below the surface is $dA = L_w\,dy$. From Eq. (9-36) the ideal jet velocity is given by

(9-36) $$U_i = \sqrt{2gy}$$

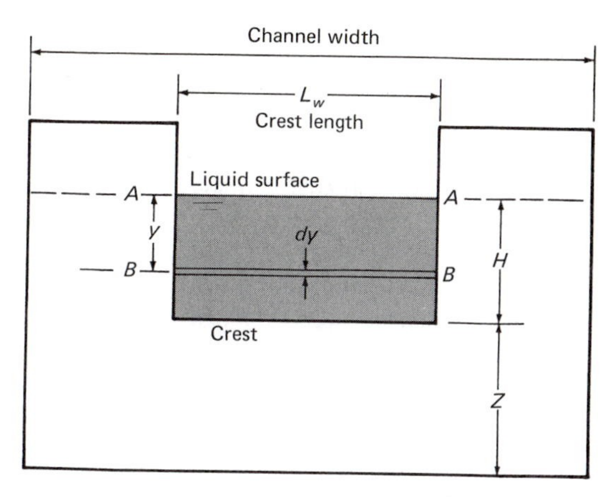

FIGURE 9-7 Notation for rectangular weir study

Substituting these values in the continuity equation (3-3) and integrating between the liquid surface (O) and the weir crest (H),

$$Q_i = \int_O^A U\,dA = \int_O^H (\sqrt{2gy})\,(L_w\,dy) = L_w\sqrt{2g}\int_O^H y^{1/2}\,dy$$

$$= L_w\sqrt{2g}\left[\frac{2y^{3/2}}{3}\right]_O^H$$

$$Q_i = (2/3)L_w\sqrt{2g}\,H^{3/2} \quad \text{(ideal volume flow rate)} \tag{9-37}$$

For water measurement, the ASME Research Committee on Fluid Meters recommends the following:

$$Q = (2/3)CL_a\sqrt{2g}\,H_a^{3/2} \tag{9-38}$$

where

Q = The actual discharge from the weir

Table 9-3 — C = Coefficient of discharge $= f\left(\dfrac{L_w}{L_c}, \dfrac{H}{Z}\right)$

Table 9-3 — L_a = Adjusted crest length $= L_w + \Delta L$

H_a = Adjusted weir head $= H + 0.003$ ft or $H + 0.9$ mm

EXAMPLE 1 Rectangular Weir Flow Measurement Water flows in a channel whose width is 10 ft. At the end of this channel is a weir whose crest height is 4 ft and whose width is the same as that of the channel. The water flows at a height of 3 ft over the crest of the weir. Estimate the flow in cubic feet per second.

TABLE 9-3 *Values of C and ΔL for Use in Equation (9-38)*

H/Z	**Crest Length/Channel Width = L_w/L_c**							
	0	0.2	0.4	0.6	0.7	0.8	0.9	1.0
	Coefficient of Discharge — C							
0	0.587	0.589	0.591	0.593	0.595	0.597	0.599	0.603
0.5	0.586	0.588	0.594	0.602	0.610	0.620	0.631	0.640
1.0	0.586	0.587	0.597	0.611	0.625	0.642	0.663	0.676
1.5	0.584	0.586	0.600	0.620	0.640	0.664	0.695	0.715
2.0	0.583	0.586	0.603	0.629	0.655	0.687	0.726	0.753
2.5	0.582	0.585	0.608	0.637	0.671	0.710	0.760	0.790
3.0	0.580	0.584	0.610	0.647	0.687	0.733	0.793	0.827
	Adjustment for Crest Length — ΔL							
ft	0.007	0.008	0.009	0.012	0.013	0.014	0.013	−0.005
mm	2.1	2.4	2.7	3.7	4.0	4.3	4.0	−1.5

Step A *Data Reduction*

(1) *gravity*

Assume standard

(B-50) $g = 32.17$ ft/sec²

(2) *weir data*

Fig. 9-7 $\frac{L_w}{L_c} = 10/10 = 1, \quad \frac{H}{Z} = 3/4 = 0.75$

Table 9-3 $C = 0.658$ (by linear interpolation)

Table 9-3 $\Delta L = -0.005$ ft

Step B *Compute Flow Rate*

(9-38) $L_a = L_w + \Delta L = 10 - 0.005 = 9.995$ ft

(9-38) $H_a = H + 0.003 = 3 + 0.003 = 3.003$ ft

(9-38) $Q = (2/3) C L_a \sqrt{2g}\, H_a^{3/2}$

$Q = (2/3) 0.658 \times 9.995 \sqrt{2 \times 32.17}\, (3.003)^{3/2}$

$Q = 183.0$ ft³/sec

Triangular Weirs

Consider the jet issuing from the triangular weir shown in Fig. 9-8. The flow area of the fluid element at a distance below the surface of y is $dA = L_x\, dy$. From geometry,

$$(L_x/L_H) = (H - y)/H \qquad \text{or} \qquad L_x = (L_H/H)(H - y),$$

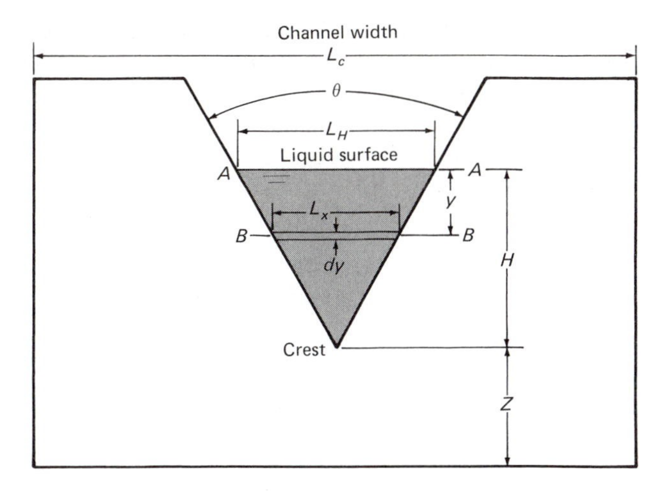

FIGURE 9-8 Notation for triangular weir study

and from trigonometry,

$$L_H = 2H \tan(\theta/2),$$

so that

$$L_x = (2H \tan(\theta/2)/H)(H - y)$$

or

$$L_x = 2 \tan(\theta/2)(H - y) \qquad \text{and} \qquad dA = 2 \tan(\theta/2)(H - y)dy.$$

From Eq. (9-36) the ideal jet velocity is $U_i = \sqrt{2gy}$. Substituting these values for dA and U_i in the equation of continuity (3-3) and integrating between the liquid surface (O) and the crest (H),

$$Q_i = \int_O^A U\,dA = \int_O^H (\sqrt{2gy})\,[2 \tan(\theta/2)]\,(H - y)\,dy$$

$$= 2\sqrt{2g} \tan\left(\frac{\theta}{2}\right) \int_O^H (H - y)y^{1/2}\,dy$$

$$Q_i = 2\sqrt{2g} \tan\left(\frac{\theta}{2}\right)\Big[(2/3)Hy^{3/2} - (2/5)y^{5/2}\Big]_O^H$$

$$Q_i = (8/15) \tan\left(\frac{\theta}{2}\right)\sqrt{2g}H^{5/2} \quad \text{(ideal volume flow rate)} \tag{9-39}$$

The value of the discharge coefficient for triangular weirs is dependent primarily on the notch angle θ, and only slightly on the head-crest ratio H/Z. The fluid viscosity and surface tension may affect the value of the discharge coefficient slightly, but experimental data are inadequate to define such effects. The ASME Research Committee on Fluid Meters recommends the following for the measurement of liquid flow through triangular weirs:

$$Q = (8/15)C \tan\left(\frac{\theta}{2}\right)\sqrt{2g}\,(H + \Delta H)^{5/2} \tag{9-40}$$

where

Q = The actual discharge from the weir

Table 9-4 C = Coefficient of discharge = $f(\theta)$

Table 9-4 ΔH = Correction for head/crest ratio = $f(\theta)$

EXAMPLE 2 Flow Over a Triangular Weir A city engineer wishes to maintain a flow of 5 m^3/s of water in an open channel whose width is 3 m at a height of 3 m by locating a triangular weir at the end of the channel. The weir has a crest height of 1.2 m. What angle of notch is required to maintain these conditions?

Step A *Data Reduction*

(1) *gravity*

Assume standard

(B-51) $g = 9.807$ m/s^2

(2) *weir data*

Fig. 9-8 If the total depth of channel $H + Z = 3$ m and the crest height $Z = 1.2$ m, then $H = 3 - Z = 3 - 1.2 = 1.8$ m

Step B *First Trial*

Assume $\theta = 60°$ (average angle, Table 9-4)

Table 9-4 $C = 0.576$, $\Delta H = 1.2$ mm

TABLE 9-4 *Values of C and ΔH for Use in Equation (9-39)*

Item		Weir Notch Angle θ (Degrees)						
		20	30	45	60	75	90	100
C		0.592	0.586	0.580	0.576	0.576	0.579	0.581
ΔH	ft	0.010	0.007	0.005	0.004	0.003	0.003	0.003
	mm	3.0	2.1	1.5	1.2	0.9	0.9	0.9

(9-40) $$Q = (8/15)C \tan\left(\frac{\theta}{2}\right)\sqrt{2g}\,(H + \Delta H)^{5/2}$$

$$5 = 8/15 \times 0.576 \tan\left(\frac{\theta}{2}\right)\sqrt{2 \times 9.807}\ (1.8 + 0.001\,2)^{5/2}$$

$$\tan\frac{\theta}{2} = 0.844\,036,\ \theta = 80.33°$$

Step C *Second Trial*

Use first trial $\theta = 80.33°$

Table 9-4 $\Delta H = 0.9$ mm, $C = 0.577$ (interpolated)

(9-40) $$Q = (8/15)C \tan\left(\frac{\theta}{2}\right)\sqrt{2g}\,(H + \Delta H)^{5/2}$$

$$5 = (8/15)0.577 \tan\left(\frac{\theta}{2}\right)\sqrt{2 \times 9.807}\ (1.8 + 0.000\,9)^{5/2}$$

$$\tan\frac{\theta}{2} = 0.842\,923,\ \theta = 80.26°$$

Further trials not necessary.

PROBLEMS

1. Water at 68°F flows at a depth of ½ D in a 12-in. 250 psi cast iron pipe. If the pipe is laid to a slope of 1/100, estimate the average velocity using (a) the Colebrook friction factor, and (b) the Manning roughness factor.
2. In a test of a rectangular channel 1.219 m wide flowing at a depth of 0.610 m, a discharge of 0.453 m^3/s was measured for a slope of 1/2500. What is the Manning roughness factor n for this channel material?
3. How deep will water flow at the rate of 120 ft^3/sec in a brick-lined rectangular channel 10 ft wide, laid on a slope of 1/10,000?
4. Estimate the minimum slope required for a flow of 40 m^3/s in an open rectangular channel made of finished concrete whose width is 15 m if the depth may not exceed 3 m.
5. What are the best proportions for a gravel-lined trapezoidal channel whose sides slope at 45° if it is to carry 200 ft^3/sec of water with a drop of 1 ft in 1000 ft?
6. What are the best proportions for a trapezoidal channel whose area is 96 m^2 and whose sides slope at 30°?
7. A rectangular channel lined with planed lumber has a width of 27 ft. When the depth of water is 3.5 ft, the flow is 370 ft^3/sec.

Determine (a) the required slope, (b) the specific energy, and (c) type of flow.

8. A rubble-lined rectangular channel 4-m wide is to carry a flow of 6 m^3/s. What slope is required for critical flow?
9. Water flows 4 ft deep in a rectangular channel with an average velocity of 3 ft/sec. Can a hydraulic jump be formed?
10. With a discharge of 11 m^3/s of water in a rectangular channel 6-m wide, determine the depth y_1 which will sustain a hydraulic jump if $y_2 = 1.2$ m. What portion of the initial energy is lost in the jump?
11. An ASME rectangular weir whose crest length is 4 ft is to be installed at the end of an 8-ft wide channel. It is desired to have a total depth of 7 ft in the channel when the flow rate is 60 ft^3/sec. Estimate the weir height required.
12. A channel 8-m wide supplies a flow of 10 m^3/s at a depth of 2 m to an ASME rectangular weir located at the end of the channel. For a crest height of 1 m, estimate the length of weir required.
13. An ASME triangular weir whose notch angle is 45° has a head of 0.880 ft. Estimate the discharge.
14. Estimate the head required to discharge 250 m^3/hr from a 90° ASME triangular weir.

REFERENCES

Binder, R. C., *Fluid Mechanics,* 4th ed. Prentice-Hall, Englewood Cliffs, N.J., 1962.

Giles, R. V., *Theory and Problems of Fluid Mechanics and Hydraulics,* 2d ed. Schaum Publishing Company, New York, 1962.

Kindsvater, C. F., and R. W. Carter *Discharge Characteristics of Rectangular Thin Plate Weirs.* Transactions American Society of Civil Engineers, vol. 124, pp. 772–822, 1959.

Vennard, J. K., *Elementary Fluid Mechanics,* 3d ed. John Wiley and Sons, New York, 1954.

Fluid Meters—Their Theory and Application, 6th ed. Edited by H. S. Bean, American Society of Mechanical Engineers, New York, 1971.

10 Fluidics

10-1 INTRODUCTION

A fluidic system is one that used fluid dynamic phenomena to perform sensing, logic, amplification, and control functions. A flueric system is one that contains no mechanical moving parts. The first known use of the term *fluidics* was in a special report appearing in the February 8, 1965, issue of *Missiles and Rockets.* The technology of fluidics dates from a public announcement made by the U.S. Army's then Diamond Ordnance Fuse Laboratories early in 1960. However, most of the fluid phenomena now being applied have been known for many years.

In 1904 L. Prandtl discovered that flow separation could be varied by applying suction at the boundary layer of a wide-angle diffuser. By installing ports on each side of the diffuser, he found that when suction was applied to one side of the diffuser, the discharge fluid would adhere to that side, as shown in Fig. 10-1(B). When suction was applied to ports on both sides, the discharge fluid expanded and filled the entire diffuser. If Prandtl had installed an output duct on each side of the diffuser, he would have invented the first fluidic logic element.

In the early 1930s H. Coanda discovered the effect that bears his name. Coanda observed that when a free jet was introduced near an adjacent curved or flat plate, the jet would adhere to the plate and follow the plate even though the flow path is highly divergent in the direction of the jet. Consider the two-dimensional flow shown in Fig. 10-2. The jet issuing from the nozzle begins to entrain ambient fluid into the jet mixing region. Near the adjacent plate the entrained fluid is not easily replaced. On the side of the jet away from

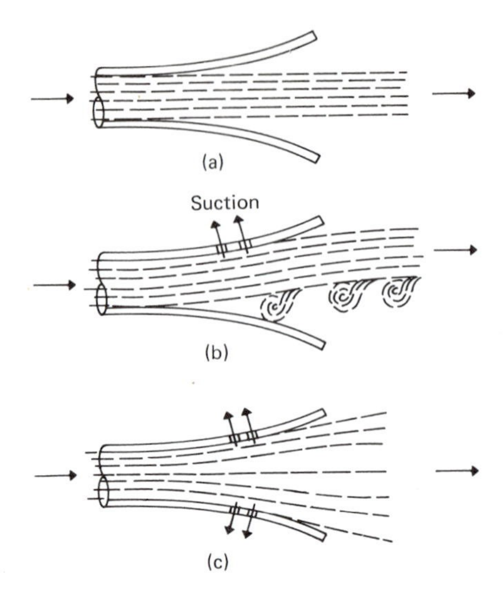

FIGURE 10-1 Prandtl diffuser

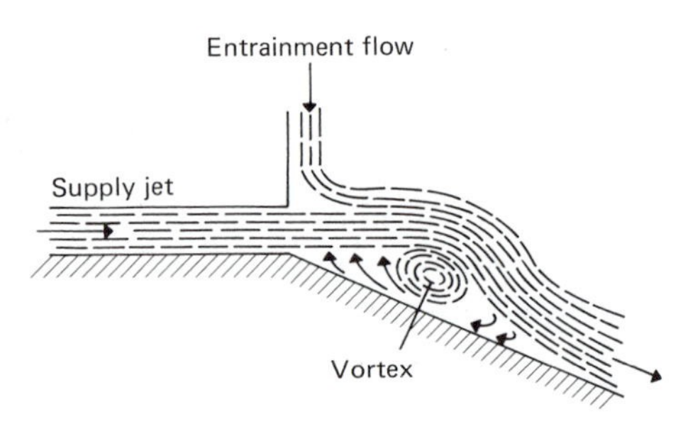

FIGURE 10-2 Coanda effect

the adjacent plate, the entrained fluid is easily replaced by the ambient fluid. The result is the rapid development of a transverse pressure gradient across the jet and the formation of a vortex which forms a region of lowest pressure. The pressure gradient bends the jet toward, and at some point downstream against, the adjacent plate. The plate can be inclined as much as 45 degrees to the axis of the jet and the jet will still become attached. The Coanda effect is of major importance to fluidic technology.

As was stated earlier, the technology of fluidics dates from 1960. The primary impetus to this technology was provided by R. Bowles, W. Horton, and R. Warren of the U.S. Army's (presently called) Harry Diamond Laboratories. These three scientists subsequently disclosed and patented a number of fundamental devices and applications in this field.

To meet the needs of a new and expanding technology, the National Fluid Power Association, the Society of Automotive Engineers, and the Department of Defense have published standards for fluidic terminology, nomenclature, graphical symbology, and definitions.

The National Fluid Power Association publication [reference (a)], which was adopted as an American National Standard in 1971, provides methods for rating the performance of fluidic devices. The Society of Automotive Engineers publication [reference (b)] was prepared by its Aerospace Fluidics Panel. The Department of Defense Document [reference (c)] MIL-STD-1306A is based on information obtained from the committees who prepared references (a) and (b). There are only minor variations between the three standards.

The scope of fluidics extends beyond material normally contained in a fluid mechanics text; for this reason, this chapter is limited to a discussion of fluid jets and various means of employing them as bi-stable or proportional amplifiers.

It is hoped that this introduction will make the student aware of fluidics and encourage him to read further, particularly in technical journals and trade publications. For the basic fluid mechanics of fluidics, references (g) and (j) are recommended, for component design, reference (i), and for applications, reference (h).

10-2 SUBMERGED JETS

The fluid jet appears in one form or another in most fluidic elements. A submerged jet is one that discharges into its own fluid, i.e., air into air, water into water, etc.

Laminar Jet

Figure 10-3 shows a fluid jet discharging from a nozzle. The velocity distribution across the nozzle exit is essentially uniform, and the jet interacts with the stationary fluid in the discharge space. Because of viscosity, viscous shear stress decelerates the outer layers of the jet while accelerating the adjacent layers of the stationary body of the fluid. A central core, moving at essentially the original velocity, is enclosed by a boundary layer in which the velocity declines to zero.

The width of the boundary layer increases as an increasing amount of the stationary fluid is entrained by the jet, spreading the

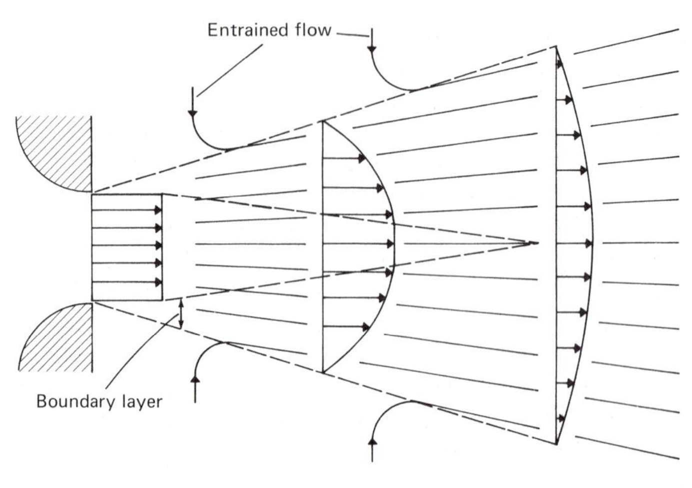

FIGURE 10-3 Laminar jet

jet and narrowing the central core until it becomes zero. The center line velocity decreases asymptotically to zero, the original kinetic energy being dissipated by viscous forces. Momentum, however, is conserved along the length of the jet, the reduction in average velocity being exactly compensated by an increase in the mass flow caused by entrainment.

The motion is completely steady, and shear stress between adjacent layers in relative motion is due only to the viscosity of the fluid. Except for the increase in jet width due to fluid entrainment, the formation of the boundary layer is similar to that shown in Fig. 8-4, Sec. 8.5, for a closed conduit.

Turbulent Flow

A laminar jet breaks down and becomes turbulent at some distance along its length, as shown in Fig. 10-4. Turbulent motion greatly speeds the spread of the jet and reduces its penetration into the stationary fluid. If conditions are such that the jet remains laminar for say 100 L/D, then transition to turbulence may be observed as a growth of periodic disturbances, which start as waves in the jet. For shorter laminar lengths, this phenomenon has not been observed and transition from laminar to turbulent is direct, as shown in Fig. 10-4.

The useful range of Reynolds numbers for fluidic applications is from about 400 to 20,000. In this range, the laminar length decreases as Reynolds number decreases. For a circular air jet discharging into still air, $L/D \approx 300$ for a Reynolds number of 500 to $L/D \approx 8$ to a Reynolds number of 1,500.

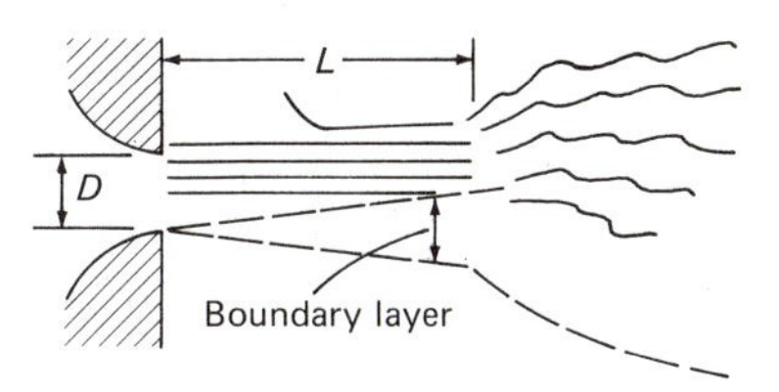

FIGURE 10-4 Transition from laminar to turbulent flow

10-3 TURBULENCE AMPLIFIER

The turbulence amplifier was invented by Raymond Auger in 1965 [reference (e)]. Figure 10-5 shows a typical one. Pressure is applied to the supply tube, producing a low-velocity jet directed toward the output tube. The apparatus is designed for the Reynolds number required to produce conditions, so that with no input the jet remains laminar from the supply to output tubes.

With no input, maximum pressure is recovered in the output tube. When input pressure is applied to the control tube, the control jet breaks up the laminar supply jet, and, as a result of the turbulence

FIGURE 10-5 Typical turbulence amplifier

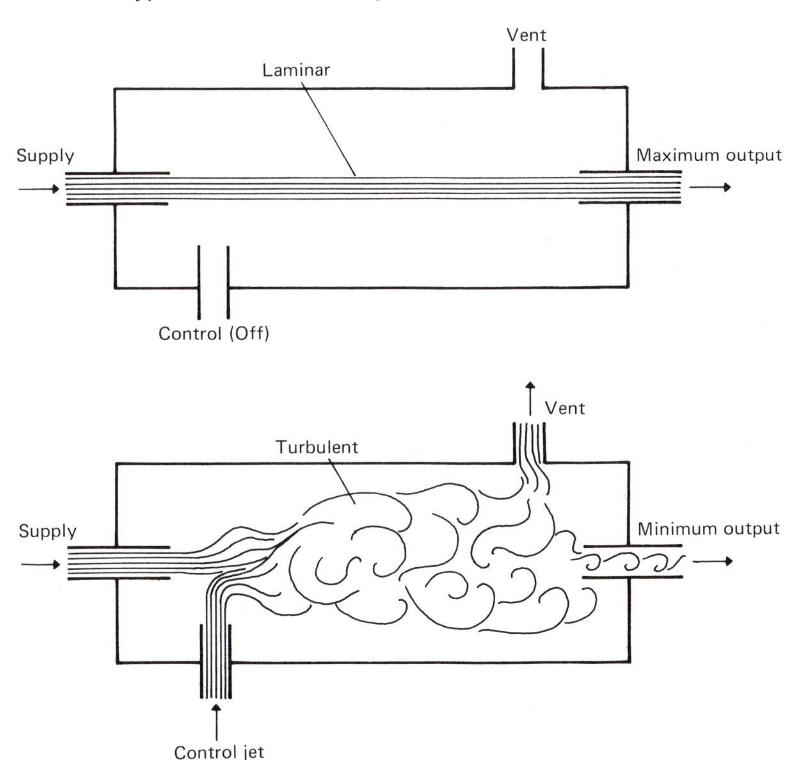

produced, the pressure recovery in the output tube is minimal. Because of the highly sensitive nature of the phenomena involved, the turbulence amplifier is used only as a digital (on-off) device.

An actual turbulence amplifier may consist of more than one control tube. In other variations of this principle, the laminar jet may be changed to turbulent by an acoustic signal or by physical interruption of the power jet.

10-4 IMPACT MODULATION AMPLIFIER

In the impact modulation amplifier, two axially opposed circular supply nozzles are used to cause impingement of two jets. If the momenta of these jets are equal, the impact point occurs halfway between the two nozzles. The location of the impact point is very sensitive to small changes in axial momentum. There are two practical versions of this amplifier, the direct impact modulator and the transverse impact modulator. These devices were described by B. G. Bjornsen in 1964 [reference (f)].

Direct Impact Modulation

As shown in Fig. 10-6, a direct impact modulation amplifier consists of two circular supply jets, *A* and *B*, with control being supplied through the annulus around supply jet *A* and output from the annulus around supply jet *B*, the point of impact being at *X*. Because this device is very sensitive to pressure differences between *A* and *B*, these supply jets are fed from the same source, the pressure in *B* being lowered by increased line resistance until the impact point *X* is just outside the output annulus. With no control jet, the output is zero. When a control pressure introduces a flow in the control annulus, it mixes with the supply jet *A* and increases its momentum, causing the impact point *X* to move into the output

FIGURE 10-6 Direct impact modulation amplifier

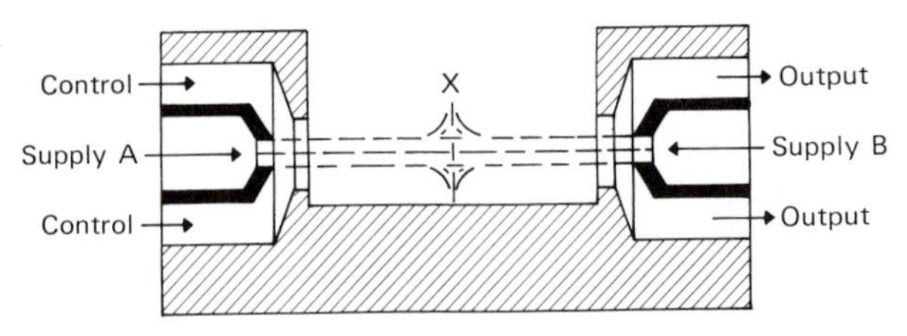

annulus and producing an output signal. The input-output characteristics of this amplifier are proportional (analog) and positive.

Transverse Impact Modulation

As shown in Fig. 10-7 the transverse modulation amplifier is similar to the direct modulator except the control jet is now at 90° from the supply stream. In the transverse modulator supply pressures, B is adjusted so that the impact point X is just inside the output annulus. With no control jet, the output is a maximum. When the control jet is in action, it bends supply jet A upward, causing the impact point X to move to the left, reducing the output pressure. The input-output characteristics of this amplifier are proportional and negative.

10-5 JET DEFLECTION AMPLIFIER

The jet deflection amplifier is also called a jet interaction amplifier because its operation depends upon the interaction of two control jets with a supply jet. There are two types of this amplifier, vented (open) and closed.

Vented Type

Figure 10-8 shows a vented jet deflection amplifier. When both control ports are off, the supply jet will strike the splitter plate, one-half going to output port A, the other to B. The resultant pressure differential across the output ports will be zero.

When a differential pressure is applied across the control ports, the supply jet will be deflected so that more will flow through one of the output ports than the other, thus creating a pressure differential across them. For example, as shown in Fig. 10-8, the left control jet caused the supply jet to deflect to the right, resulting in more flow through output B than through output A.

FIGURE 10-7 Transverse impact modulation amplifier

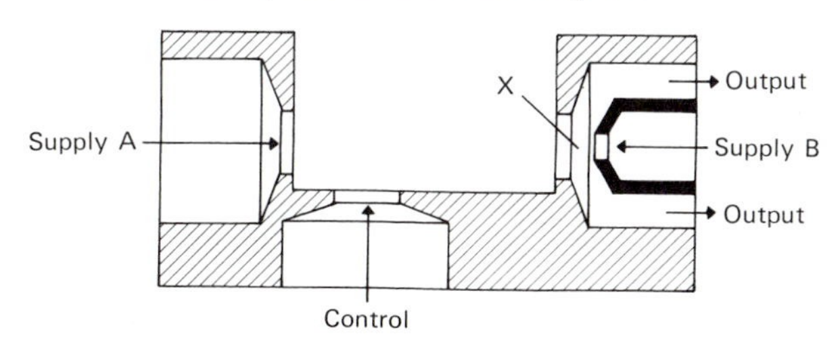

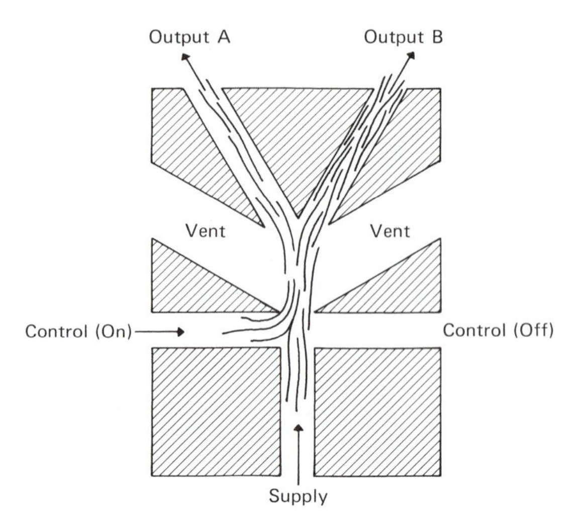

FIGURE 10-8 Vented jet deflection amplifier

Closed Type

The closed jet deflection amplifier operates in the same way as the vented. The primary difference between the two is that, since the closed jet is not vented to the atmosphere, the effect of loading the output ports is reflected back to the interaction region, affecting the interactions between the supply and control jets. When both output ports are restricted, the pressure backs up in the interaction region, raising the internal pressure of the amplifier.

10-6 WALL ATTACHMENT AMPLIFIER

The wall attachment amplifier operates on the Coanda effect described in Sec. 10-1. The configuration of the basic device is shown in Fig. 10-9. Pressure applied at the supply port produces a jet that issues into the interaction region. Initially some of the flow goes out both output ports, but this is an unstable condition. Owing to the entrainment caused by the power jet, a low pressure region is created in the vicinity of the two walls. The jet is attracted to one or the other as it deflects closer and closer to the wall of its choice. The pressure becomes lower and lower until the supply stream attaches itself to that wall.

When control flow is injected into the lower pressure region, allowing the supply jet to detach from that wall, there is still a low pressure region due to entrainment on the other side, so that the

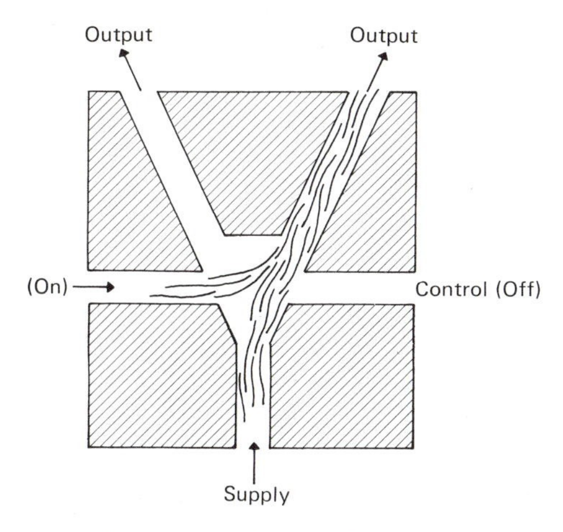

FIGURE 10-9 Typical wall attachment amplifier

supply stream is attracted to the opposite wall. The jet will then switch over and attach itself to that wall. The wall attachment amplifier output is digital and has a memory.

10-7 BOUNDARY LAYER AMPLIFIER

This two-dimensional device uses the principle of forced separation of a stream flowing over a curved surface, as shown in Fig. 10-10. With the control off, the supply jet clings to the curved surface and discharges through the lower output port. With the

FIGURE 10-10 Principle of the boundary layer amplifier

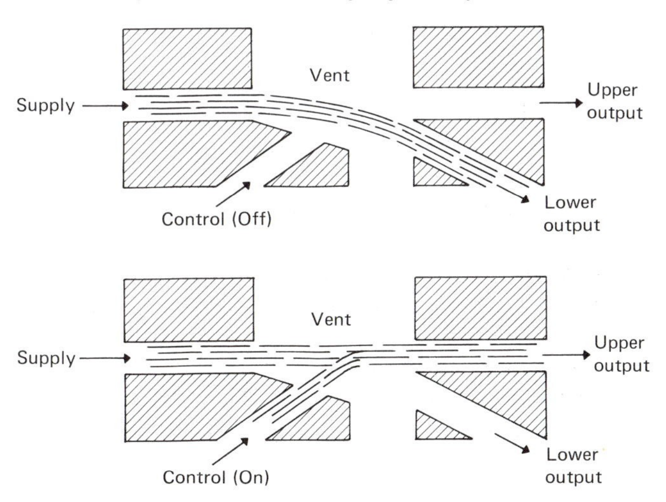

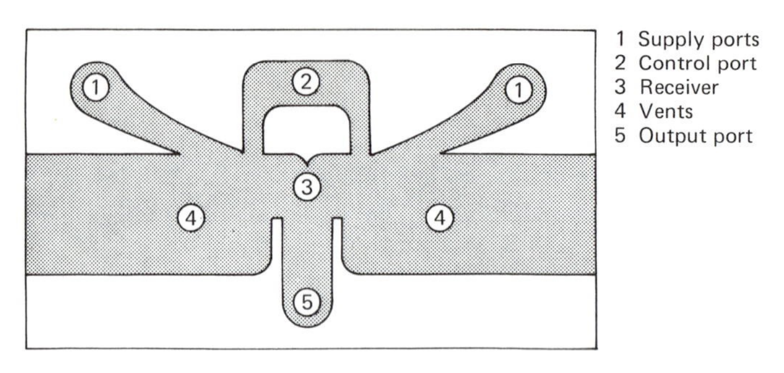

FIGURE 10-11 Focused-jet amplifier

control on, the point of separation will change so that some of the flow is diverted to the upper output port. The point of separation and thus the amount of flow diverted is proportional to the amount of flow injected into the boundary layer.

10-8 FOCUSED-JET AMPLIFIER

In the focused-jet amplifier, two supply jets (1) are focused on the receiver (3), as shown in Fig. 10-11. With no control input, the combined supply jets leave via the output port, so that maximum output is obtained with no control flow. When pressure is applied to the control port (2), the resultant control jets force the supply jets to discharge into the vents (4), so that minimum output is obtained with control flow activated. The device is digital (stable in either of two states).

10-9 VORTEX AMPLIFIER

The basic vortex amplifier is shown in Fig. 10-12. Pressure is supplied to the supply port, producing a jet which enters the cylindrical chamber radially. If the supply jet is not deflected, it will exit through the center output port. When pressure is applied at the control port, the tangential jet produced deflects the supply jet and imparts a momentum component in the tangential direction and a vortex is formed. As a result of the vortex formation, the pressure drop across the cylindrical chamber is increased in proportion to the control flow.

The pressure at the outlet port is inversely proportional to the pressure applied at the control port. The vortex amplifier is essentially a throttling valve with no moving parts.

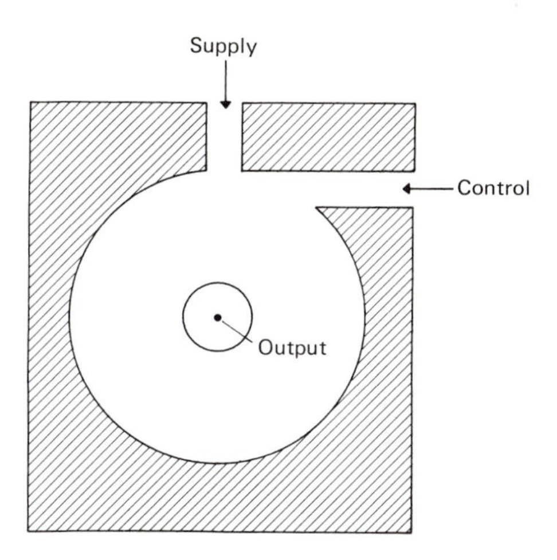

FIGURE 10-12 Vortex amplifier

REFERENCES

(a) "Method of Rating Performance of Fluidic Devices," National Fluid Power Association, T3.70.3-70, American National Standard, ANSI B93.14, 1971.

(b) "Fluidic Technology," Aerospace Recommended Practice, Society of Automotive Engineers, A RP 993A, revised February 15, 1969.

(c) "Fluerics—Terminology and Symbols," Military Standard, MIL-STD-1306A, 8 December 1972.

(d) Elliott Mendelson, *Theory and Problems of Boolean Algebra and Switching Circuits*. McGraw-Hill, New York, 1970.

(e) R. N. Auger, "Turbulence Amplifier Design and Application," Proceedings, Second Harry Diamond Laboratory Symposium, Washington, D.C., vol. 1, May 1964, pp. 357–366.

(f) B. G. Bjornsen, "The Impact Modulator," Proceedings, Second Harry Diamond Laboratories Symposium, Washington, D.C., vol. 2, May 1964, pp. 5–12.

(g) *Fluid Amplifiers*. Edited by J. M. Kirshner, McGraw-Hill, New York, 1966.

(h) *A Guide to Fluidics*. Edited by A. Conway. MacDonald and Co., London, 1971.

(i) C. A. Belsterling, *Fluidic Systems Design*. John Wiley, New York, 1971.

(j) Foster, K., and G. A. Parker, *Fluidics—Components and Circuits*. John Wiley, New York, 1970.

Appendix A
Selected Data

Table A-1 *Density of Selected Liquids*

TEMPERATURE °C		0	10	20	30	40	50
TEMPERATURE °F		32	50	68	86	104	122
LIQUID	**Pressure, psia**	ρ in slugs ft^{-3} or lbf sec^2 ft^{-4}					
Alcohol, ethyl[f]	14.7	1.564	1.548	1.532	1.515	1.498	1.481
Benzene[f]	14.7	1.746	1.726	1.705	1.684	1.663	1.642
Carbon tetrachloride[f]	14.7	3.168	3.130	3.093	3.055	3.017	2.979
Glycerin[f]	14.7	2.469	2.459	2.447	2.436	2.424	2.416
Mercury[f]	14.7	26.38	26.33	26.28	26.24	26.19	26.14
Octane, normal[f]	14.7	1.394	1.378	1.362	1.347	1.331	1.315
Oil, castor[k]	14.7	1.889	1.876	1.863	1.850	1.836	1.832
Oil, linseed[f]	14.7	1.853	1.840	1.828	1.815	1.803	1.791
Turpentine[f]	14.7	1.705	1.689	1.673	1.657	1.641	1.625
Water, fresh[d]	14.7	1.940	1.940	1.937	1.932	1.925	1.917
	10,000	2.005	2.005	1.992	1.986	1.980	1.967
	15,000	2.031	2.031	2.018	2.012	2.005	1.992
Water, sea[h]	14.7	1.995	1.992	1.988	1.982	—	—
LIQUID	**Pressure, 10^5 N · m^{-2}**	ρ in kg · m^{-3}					
Alcohol, ethyl[f]	1.013	806.0	797.8	789.3	780.7	772.0	763.3
Benzene[f]	1.013	900.0	898.4	878.7	868.0	857.2	846.5
Carbon tetrachloride[f]	1.013	1 633	1 613	1 594	1 575	1 555	1 535
Glycerin[f]	1.013	1 273	1 267	1 261	1 255	1 249	1 243
Mercury[f]	1.013	13 600	13 570	13 550	13 520	13 500	13 470
Octane, normal[f]	1.013	718.5	710.3	702.1	694.0	685.9	677.6
Oil, castor[k]	1.013	973.7	967.0	960.2	953.2	946.2	939.0
Oil, linseed[f]	1.013	955.0	948.5	942.0	935.5	929.0	922.6
Turpentine[f]	1.013	879.0	870.7	862.4	854.1	845.8	837.5
Water, fresh[d]	1.013	999.8	999.7	998.3	995.7	992.2	998.0
	689.5	1 033	1 033	1 026	1 024	1 020	1 014
	1 034	1 047	1 047	1 040	1 036	1 033	1 026
Water, sea[h]	1.013	1 028	1 027	1 025	1 021	—	—

Table A-2 *Isothermal Bulk Modulus of Elasticity of Selected Liquids*

TEMPERATURE °C	0	10	20	30	40	50
°F	32	50	68	86	104	122
LIQUID	**E_T in 10^6 lbf ft^{-2}**					
Alcohol, ethyl[i]	21.16	19.91	18.71	17.54	16.39	15.24
Benzene[i]	25.82	23.92	22.12	20.54	18.99	17.57
Carbon tetrachloride[i]	23.26	21.53	19.97	18.50	17.08	15.68
Glycerin[m]	93.84	91.20	88.47	85.93	83.40	80.94
Mercury[c]	526.1	520.1	516.3	510.5	504.6	499.0
Octane, normal[l,u]	20.94	19.47	18.10	16.83	15.66	14.60
Oil, castor[k]	43.54	41.30	39.19	37.21	35.33	33.52
Oil, linseed[o]	39.83	38.31	36.80	35.28	33.90	32.10
Turpentine[a,b,f]	26.84	25.28	23.75	22.25	20.92	19.59
Water, fresh[a]	40.96	43.61	45.49	46.67	47.23	47.28
Water, sea[p]	44.83	47.17	48.79	49.76	—	—
LIQUID	**E_T in 10^8 N · m^{-2} or 10^8 Pa**					
Alcohol, ethyl[i]	10.13	9.533	8.958	8.398	7.848	7.302
Benzene[i]	12.36	11.45	10.59	9.825	9.092	8.407
Carbon tetrachloride[i]	11.14	10.31	9.562	8.858	8.178	7.508
Glycerin[m]	44.93	43.67	42.36	41.14	39.93	38.75
Mercury[e]	251.8	249.0	247.2	244.4	241.6	238.9
Octane, normal[l,u]	10.03	9.322	8.666	8.058	7.498	6.991
Oil, castor[k]	20.84	19.77	18.76	17.82	16.92	16.05
Oil, linseed[o]	19.07	18.34	17.62	16.89	16.23	15.23
Turpentine[a,b,f]	12.85	12.10	11.37	10.65	10.02	9.379
Water, fresh[a]	19.62	20.88	21.78	22.46	22.61	22.64
Water, sea[p]	21.46	22.59	23.36	23.83	—	—

Table A-3 *Isentropic Bulk Modulus of Elasticity of Selected Liquids*

TEMPERATURE °C	0	10	20	30	40	50
TEMPERATURE °F	32	50	68	86	104	122
LIQUID	**E_s in 10^6 lbf ft^{-2}**					
Alcohol, ethyl[t]	25.56	23.86	22.26	20.73	19.30	17.96
Benzene[j]	37.87	34.89	32.09	29.46	27.04	24.76
Carbon tetrachloride[k]	34.18	31.69	29.34	27.08	24.94	22.90
Glycerin[m]	102.1	99.82	97.40	95.16	92.91	90.65
Mercury[e]	606.6	601.8	596.9	592.2	587.4	587.8
Octane, normal[m]	24.47	22.62	20.83	19.17	17.59	16.08
Oil, castor[k]	50.11	47.70	45.42	43.27	41.27	39.32
Oil, linseed[o]	46.08	44.33	42.58	40.85	39.19	37.01
Turpentine[a,b,f]	33.89	31.73	29.68	27.75	25.93	24.21
Water, fresh[e]	41.06	43.67	45.81	47.35	48.42	49.10
Water, sea[l]	45.11	47.63	49.58	50.98	–	–
LIQUID	**E_s in 10^8 N · m^{-2} or 10^8 Pa**					
Alcohol, ethyl[t]	12.26	11.42	10.66	9.926	9.241	8.600
Benzene[j]	18.13	16.71	15.36	14.11	12.95	11.86
Carbon tetrachloride[k]	16.37	15.17	14.05	12.97	11.94	10.96
Glycerin[m]	48.89	47.79	46.64	45.57	44.49	43.40
Mercury[e]	290.4	288.1	285.7	283.5	281.2	278.9
Octane, normal[m]	11.71	10.83	9.973	9.179	8.422	7.699
Oil, castor[k]	23.99	22.84	21.75	20.72	19.76	18.83
Oil, linseed[o]	22.06	21.23	20.39	19.56	18.76	17.72
Turpentine[a,b,f]	16.22	15.19	14.21	13.29	12.42	11.59
Water, fresh[e]	19.66	20.91	21.93	22.67	23.18	23.51
Water, sea[e]	21.60	22.81	23.74	24.41	–	–

Table A-4 *Dynamic Viscosity of Selected Liquids at Atmospheric Pressure*

TEMPERATURE °C	0	10	20	30	40	50
TEMPERATURE °F	32	50	68	86	104	122
LIQUID	μ in 10^{-6} lbf sec ft^{-2}					
Alcohol, ethyl[f]	37.10	30.22	24.85	20.63	17.25	14.52
Benzene[f]	18.79	15.81	13.52	11.72	10.28	9.103
Carbon tetrachloride[f]	28.12	23.69	20.28	17.58	15.41	13.64
Glycerin[q]	252,100	81,450	29,500	12,780	5,931	2,966
Mercury[a]	35.19	33.73	32.46	31.31	30.28	29.39
Octane, normal[g]	14.88	12.96	11.42	10.16	9.123	8.243
Oil, castor[r]	133,800	49,780	20,070	9,526	4,648	2,517
Oil, linseed[a]	1,921	1,337	950.2	691.3	457.4	367.6
Turpentine[a]	46.95	37.24	31.06	25.57	22.36	19.13
Water, fresh[d]	36.61	27.14	20.92	16.65	13.61	11.37
Water, sea[h]	39.40	29.08	22.61	18.20	–	–
LIQUID	μ in 10^{-4} N · s · m^{-2} or 10^{-4} Pa · s					
Alcohol, ethyl[f]	17.76	14.47	11.90	9.877	8.259	6.952
Benzene[f]	8.995	7.568	6.473	5.612	4.922	4.358
Carbon tetrachloride[f]	13.46	11.34	9.708	8.418	7.379	6.529
Glycerin[q]	120 700	39 000	14 120	6 120	2 840	1 420
Mercury[a]	16.85	16.16	15.54	14.49	14.50	14.07
Octane, normal[g]	7.125	6.203	5.466	4.865	4.368	3.947
Oil, castor[r]	64 060	23 830	9 610	4 565	2 242	1 205
Oil, linseed[a]	920.0	640.0	455.0	331.0	219.0	176.0
Turpentine[a]	22.48	17.83	14.87	12.72	10.71	9.160
Water, fresh[d]	17.53	13.00	10.02	7.972	6.514	5.542
Water, sea[h]	18.86	12.99	10.83	8.714	–	–

Table A-5 *Surface Tension of Selected Liquids in Contact with Air at Atmospheric Pressure*

TEMPERATURE °C	0	10	20	30	40	50
TEMPERATURE °F	32	50	68	86	104	122
LIQUID	σ **in** 10^{-4} **lbf ft**$^{-1}$					
Alcohol, ethyl[f]	16.48	15.86	15.26	14.68	14.12	13.57
Benzene[f]	21.64	20.71	19.79	18.89	18.00	17.12
Carbon tetrachloride[f]	20.03	19.19	18.34	17.49	16.67	15.86
Glycerin[f]	43.98	43.73	43.45	43.11	42.73	42.31
Mercury[a,f]	329.2	327.5	326.0	323.6	322.9	321.4
Octane, normal[f]	16.31	15.62	14.94	14.26	13.58	12.92
Turpentine[f]	20.02	19.35	18.69	18.03	17.36	16.70
Water, fresh[f]	51.83	50.86	49.85	48.77	47.66	46.53
Water, sea[p]	52,36	51.37	50.38	49.40	—	—
LIQUID	σ **in** 10^{-3} **N · m**$^{-1}$					
Alcohol, ethyl[f]	24.05	23.14	22.27	21.43	20.60	19.80
Benzene[f]	31.58	30.22	28.88	27.56	26.26	24.98
Carbon tetrachloride[f]	29.23	28.00	26.77	25.53	24.33	23.14
Glycerin[f]	64.18	63.82	63.41	62.92	62.36	61.75
Mercury[a,f]	480.4	477.9	475.7	472.4	471.2	469.0
Octane, normal[f]	23.80	22.79	21.80	20.81	19.82	18.85
Turpentine[f]	29.22	28.25	27.28	26.31	25.34	24.37
Water, fresh[f]	75.64	74.22	72.75	71.18	69.56	67.91
Water, sea[p]	76.71	74.97	73.53	72.09	—	—

Table A-6 *Vapor Pressure of Selected Liquids*

TEMPERATURE °C		0	10	20	30	40	50
TEMPERATURE °F		32	50	68	86	104	122
LIQUID	a	p_v in 10^a lbf ft^{-2}					
Alcohol, ethyl[e]	1	3.408	6.621	12.25	21.73	37.22	61.26
Benzene[c]	1	7.336	12.68	20.94	32.13	50.89	75.54
Carbon tetrachloride[c]	1	9.479	15.91	25.65	39.82	59.85	87.38
Glycerin[a]	−5	8.142	26.58	80.05	224.2	587.8	1,042
Mercury[f]	−4	5.151	13.64	33.44	81.65	169.2	352.8
Octane, normal[g]	0	7.956	15.68	29.13	51.37	86.60	140.2
Turpentine[a]	0	3.094	5.861	10.62	18.52	31.16	50.77
Water, fresh[d]	1	1.276	2.563	4.880	8.859	15.41	25.76
Water, sea[e]	1	1.252	2.514	4.788	8.692	–	–
LIQUID	b	p_v in 10^b N · m^{-2} or 10^b Pa					
Alcohol, ethyl[e]	3	1.632	3.170	5.866	10.41	17.83	29.33
Benzene[c]	3	3.512	6.070	10.00	15.38	24.37	36.17
Carbon tetrachloride[c]	3	4.538	7.621	12.28	19.07	28.65	41.84
Glycerin[a]	−3	3.898	12.73	38.33	107.3	281.4	695.3
Mercury[f]	−2	2.466	6.532	16.00	36.96	81.05	168.9
Octane, normal[g]	2	3.809	7.507	13.95	24.60	41.46	67.13
Turpentine[a]	2	1.481	2.806	5.088	8.869	14.92	24.30
Water, fresh[d]	3	0.6108	1.227	2.337	4.242	7.377	12.33
Water, sea[e]	3	0.5928	1.203	2.292	4.162	–	–

Table A-7 *Properties of Some Common Gases*[c]

	GAS			Air	Carbon dioxide	Carbon monoxide	Helium	Hydrogen	Methane	Nitrogen	Oxygen
	SYMBOL			–	CO_2	CO	He	H_2	CH_4	N_2	O_2
M	Molecular weight			28.96	44.01	28.01	4.003	2.016	16.04	28.01	32.00
T_c	Critical temperature		°R	238.7	547.7	239.8	9.45	59.85	344.0	227.2	277.9
			K	132.6	304.2	133.2	5.25	33.25	191.9	126.2	154.4
p_c	Critical pressure		psia	546.7	1,071	516.4	33.83	191.6	685.6	492.3	730.4
			$10^5\ N \cdot m^{-2}$	37.69	73.84	35.60	2.322	13.21	47.27	33.94	50.36
R	Gas constant		ft-lbf/lbm°R	53.36	35.11	55.17	386.0	766.5	96.34	55.17	48.29
			J/(kg · K)	287.1	188.9	296.5	2 077	4 124	518.3	296.2	259.8
c_p	Specific heat	Constant pressure	Btu/(lbm°R)	0.2400	0.1935	0.2463	1.240	3.387	0.5232	0.2464	0.2168
			J/(kg · K)	1 005	814.7	1 032	5 192	14 180	2 191	1 032	660.0
c_v		Constant volume	Btu/(lbm°R)	0.1714	0.1484	0.1755	0.7441	2.402	0.3994	0.1755	0.1548
			J/(kg · K)	717.2	621.1	734.7	3 115	10 060	1 672	734.8	471.1
k		Ratio	c_p/c_v	1.400	1.304	1.404	1.667	1.410	1.310	1.404	1.401

Table A-8 *Dynamic Viscosity of Selected Gases*[a]

TEMPERATURE °C	0	10	20	30	40	50
TEMPERATURE °F	32	50	68	86	104	122
GAS	μ in 10^{-8} lbf sec ft^{-2}					
Air	35.67	37.06	38.31	38.94	39.92	40.58
Carbon dioxide	29.03	29.97	30.91	31.95	32.79	33.84
Carbon monoxide	34.67	35.51	36.44	37.38	38.28	39.17
Helium	38.85	39.71	40.54	41.58	42.59	43.59
Hydrogen	17.44	17.86	18.27	18.71	19.15	19.58
Methane	21.43	22.28	22.70	23.37	24.02	24.67
Nitrogen	34.65	35.57	36.51	37.43	38.37	39.29
Oxygen	39.47	40.89	42.25	43.40	44.51	45.61
GAS	μ in 10^{-6} N · s · m^{-2} or 10^{-6} Pa · s					
Air	17.08	17.74	18.34	18.64	19.04	19.43
Carbon dioxide	13.90	14.35	14.80	15.30	15.70	16.20
Carbon monoxide	16.60	17.00	17.45	17.90	18.33	18.75
Helium	18.60	19.01	19.41	19.91	20.93	20.87
Hydrogen	8.350	8.550	8.746	8.960	9.169	9.375
Methane	10.26	10.67	10.87	11.19	11.50	11.81
Nitrogen	16.59	17.03	17.48	17.92	18.37	18.81
Oxygen	18.90	19.58	20.23	20.78	21.31	21.84

Table A-9 *1962 U.S. Standard Atmosphere*[v]

Altitude, z	Temperature, T	Pressure, p	Viscosity, μ	Acoustic velocity, c
U.S. UNITS				
10^3 ft	°R	lbf/ft^2	$10^8 \frac{\text{lbf-sec}}{\text{ft}^2}$	ft/sec
0	518.67	2,116	37.36	1,117
10	483.03	1,456	35.34	1,077
20	447.42	973.2	33.33	1,037
30	411.84	629.7	31.07	994.8
40	389.97	393.1	29.69	968.1
50	389.97	243.6	29.69	968.1
60	389.97	151.0	29.69	968.1
70	392.25	93.72	29.83	970.9
80	397.69	58.51	30.18	977.6
90	403.14	36.78	30.52	984.3
100	408.57	23.27	28.91	990.9
SI UNITS				
km	K	10^3 N/m^2	$10^{-6} \frac{\text{N} \cdot \text{s}}{\text{m}^2}$	m/s
0	288.15	101.325	17.89	340.3
5	255.68	54.05	16.28	320.5
10	223.25	26.50	14.58	299.5
15	216.65	12.11	14.22	295.1
20	216.65	5.529	14.22	295.1
25	221.25	2.549	14.48	298.4
30	226.51	1.390	14.75	301.7

Table A-10 *Properties of Areas and Volumes*

RECTANGLE

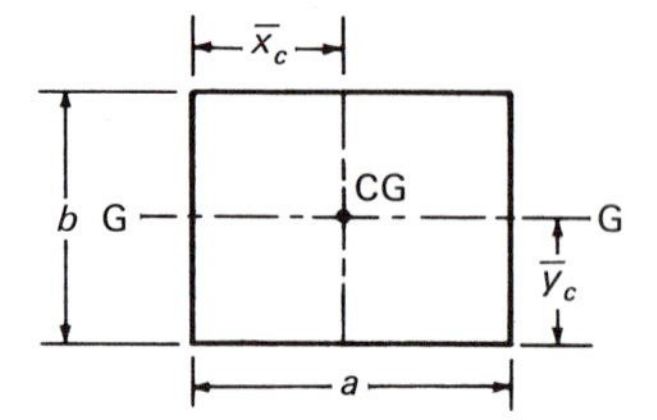

$$A = ab$$
$$\bar{x}_c = a/2$$
$$\bar{y}_c = b/2$$
$$I_G = ab^3/12$$
$$I_G/A = b^2/12$$

TRIANGLE

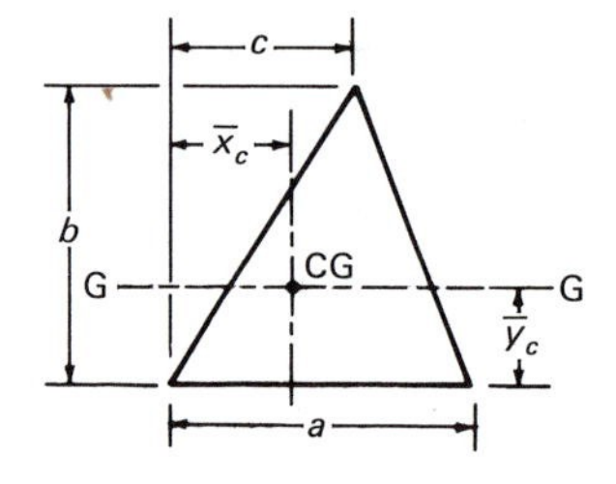

$$A = ab/2$$
$$\bar{x}_c = (a + c)/3$$
$$\bar{y}_c = b/3$$
$$I_G = ab^3/36$$
$$I_G/A = b^2/18$$

TRAPEZOID

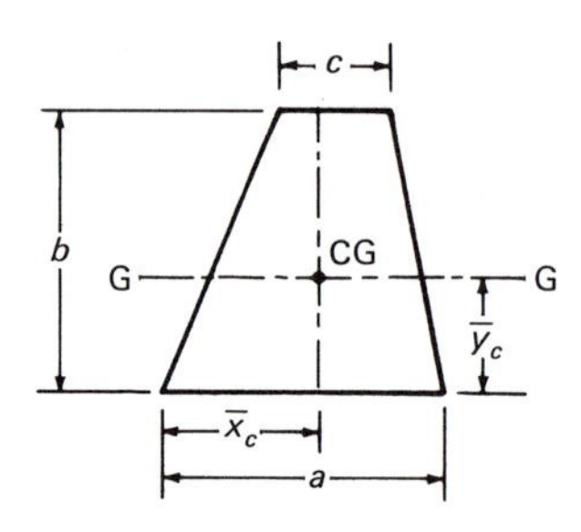

$$A = (b/2)(a + c)$$

$\bar{x}_c$ is on a line connecting the midpoints of sides a and b

$$\bar{y}_c = (b/3)(a + 2c)/(a + c)$$
$$I_G = (b^3/36)(a^2 + 4ac + c^2)/(a + c)$$
$$I_G/A = (b^2/12)(a^2 + 4ac + c^2)/(a + c)^2$$

CIRCLE

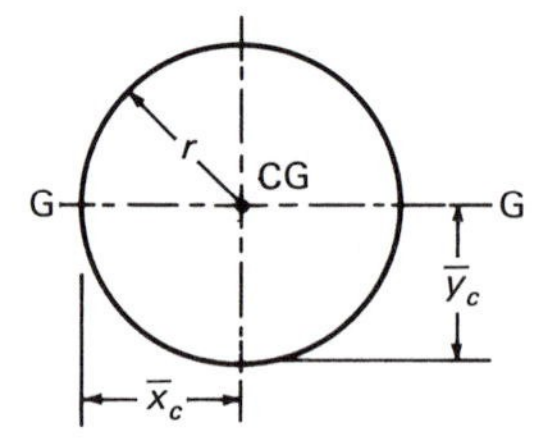

$$r^2 = x^2 + y^2$$
$$A = \pi r^2$$
$$\bar{x}_c = r$$
$$\bar{y}_c = r$$
$$I_G = \pi r^2/4$$
$$I_G/A = r^2/4$$

Table A-10 (*continued*)

HALF CIRCLE

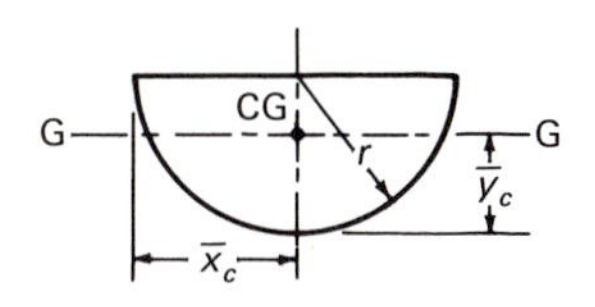

$$A = \pi r^2/2$$
$$\bar{x}_c = r$$
$$\bar{y}_c = r(1 - 4/3\pi)$$
$$I_G = r^4(\pi/8 - 8/9\pi)$$
$$I_G/A = (r/6\pi)^2(9\pi^2 - 64)$$

QUARTER CIRCLE

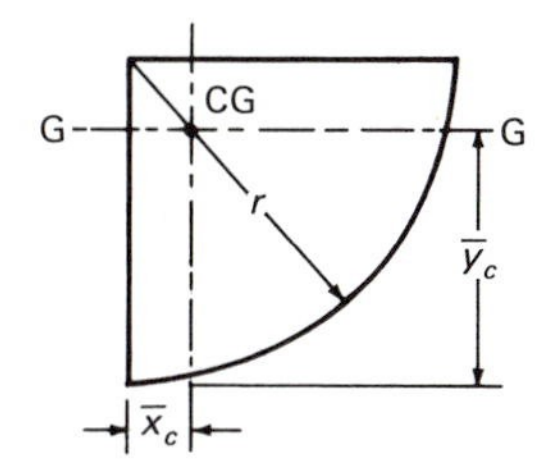

$$A = \pi r^2/4$$
$$\bar{x}_c = 4r/3\pi$$
$$\bar{y}_c = r(1 - 4/3\pi)$$
$$I_G = (r^4/2)(\pi/8 - 8/9\pi)$$
$$I_G/A = (r/6\pi)^2(9\pi^2 - 64)$$

ELLIPSE

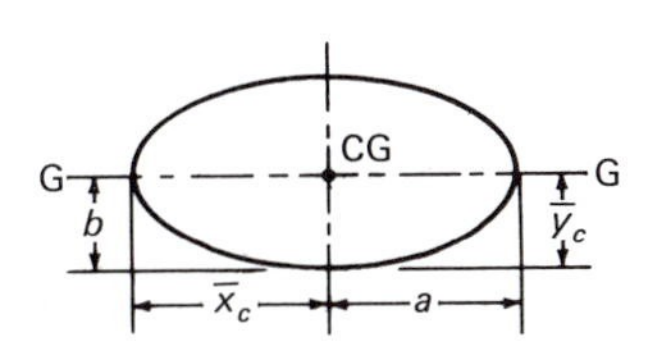

$$(x/a)^2 + (y/b)^2 = 1$$
$$A = \pi ab$$
$$\bar{x}_c = a$$
$$\bar{y}_c = b$$
$$I_G = \pi ab^3/4$$
$$I_G/A = b^2/4$$

HALF ELLIPSE

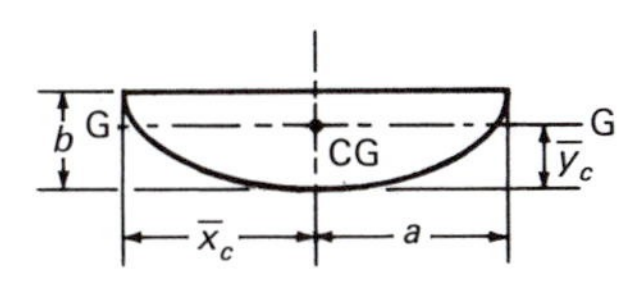

$$A = \pi ab/2$$
$$\bar{x}_c = a$$
$$\bar{y}_c = b(1 - 4/3\pi)$$
$$I_G = ab^3(\pi/8 - 8/9\pi)$$
$$I_G/A = (b/6\pi)^2(9\pi^2 - 64)$$

Table A-10 (*continued*)

QUARTER ELLIPSE

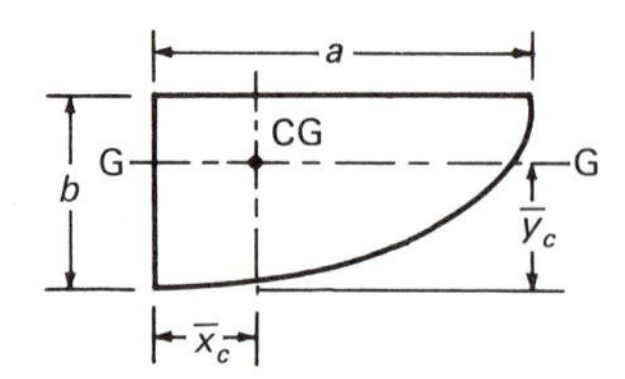

$$A = \pi ab/4$$
$$\bar{x}_c = 4a/3\pi$$
$$\bar{y}_c = b(1 - 4/3\pi)$$
$$I_G = (ab^3/2)(\pi/8 - 8/9\pi)$$
$$I_G/A = (b/6\pi)^2(9\pi^2 - 64)$$

PARABOLA

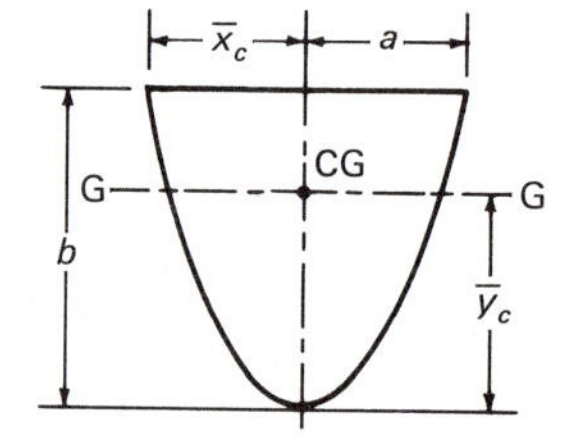

$$x^2 = (a^2/b)y$$
$$A = 4ab/3$$
$$\bar{x}_c = a$$
$$\bar{y}_c = 3b/5$$
$$I_G = 16ab^3/175$$
$$I_G/A = 4b^2/55$$

HALF PARABOLA

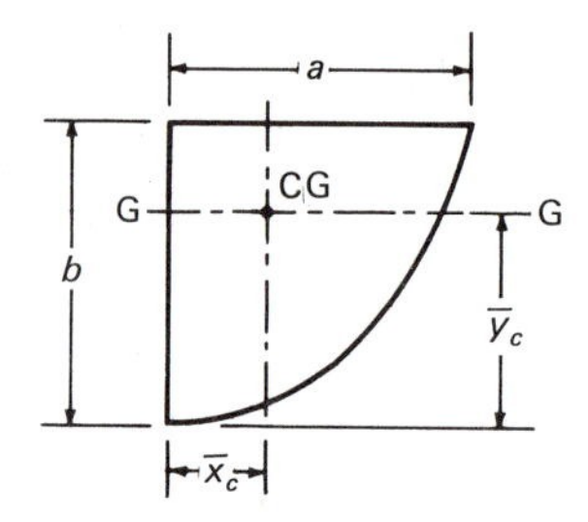

$$A = 2ab/3$$
$$\bar{x}_c = 3a/8$$
$$\bar{y}_c = 3b/5$$
$$I_G = 8ab^3/175$$
$$I_G/A = 4b^2/55$$

CYLINDER

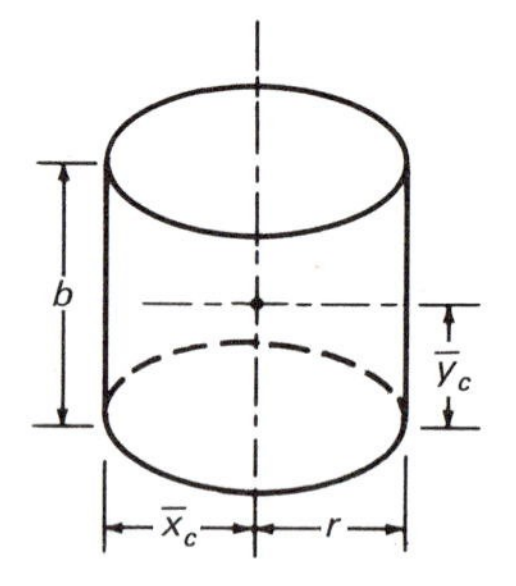

$$\text{Vol} = \pi b r^2$$
$$A = 2\pi r(b + r)$$
$$\bar{x}_c = r$$
$$\bar{y}_c = b/2$$

Table A-10 *(continued)*

CONE

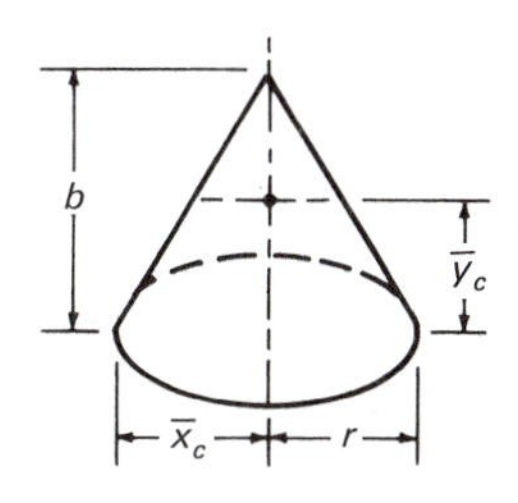

$$\text{Vol} = \pi r^2 b/3$$
$$A = \pi r(r + \sqrt{r^2 + b^2})$$
$$\bar{x}_c = r$$
$$\bar{y}_c = b/4$$

SPHERE

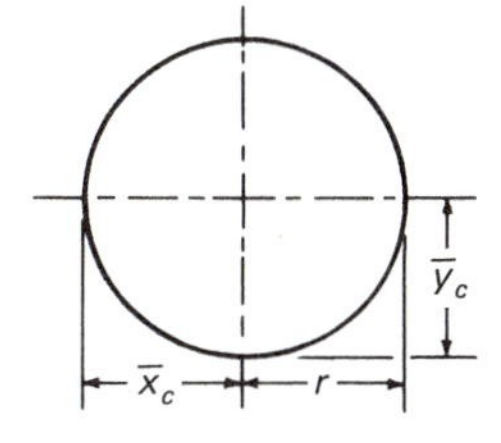

$$\text{Vol} = 4\pi r^3/3$$
$$A = 4\pi r^2$$
$$\bar{x}_c = r$$
$$\bar{y}_c = r$$

HEMISPHERE

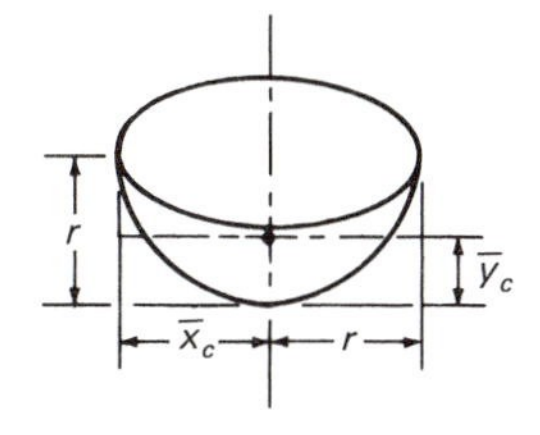

$$\text{Vol} = 2\pi r^3/3$$
$$A = 3\pi r^2$$
$$\bar{x}_c = r$$
$$\bar{y}_c = 5r/8$$

PARABOLOID OF REVOLUTION

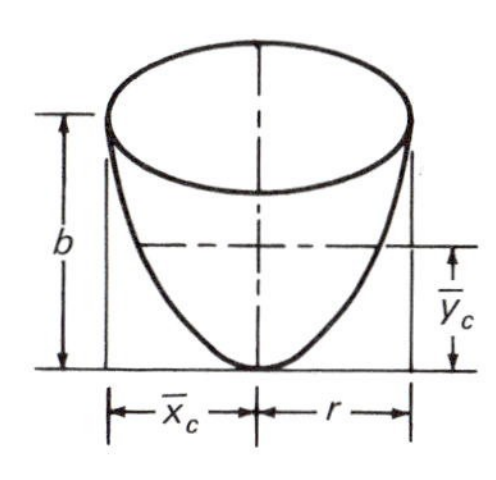

$$\text{Vol} = \pi b r^2/2$$
$$A = \pi r^2 + \frac{\pi r}{3b^2}\left[\left(\frac{r^2}{4} + b^2\right)^{3/2} - \left(\frac{b}{4}\right)^3\right]$$
$$\bar{x}_c = r$$
$$\bar{y}_c = 2b/3$$

ELLIPSOIDS AND SPHEROIDS

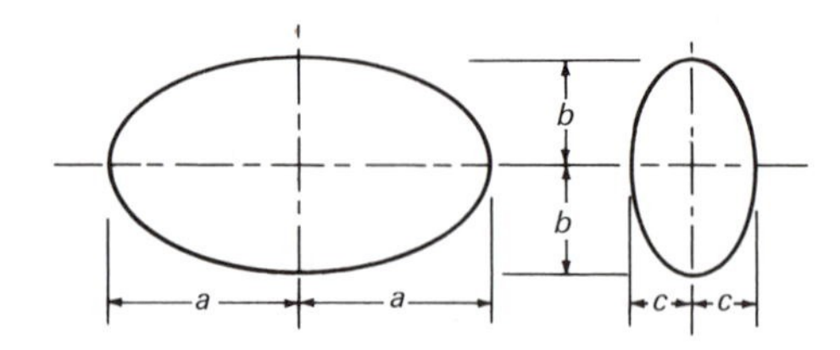

For Ellipsoid:

$\text{Vol} = 4\pi abc/3$

For Prolate Spheroid: $c = b = r$, let $f = \sqrt{1 - (r/a)^2}$

$\text{Vol} = 4\pi ar^2/3$

$A = 2\pi r^2 + (2\pi ar/f) \sin^{-1} f$

For Oblate Spheroid: $a = c = r$, let $f = \sqrt{1 - (b/r)^2}$

$\text{Vol} = 4\pi br^2/3$

$A = 2\pi r^2 + (\pi b^2/f) \log_e (1 + f)/(1 - f)$

Table A-11 *Properties of Wrought Steel and Wrought Iron Pipe*

Pipe size, in.	Outside diameter		Schedule	Wall thickness		Internal diameter		Flow area		$10^6\epsilon/D$
	in.	mm		in.	mm	ft	mm	ft²	mm²	
1/8	0.405	10.29	40 (STD)	0.068	1.73	0.02242	6.83	0.0003947	36.64	6,690
			80 (XS)	0.095	2.41	0.01792	5.47	0.0002522	23.50	8,370
1/4	0.540	13.72	40 (STD)	0.088	2.24	0.03033	9.24	0.0007227	67.06	4,946
			80 (XS)	0.119	3.02	0.02517	7.68	0.0004974	46.32	5,959
3/8	0.675	17.14	40 (STD)	0.091	2.31	0.04108	12.52	0.001326	123.3	3,651
			80 (XS)	0.126	3.20	0.03525	10.74	0.0009759	90.59	4,255
1/2	0.840	21.34	40 (STD)	0.109	2.77	0.05183	15.80	0.002110	196.1	2,894
			80 (XS)	0.147	3.74	0.04550	13.86	0.001626	150.8	3,297
			160	0.188	4.78	0.03867	11.78	0.001174	109.0	3,879
			(XXS)	0.294	7.47	0.02100	6.40	0.0003464	32.17	7,143
3/4	1.050	26.67	40 (STD)	0.113	2.87	0.06867	20.93	0.003703	344.1	2,184
			80 (XS)	0.154	3.92	0.06183	18.83	0.003003	278.5	2,426
			160	0.219	5.56	0.05100	15.55	0.002043	189.8	2,941
			(XXS)	0.308	7.82	0.03617	11.03	0.001027	95.55	4,147
1	1.315	33.40	40 (STD)	0.133	3.38	0.08742	26.64	0.006002	557.4	1.715
			80 (XS)	0.179	4.55	0.07975	24.30	0.004995	508.3	1,887
			160	0.250	6.35	0.06792	20.70	0.003623	336.5	2,208
			(XXS)	0.358	9.09	0.04992	15.22	0.001957	181.5	3,005

Table A-11 (*continued*)

Pipe size, in.	Outside diameter		Schedule	Wall thickness		Internal diameter		Flow area		$10^6\epsilon/D$
	in.	mm		in.	mm	ft	mm	ft²	mm²	
1¼	1.660	42.16	40 (STD)	0.140	3.56	0.1150	35.04	0.01039	964.3	1,304
			80 (XS)	0.191	4.85	0.1065	32.46	0.008908	827.5	1,408
			160	0.250	6.35	0.09667	29.46	0.007339	681.6	1,552
			(XXS)	0.382	9.70	0.07467	22.76	0.004379	406.9	2,008
1½	1.900	48.26	40 (STD)	0.145	3.68	0.1342	40.90	0.01414	1 313	1,117
			80 (XS)	0.200	5.08	0.1250	38.10	0.01227	1 140	1,200
			160	0.281	7.14	0.1115	33.98	0.009764	906.8	1,345
			(XXS)	0.400	10.16	0.09167	27.94	0.006600	613.1	1,636
2	2.375	60.34	40 (STD)	0.154	3.91	0.1723	52.52	0.02330	2 166	871.6
			80 (XS)	0.218	5.54	0.1616	49.26	0.02051	1 906	928.2
			160	0.344	8.74	0.1406	42.86	0.01552	1 443	1,067
			(XXS)	0.436	11.07	0.1253	38.20	0.01232	1 146	1,197
2½	2.875	73.03	40 (STD)	0.203	5.16	0.2058	62.71	0.03325	3 089	728.9
			80 (XS)	0.276	7.01	0.1936	59.01	0.02943	2 735	774.8
			160	0.375	9.53	0.1771	53.97	0.02463	2 288	847.0
			(XXS)	0.552	14.02	0.1476	44.99	0.01711	1 590	1,016
3	3.500	88.90	40 (STD)	0.216	5.49	0.2557	77.92	0.05134	4 769	586.6
			80 (XS)	0.300	7.62	0.2417	73.66	0.04587	4 261	620.6
			160	0.438	11.13	0.2187	66.64	0.03755	3 488	685.9
			(XXS)	0.600	15.24	0.1917	58.42	0.02885	2 680	782.5

3½	4.000	101.6	40 (STD)	0.226	5.74	0.2957	90.12	0.06866	6 379	507.3
			80 (XS)	0.318	8.08	0.2803	85.44	0.06172	5 733	535.1
4	4.500	114.3	40 (STD)	0.237	6.02	0.3355	102.3	0.08841	8 219	447.1
			80 (XS)	0.337	8.56	0.3198	97.18	0.07984	7 417	469.0
			120	0.438	11.13	0.3020	92.04	0.07163	6 654	496.7
			160	0.531	13.49	0.2865	87.32	0.06447	5 988	523.6
			(XXS)	0.674	17.12	0.2627	80.06	0.05419	5 034	571.0
5	5.563	141.3	40 (STD)	0.258	6.55	0.4206	128.2	0.1389	12 910	356.6
			80 (XS)	0.375	9.53	0.4011	122.2	0.1263	11 730	374.0
			120	0.500	12.70	0.3803	115.9	0.1136	10 550	394.4
			160	0.625	15.88	0.3594	109.5	0.1015	9 417	417.3
			(XXS)	0.750	19.05	0.3386	103.2	0.09004	8 365	443.0
6	6.625	168.3	40 (STD)	0.280	7.11	0.5054	154.1	0.2006	18 650	296.8
			80 (XS)	0.432	10.97	0.4801	146.4	0.1810	16 830	311.9
			120	0.562	14.27	0.4584	139.8	0.1650	15 350	327.1
			160	0.719	18.26	0.4823	131.8	0.1467	13 640	311.0
			(XXS)	0.864	21.95	0.4081	124.4	0.1308	12 150	367.6
8	8.625	219.1	20	0.250	6.35	0.6771	206.4	0.3601	33 460	221.5
			30	0.277	7.04	0.6726	205.0	0.3553	33 010	223.0
			40 (STD)	0.322	8.18	0.6651	202.7	0.3474	32 270	225.5
			60	0.406	10.31	0.6511	198.5	0.3329	30 950	230.7
			80 (XS)	0.500	12.70	0.6354	193.7	0.3171	29 470	236.1
			100	0.594	15.09	0.6198	188.9	0.3017	28 030	242.0
			120	0.719	18.26	0.5989	182.6	0.2817	26 190	250.5
			140	0.812	20.62	0.5834	177.9	0.2673	24 860	257.1
			(XXS)	0.875	22.23	0.5729	174.6	0.2578	23 940	261.8

Table A-11 *(continued)*

Pipe size, in.	Outside diameter		Schedule	Wall thickness		Internal diameter		Flow area		$10^6\epsilon/D$
	in.	mm		in.	mm	ft	mm	ft²	mm²	
8	8.625	219.1	160	0.906	23.01	0.5678	173.1	0.2532	23 530	264.2
10	10.750	273.1	20	0.250	6.35	0.8542	260.4	0.5730	33 260	175.6
			30	0.307	7.80	0.8447	257.5	0.5604	52 080	177.6
			40 (STD)	0.365	9.27	0.8350	254.6	0.5476	50 910	179.6
			60 (XS)	0.500	12.70	0.8125	247.7	0.5185	48 190	184.6
			80	0.594	15.09	0.7968	242.9	0.4987	46 340	188.3
			100	0.719	18.27	0.7760	236.6	0.4730	43 970	193.3
			120	0.844	21.44	0.7552	230.2	0.4470	41 620	198.6
			140 (XXS)	1.000	25.40	0.7292	222.3	0.4176	38 810	205.7
			160	1.125	28.58	0.7083	215.9	0.3941	36 610	211.8
12	12.750	323.9	20	0.250	6.35	1.021	311.2	0.8185	76 060	147.1
			30	0.330	8.38	1.008	307.1	0.7972	74 071	148.8
			(STD)	0.375	9.53	1.000	304.8	0.7854	72 970	150.0
			40	0.406	10.31	0.9948	303.3	0.7773	72 250	150.8
			(XS)	0.500	12.70	0.9792	298.5	0.7530	69 980	153.2
			60	0.562	14.27	0.9688	295.3	0.7372	68 490	154.8
			80	0.688	17.48	0.9478	288.9	0.7056	65 550	158.3
			100	0.844	21.44	0.9218	281.0	0.6674	62 020	162.7
			120 (XX)	1.000	25.40	0.8958	273.1	0.6303	58 580	167.5
			140	1.125	28.58	0.8750	266.7	0.6013	55 860	171.4
			160	1.312	33.32	0.8438	257.2	0.5592	51 960	177.8

14	14.000	355.6	30 (STD)	0.375	9.55	1.104	336.5	0.9575	88 930	135.9
			160	1.406	35.71	0.9323	284.2	0.6827	63 440	160.9
16	16.000	406.4	30 (STD)	0.375	9.53	1.271	387.3	1.268	117 800	118.3
			160	1.594	40.49	1.068	325.4	0.8953	83 160	140.4
18	18.000	457.2	(STD)	0.375	9.53	1.438	438.1	1.623	150 700	104.3
			160	1.781	45.24	1.203	366.7	1.137	105 600	124.7
20	20.000	508.0	20 (STD)	0.375	9.53	1.604	488.9	2.021	187 700	93.52
			160	1.969	50.01	1.339	408.0	1.407	130 700	112.0
22	22.000	558.8	20 (STD)	0.375	9.53	1.771	539.7	2.463	228 800	84.70
			160	2.125	53.98	1.479	450.8	1.718	159 600	101.4
24	24.000	609.6	20 (STD)	0.375	9.53	1.938	590.5	2.948	273 900	77.40
			160	2.344	59.54	1.609	490.5	2.034	189 000	93.22
26	26.000	660.4	(STD)	0.375	9.53	2.104	641.3	3.477	323 000	71.29
28	28.000	711.2	(STD)	0.375	9.53	2.271	692.1	4.050	376 200	66.05
30	30.000	762.0	(STD)	0.375	9.53	2.438	742.9	4.666	433 500	61.53
32	32.000	812.8	(STD)	0.375	9.53	2.604	793.4	5.326	494 400	57.60
34	34.000	863.6	(STD)	0.375	9.53	2.771	844.5	6.030	560 100	54.13
36	36.000	914.4	(STD)	0.375	9.53	2.938	895.3	6.777	629 500	51.06
38	38.000	965.2	—	0.375	9.53	3.104	946.1	7.568	703 000	48.32
40	40.000	1 016	—	0.375	9.53	3.271	996.9	8.403	780 500	45.86

NOTES

1. Outside diameter and wall thickness in inches extracted from *American National Standard Wrought Steel and Wrought Iron Pipe,* ANSI B36.10–1970, with permission of the publisher, American Society of Mechanical Engineers.
2. Metrication by author.
3. Data for sizes above 12 in. were selected for purposes of illustration only. ANSI B36.10 should be consulted for full range of wall thicknesses.
4. Letter schedules: (STD) Standard, (XS) Extra Strong, (XXS) Double Extra Strong.
5. Values of relative roughness, ϵ/D, computed using an absolute roughness $\epsilon = 150 \times 10^{-6}$ ft (46 μm).

Table A-12 *Properties of 250-psi Cast Iron Pipe*

Pipe size, in.	Outside diameter		Thickness class	Wall thickness		Internal diameter		Flow area		$10^6\epsilon/D$
	in.	mm		in.	mm	ft	mm	ft^2	mm^2	
3	3.96	100.6	22	0.32	8.1	0.2767	84.8	0.06012	5 595	3,072
4	4.80	121.9	22	0.35	8.9	0.3417	104.1	0.09168	8 511	2,488
6	6.90	175.3	22	0.38	9.7	0.5117	155.9	0.2056	19 090	1,661
8	9.05	229.9	22	0.41	10.4	0.6858	209.11	0.3694	34 340	1,239
10	11.10	281.9	22	0.44	11.2	0.8517	259.5	0.5696	52 890	998.0
12	13.20	335.3	23	0.52	13.2	1.013	308.9	0.8065	74 940	838.8
14	15.30	388.6	24	0.59	15.0	1.177	358.6	1.087	101 000	722.4
16	17.40	442.0	24	0.63	16.0	1.345	410.0	1.421	132 000	632.0
18	19.50	495.3	24	0.68	17.3	1.512	460.7	1.795	166 700	562.3
20	21.60	548.6	24	0.72	18.3	1.680	512.0	2.217	205 900	506.0
24	25.80	655.3	24	0.79	20.1	2.018	615.1	3.199	297 200	421.1
30	32.00	812.8	25	0.99	25.1	2.501	762.6	4.915	456 800	399.8
36	38.30	972.9	25	1.10	27.9	3.008	917.1	7.108	660 600	282.5
42	44.50	1 130	25	1.22	31.0	3.505	1 068	9.649	895 800	242.5
48	50.80	1 290	25	1.33	33.8	4.012	1 222	12.64	1 174 000	211.9

NOTES

1. Outside diameter and wall thickness extracted from *American National Standard Cast Iron Pipe Centrifugally Cast in Metal Molds for Water or Other Liquids*, ANSI A21.6–1970
2. Metrication by author.
3. Values of relative roughness, ϵ/D, were computed using $\epsilon = 850 \times 10^{-6}$ ft (260 μm).

Table A-13 *Properties of Seamless Copper Water Tube*

Standard size, in.	Outside diameter		Type	Wall thickness		Internal diameter		Flow area		$10^6\epsilon/D$
	in.	mm		in.	mm	ft	mm	ft^2	mm^2	
1/4	0.375	9.53	K	0.035	0.89	0.02542	7.75	0.0005074	47.17	196.7
			L	0.030	0.76	0.02625	8.01	0.0005412	50.39	190.5
3/8	0.500	12.70	K	0.049	1.25	0.03350	10.22	0.0008814	82.03	149.3
			L	0.035	0.89	0.03583	10.92	0.001008	93.66	141.6
			M	0.025	0.64	0.03750	11.42	0.001104	102.4	133.3
1/2	0.625	15.88	K	0.049	1.24	0.04392	13.40	0.001515	141.0	113.8
			L	0.040	1.02	0.04542	13.84	0.001620	150.5	110.1
			M	0.028	0.71	0.04742	14.46	0.001766	164.2	103.4
5/8	0.750	19.05	K	0.049	1.24	0.05433	16.57	0.002319	215.6	92.02
			L	0.042	1.07	0.05550	16.91	0.002419	224.6	90.09
3/4	0.875	22.22	K	0.065	1.65	0.06208	18.92	0.003027	281.1	80.54
			L	0.045	1.14	0.06542	19.94	0.003361	312.3	76.43
			M	0.032	0.81	0.06758	20.60	0.003587	333.3	73.98
1	1.125	28.58	K	0.065	1.65	0.08292	25.28	0.005400	501.9	60.30
			L	0.050	1.27	0.08542	26.04	0.005730	532.6	58.54
			M	0.035	0.89	0.08792	26.80	0.006071	564.1	56.87
1 1/4	1.375	34.93	K	0.065	1.65	0.1038	31.63	0.008454	785.8	48.19
			L	0.055	1.40	0.1054	32.13	0.008728	810.8	47.93
			M	0.042	1.07	0.1076	32.79	0.009090	844.4	46.48

Table A-13 *(continued)*

Standard size, in.	Outside diameter		Type	Wall thickness		Internal diameter		Flow area		$10^6\epsilon/D$
	in.	mm		in.	mm	ft	mm	ft^2	mm^2	
$1\frac{1}{2}$	1.625	41.28	K	0.072	1.83	0.1234	37.62	0.01196	1 112	40.51
			L	0.060	1.52	0.1254	38.24	0.01235	1 148	39.87
			M	0.049	1.24	0.1273	38.80	0.01272	1 182	39.29
2	2.125	53.98	K	0.083	2.11	0.1633	49.76	0.02093	1 195	30.62
			L	0.070	1.78	0.1654	50.42	0.02149	1 997	30.22
			M	0.058	1.48	0.1674	51.02	0.02701	2 044	29.87
$2\frac{1}{2}$	2.625	66.68	K	0.095	2.41	0.2029	61.86	0.03234	3 005	24.64
			L	0.080	2.03	0.2054	62.62	0.03314	3 080	24.34
			M	0.065	1.65	0.2079	63.38	0.03395	4 017	24.05
3	3.125	79.38	K	0.109	2.77	0.2423	73.84	0.04609	4 282	20.64
			L	0.090	2.29	0.2454	74.80	0.04730	4 394	20.37
			M	0.072	1.83	0.2484	75.72	0.04847	4 503	20.13
$3\frac{1}{2}$	3.625	92.08	K	0.120	3.05	0.2821	85.98	0.06249	5 806	17.72
			L	0.100	2.54	0.2854	87.00	0.06398	5 945	17.52
			M	0.083	2.11	0.2883	87.86	0.06523	6 063	17.35
4	4.125	104.8	K	0.134	3.40	0.3214	98.00	0.08114	7 543	15.56
			L	0.110	2.79	0.3254	99.22	0.08317	7 732	15.36
			M	0.095	2.41	0.3279	99.98	0.08445	7 851	15.25

Table A-13 *(continued)*

Standard size, in.	Outside diameter		Type	Wall thickness		Internal diameter		Flow area		$10^6\epsilon/D$
	in.	mm		in.	mm	ft	mm	ft^2	mm^2	
5	5.125	130.2	K	0.160	4.06	0.4004	122.1	0.1259	11 710	12.49
			L	0.125	3.18	0.4063	123.8	0.1296	12 050	12.31
			M	0.109	2.77	0.4089	124.7	0.1313	12 210	12.23
6	6.125	155.6	K	0.192	4.88	0.4784	145.8	0.1798	16 700	10.45
			L	0.140	3.56	0.4871	148.5	0.1863	17 320	10.27
			M	0.122	3.10	0.4901	143.9	0.1886	17 530	10.20
8	8.125	206.4	K	0.271	6.88	0.6319	192.6	0.3136	29 150	7.912
			L	0.200	5.08	0.6438	196.2	0.3255	30 250	7.767
			M	0.170	4.32	0.6488	197.8	0.3306	30 720	7.707
10	10.125	257.2	K	0.388	8.59	0.7874	240.0	0.4870	45 250	6.350
			L	0.250	6.35	0.8021	244.5	0.5053	46 950	6.234
			M	0.212	5.38	0.8084	246.4	0.5133	47 680	6.185
12	12.125	308.0	K	0.405	10.29	0.9429	287.4	0.6983	64 880	5.303
			L	0.280	7.11	0.9638	293.8	0.7295	67 790	5.188
			M	0.254	6.45	0.9681	295.1	0.7361	68 400	5.165

NOTES

1. Outside diameter and wall thickness in inches extracted from *American National Standard Seamless Copper Water Tube,* ANSI H 23.1–1970.
2. Metrication by author.
3. Types: K, for underground service and general plumbing; L, for interior plumbing; M, for use with soldered fittings only.
4. Values of relative roughness, ϵ/D, were computed using $\epsilon = 5 \times 10^{-6}$ ft (1.5 μm).

Table A-14 *Values of Flow Areas* A *and Hydraulic Radius* R_h *for Various Cross Sections*

CROSS SECTIONS	CONDITION		EQUATIONS
	Flowing Full	$h/D = 1$	$A = \pi D^2/4$ $R_h = D/4$
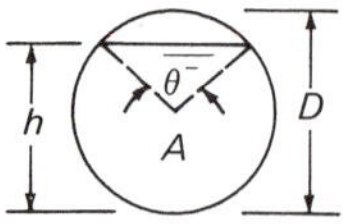	Upper Half Partly Full	$0.5 < h/D < 1$	$\cos^{-1}(\theta/2) = (2h/D - 1)$ $A = [\pi(360 - \theta) + 180 \sin\theta](D^2/1440)$ $R_h = [1 + (180 \sin\theta)/(\pi\theta)](D/4)$
		$h/D = 0\ 8128$	$A = 0\ 6839\ D^2$ R_h max $= 0\ 30430$
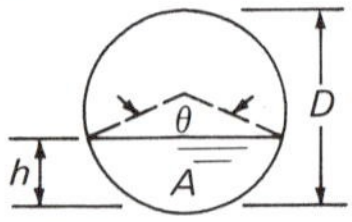	Lower Half Partly Full	$h/D = 0.5$	$A = \pi D^2/8$ R_h max $= h/2$
		$0 < h/D < 0.5$	$\cos^{-1}(\theta/2) = (1 - 2h/D)$ $A = [\pi\theta - 180 \sin\theta](D^2/1440)$ $R_h = [1 - (180 \sin\theta)/(\pi\theta)](D/4)$
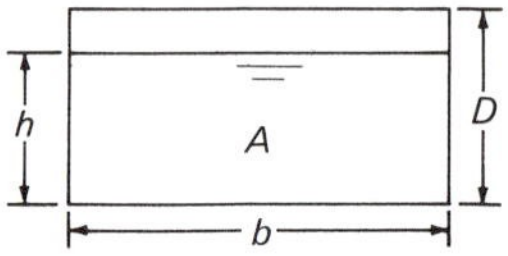	Flowing Full	$h/D = 1$	$A = bD$ $R_h = bD/2(b + D)$
		Square $b = D$	$A = D^2$ $R_h = D/4$
	Partly Full	$h/D < 1$ $h/b = 0.5$ $b \to \infty,\ h \to 0$	$A = bh$ $R_h = bh/(2h + b)$ $A = b^2/2$ R_h max $= h/2$ $R_h \to h$ (wide shallow stream)
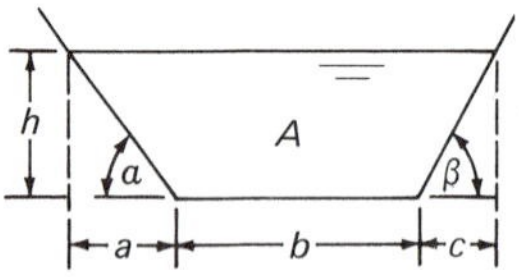	$\alpha \neq \beta$		R_h max $= h/2$ $A = [b + \frac{1}{2}h\ (\cot\alpha + \cot\beta)]h$ $R_h = A/[b + h(\operatorname{cosec}\alpha + \operatorname{cosec}\beta)]$
	$\alpha = \beta$	$\frac{h}{a} = \frac{1}{2}$ $\alpha = 26°\ 34'$	$A = (b + 2h)h$ $R_h = (b + 2h)h/(b + 4.472h)$
		$\frac{h}{a} = \frac{1}{\sqrt{3}}$ $\alpha = 30°$	$A = (b + 1.732h)h$ $R_h = (b + 1.732h)h/(b + 4h)$
		$\frac{h}{a} = \frac{2}{3}$ $\alpha = 33°\ 41'$	$A = (b + 1.5h)h$ $R_h = (b + 1.5h)h/(b + 3.606h)$
		$\frac{h}{a} = 1$ $\alpha = 45°$	$A = (b + h)h$ $R_h = (b + h)h/(b + 2.828h)$
		$\frac{h}{a} = \frac{3}{2}$ $\alpha = 56°\ 19'$	$A = (b + 0.6667h)h$ $R_h = (b + 0.6667h)h/(b + 2.404/h)$
		$\frac{h}{a} = \sqrt{3}$ $\alpha = 60°$	$A = (b + 0.5774h)h$ $R_h = (b + 0.5774h)h/(b + 2.309h)$
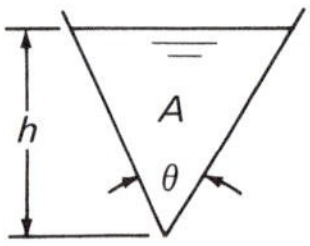	θ = any angle		$A = \tan(\theta/2)h^2$ $R_h = \sin(\theta/2)h/2$
	$\theta = 30$		$A = 0.2679h^2$ $R_h = 0.1294h$
	$\theta = 45$		$A = 0.4142h^2$ $R_h = 0.1913h$
	$\theta = 60$		$A = 0.5774h^2$ $R_h = 0.2500h$
	$\theta = 90$		$A = h^2$ $R_h = 0.3536h$

Table A-15 *Isentropic Expansion Factor* Y

β	k	Critical Values		Expansion Factor Y			
		r_c	Y_c	$r = 0.60$	$r = 0.70$	$r = 0.80$	$r = 0.90$
0	1.10	0.5846	0.6894	0.7021	0.7820	0.8579	0.9304
	1.20	0.5644	0.6948	0.7228	0.7981	0.8689	0.9360
	1.30	0.5457	0.7000	0.7409	0.8119	0.8783	0.9408
	1.40	0.5282	0.7049	0.7568	0.8240	0.8864	0.9449
0.20	1.10	0.5848	0.6892	0.7017	0.7817	0.8577	0.9303
	1.20	0.5546	0.6946	0.7225	0.7978	0.8687	0.9359
	1.30	0.5459	0.6998	0.7406	0.8117	0.8781	0.9407
	1.40	0.5284	0.7047	0.7576	0.8237	0.8862	0.9448
0.50	1.10	0.5921	0.6817	0.6883	0.7699	0.8485	0.9250
	1.20	0.5721	0.6872	0.7094	0.7864	0.8600	0.9310
	1.30	0.5535	0.6923	0.7278	0.8007	0.8699	0.9361
	1.40	0.5362	0.6973	0.7440	0.8133	0.8785	0.9405
0.60	1.10	0.6006	0.6729	—	0.7556	0.8374	0.9186
	1.20	0.5808	0.6784	0.6939	0.7727	0.8495	0.9250
	1.30	0.5625	0.6836	0.7126	0.7875	0.8599	0.9305
	1.40	0.5454	0.6885	0.7292	0.8006	0.8689	0.9352
0.70	1.10	0.6160	0.6570	—	0.7290	0.8160	0.9058
	1.20	0.5967	0.6624	0.6651	0.7469	0.8292	0.9131
	1.30	0.5788	0.6676	0.6844	0.7626	0.8405	0.9193
	1.40	0.5621	0.6726	0.7015	0.7765	0.8505	0.9247
0.80	1.10	0.6441	0.6277	—	0.6778	0.7731	0.8788
	1.20	0.6238	0.6331	—	0.6970	0.7881	0.8877
	1.30	0.6087	0.6383	—	0.7140	0.8012	0.8954
	1.40	0.5926	0.6433	0.6491	0.7292	0.8182	0.9021

Table A-16 *Dimensions and Units of Common Variables*

Type	Symbol	Variable	Dimensions		Units	
			MLT	FLT	US	SI
Geometric	L	Length	L		ft	m
	A	Area	L^2		ft^2	m^2
	Vol	Volume	L^3		ft^3	m^3
Kinematic	t	Time	T		sec	s
	ω	Angular velocity	T^{-1}		sec^{-1}	s^{-1}
	f	Frequency				
	V	Velocity	LT^{-1}		ft sec^{-1}	$m \cdot s^{-1}$
	ν	Kinematic viscosity	L^2T^{-1}		ft^2 sec^{-1}	$m^2 \cdot s^{-1}$
	Q	Volume flow rate	L^3T^{-1}		ft^3 sec^{-1}	$m^3 \cdot s^{-1}$
	α	Angular acceleration	T^{-2}		sec^{-2}	s^{-2}
	a	Acceleration	LT^{-2}		ft sec^{-2}	$m \cdot s^{-2}$
Dynamic	ρ	Density	ML^{-3}	$FL^{-4}T^2$	slug ft^{-3}	$kg \cdot m^{-3}$
	M	Mass	M	$FL^{-1}T^2$	slugs	kg
	I	Moment of inertia	ML^2	FLT^2	slug ft^2	$kg \cdot m^2$
	μ	Dynamic viscosity	$ML^{-1}T^{-1}$	$FL^{-2}T$	slug ft^{-1} sec^{-1}	$kg \cdot m^{-1} \cdot s^{-1}$
	$\dot{M}$	Mass flow rate	MT^{-1}	$FL^{-1}T^{-1}$	slug sec^{-1}	$kg \cdot s^{-1}$

Table A-16 *(continued)*

Type	Symbol	Variable	Dimensions		Units	
			MLT	FLT	US	SI
Dynamic	MV	Momentum	MLT^{-1}	FT	lbf sec	$N \cdot s$
	Ft	Impulse				
	$I\omega$	Angular momentum	ML^2T^{-1}	FLT	slug ft^2 sec^{-1}	$kg \cdot m^2 \cdot \text{sec}^{-1}$
	γ	Specific weight	$ML^{-2}T^{-2}$	FL^{-3}	lbf ft^{-3}	$N \cdot m^{-3}$
	p	Pressure	$ML^{-1}T^{-2}$	FL^{-2}	lbf ft^{-2}	$N \cdot m^{-2}$
	τ	Unit shear stress				
	E	Modulus of elasticity				
	σ	Surface tension	MT^{-2}	FL^{-1}	lbf ft^{-1}	$N \cdot m^{-1}$
	F	Force	MLT^{-2}	F	lbf	N
	E	Energy	ML^2T^{-2}	FL	lbf ft	J
	W	Work				
	FL	Torque				
	P	Power	ML^2T^{-3}	FLT^{-1}	lbf ft sec^{-1}	W
	v	Specific volume	$M^{-1}L^3$	$F^{-1}L^4T^{-2}$	ft^3 lbm^{-1}	$kg^{-1} \cdot m^3$

REFERENCES

a. *Handbook of Chemistry and Physics,* 52d ed. Edited by Robert C. Weast. The Chemical Rubber Company, Cleveland, Ohio, 1971–72.

b. *Smithsonian Physical Tables,* 9th rev. ed. Prepared by William E. Forsythe. Smithsonian Institution, Washington, D.C., 1954.

c. *Handbook of Chemistry,* 10th rev. ed. Edited by Norbert A. Lange. McGraw-Hill Book Company, New York, 1967.

d. Meyer, C. A., McClintock, R. B., Silvestri, G. J., and R. C. Spencer, Jr. *Thermodynamic and Transport Properties of Steam,* American Society of Mechanical Engineers, New York, 1967.

e. *American Institute of Physics Handbook,* 3d ed. McGraw-Hill Book Company, New York, 1972.

f. *International Critical Tables of Numerical Data, Physics, Chemistry and Technology.* Published for the National Academy of Sciences–National Research Council by McGraw-Hill Book Company. Vols. 1 through 7, 1926–1930.

g. *Selected Values of Physical and Thermodynamic Properties of Hydrocarbons and Related Compounds.* American Petroleum Institute Project 44, Carnegie Press, Pittsburgh, Pa., 1943.

h. Sanders, H. E. *Hydrodynamics of Ship Design,* Society of Naval Architects and Marine Engineers, Washington, D.C., 1964.

i. Tyer, D. *Journal of the Chemical Society,* vol. 105, pp. 2534–2553, 1914.

j. Coppens, A. B., et al. *Journal of the Acoustical Society of America,* vol. 38, pp. 797–804, 1965.

k. Timme, R. *Journal of the Acoustical Society of America,* vol. 52, pp. 989–992, 1972.

l. Bridgman, P. W. *Proceedings of the American Academy of Arts and Sciences,* vol. 67, pp. 185–223, 1931.

m. Freyer, E. B., et al., *Journal of the American Chemical Society,* vol. 51, pp. 759–779, 1929.

n. Bhagavantam, S. and Joga Roa. *Proceedings of the Indian Academy of Arts and Sciences,* vol. 9a, pp. 312–315, 1939.

o. Hunter, J. L. *Journal of the Acoustical Society of America,* vol. 13, pp. 36–40, 1941.

p. Sverdrup, H. U., Johnson, M. W., and R. W. Fleming. *The Oceans.* Prentice-Hall, Englewood Cliffs, N.J., 1942.

q. Segur, J. B., and H. E. Oberstar. *Industrial and Engineering Chemistry,* vol. 43, pp. 2117–2120, 1951.

r. *American Institute of Physics Handbook,* 2d ed. McGraw-Hill Book Company, New York, 1963.

s. Reid, R. C., and T. K. Sherwood, *The Properties of Gases and Liquids,* 2d ed. McGraw-Hill Book Company, New York, 1966.

t. Wilson, W. and D. Bradley. *Journal of the Acoustical Society of America,* vol. 36, pp. 333–338, 1964.
u. Eduljee, H. E., et al. *Journal of the Chemical Society,* pp. 3086–3091, 1951.
v. *U.S. Standard Atmosphere Supplements, 1966.* Prepared under sponsorship of ESSA, NASA, USAF, U.S. Government Printing Office, Washington, D.C., 1967.

Appendix B

Dimensions, Unit Systems, and Conversion Factors

B-1 INTRODUCTION

The ancient Sumerians (6500–3000 B.C.) devised a numerical system in order to keep accounts for their temple communities. Their system was partly decimal and partly sexagesimal, with 10 and 6 used in an alternating fashion. This survives today in our division of the circle into 360 parts, and we still use the sexagesimal basis for angular measurement. The Sumerians were the first to use the notion of 12 subdivisions by dividing the day into 30 smaller units to give a total of 360 divisions for one 24-hour cycle.

Ancient linear units were derived from proportions of the human body. The most important of these were the cubit (length of the forearm), the digit (width of the finger), the foot, and the fathom (the distance between a man's outstretched arms). Sixteen Roman digits made a Roman foot, but the foot was also subdivided into 12 parts called *unciae*, which later became inches. The Romans retained the cubit but rated it at 24 digits. In the course of time, this became the English yard, which is really a double cubit. For longer distances, the Romans used a unit of 5000 ft, which they called *mille passus* (1000 paces), or a mile.

The oldest weighing apparatus known is a prehistoric Egyptian balance with limestone weights dating back to 5000 B.C. For several thousand years, weighing seems to have been restricted to gold and silver and other items of great value, while for ordinary commercial purposes, goods were either counted or measured by volume. We still buy oranges by the dozen instead of the pound. The first coins were nothing more than pieces of precious metal stamped with a mark of some kind to indicate their weight and fineness. The pound is still both a monetary unit and a unit of mass measure. A treatise written at the beginning of the fourteenth century on English weights and measures begins:[a]

> By consent of the whole realm the King's measure was made so that an English penny which is called Sterling, round without clipping, shall weigh thirty-two grains of dry wheat from the middle of the ear; twenty pence make an ounce and twelve ounces make a pound and eight pounds make a gallon of wine and eight gallons of wine make a bushel of London.

As technology grew in the nineteenth century, there was a great need for international standardization. In 1872 an international meeting was held in France and was attended by representatives of 26 countries, including the United States. Out of this meeting came the international treaty, the Metric Convention, which was signed by 17 countries, including the United States, in 1875. The treaty: (a) set up metric standards for length and mass; (b) established the International Bureau of Weights and Measures (abbreviated from the French as BIPM: PM for the French "Poids et Mesures," meaning weights and measures); (c) established the General Conference of Weights and Measures (CGPM) which meets every six years; and (d) set up an International Committee of Weights and Measures (CIPM), which meets every two years and which implements the recommendations of the General Conference and directs the activities of the International Bureau.

B-2 DIMENSIONS

Dimensions represent physical quantities, and units describe their magnitudes. The inch, foot, cubit, yard, fathom, rod, chain, mile, and metre all describe different magnitudes of the physical quantity whose dimension is length. In the study of fluid mechanics, interest centers on the following dimensions:

Dimension	Symbol
Length	L
Time	T
Mass	M
Force	F

Dimensions for other physical quantities may be established by application of the above dimensions to the definition of the physical quantity, as shown by the following examples.

Physical quantity	Definition	Derivation
Velocity	Length/Time	$L/T = LT^{-1}$
Acceleration	Velocity/Time	$LT^{-1}/T = LT^{-2}$
Force	Mass · Acceleration	$M \cdot LT^{-2} = MLT^{-2} = F$
Mass	Force/Acceleration	$F/LT^{-2} = FL^{-1}T^{2} = M$

From the above table it is evident that force and mass are related by Newton's second law of motion, so that, in any consistent dimensional system, if one is chosen as a fundamental dimension, the other is a derived dimension. Two dimensional systems are used in fluid mechanics, the *force* system, FLT, and the *mass* system, MLT. Again, one may derive dimensions for physical quantities by applying them to the definition of the physical quantity, as shown in the following examples.

Physical quantity	Definition	Force system	Mass system
Pressure	Force/Area	$F/L^2 = FL^{-2}$	$MLT^{-2}/L^2 = ML^{-1}T^{-2}$
Work	Force · Length	$F \cdot L = FL$	$MLT^{-2} \cdot L = ML^2T^{-2}$
Power	Work/Time	$FL/T = FLT^{-1}$	$ML^2T^{-2}/T = ML^2T^{-3}$
Density	Mass/Volume	$FL^{-1}T^2/L^3 = FL^{-4}T^2$	$M/L^3 = ML^{-3}$
Mass Flow	Mass/Time	$FL^{-1}T^2/T = FL^{-1}T$	$M/T = MT^{-1}$

B-3 SI UNITS

The 1960 Eleventh General Conference on Weights and Measures defined an international system of units, the Système Internationale d'Unites (designated as SI in all languages). This system, with six base units, was adopted by the official representatives of the 36 participating nations, including the United States. The seventh base unit, the mole, was adopted by the fourteenth CIPM in 1972. Since 1964, it has been the policy of the U.S. National Bureau of Standards to use these SI units in its publications, except where communications might be impaired. At present, the American National Standards Institute, the American Society of Mechanical Engineers, the American Society for Testing and Materials, and most other American professional engineering societies are requiring that SI units be included in their codes and standards along with the U.S. customary units as new documents are being prepared or old ones revised.

The SI system includes three classes of units: base units, supplementary units, and derived units. The seven base units are as follows:

Physical quantity	Name of unit	Symbol
Length	metre	m
Mass	kilogram	kg
Time	second	s
Electric current	ampere	A
Temperature	kelvin	K
Luminous intensity	candela	cd
Amount of substance	mole	mol

The base units are defined as follows:

(a) Unit of length is the metre, which is the length equal to 1 650 763.73 wavelengths in vacuum of the radiation corresponding to the transition between levels $2p_{10}$ and $5d_5$ of the krypton-86 atom. *Note*: In conformance with SI practice, one writes the number of wavelengths in groups of three digits *without* commas.
(b) Unit of mass is the kilogram, which is equal to the mass of the international prototype of the kilogram, located at the BIPM headquarters.
(c) Unit of time is the second, which is the duration of 9 192 631 770 periods of the radiation corresponding to the transition between the two hyperfine levels of the ground state of the cesium-133 atom.
(d) Unit of electric current is the ampere, which is that constant current which, if maintained in two straight parallel conductors of infinite length, of negligible circular cross section, and placed 1 metre apart in vacuum, would produce between these conductors a force equal to 2×10^{-7} newtons per metre of length (newton is a derived unit).
(e) Unit of thermodynamic temperature is the kelvin, which is the fraction 1/273.16 of the thermodynamic temperature of the triple point of water.
(f) Unit of luminous intensity is the candela, which is the luminous intensity, in the perpendicular direction, of a surface of 1/600 000 square metre of a blackbody at the temperature of freezing platinum under a pressure 101 325 newtons per square metre.
(g) Unit of substance is the mole, which is the amount of substance of a system which contains as many elementary entities as there are atoms in 0.012 kilogram of carbon-12.

The two supplementary units are

Physical quantity	Name of unit	Symbol
Plane angle	radian	rad
Solid angle	steradian	sr

There are 15 derived units with special names. Those of interest in the field of fluid mechanics are

Frequency	hertz	$1\ \text{Hz} = 1\ \text{s}^{-1}$
Force	newton	$1\ \text{N} = 1\ \text{kg} \cdot \text{m/s}^2$
Pressure and stress	pascal	$1\ \text{Pa} = 1\ \text{N/m}^2$
Work, energy, quantity of heat	joule	$1\ \text{J} = 1\ \text{N} \cdot \text{m}$
Power	watt	$1\ \text{W} = 1\ \text{J/s}$

Unnamed SI units of special interest in fluid mechanics have been derived.

Physical quantity	Unit combination	Formula
Area	square metre	m^2
Volume	cubic metre	m^3
Angular velocity	radians per second	rad/s
Velocity	metres per second	m/s
Acceleration	metres/second squared	m/s^2
Rotational frequency	revolutions per second	s^{-1}
Momentum	kilogram-metre per second	$kg \cdot m/s$
Density	kilogram per cubic metre	kg/m^2
Dynamic viscosity	pascal-second	$Pa \cdot s$
Kinematic viscosity	square metres per second	m^2/s
Surface tension	newtons/metre	N/m
Specific heat capacity	joules per kilogram-kelvin	$J/(kg \cdot K)$
Specific energy	joules per kilogram	J/kg

The SI system requires no conversion factors, since all physical quantities are described in terms of the base units. Decimal multiples and sub-multiples of SI units are formed by means of the prefixes given below.

Factor by Which Unit is Multiplied	Prefix	Symbol	Means
$1\ 000\ 000\ 000\ 000 = 10^{12}$	tera	T	One trillion times
$1\ 000\ 000\ 000 = 10^{9}$	giga	G	One billion times
$1\ 000\ 000 = 10^{6}$	mega	M	One million times
$1\ 000 = 10^{3}$	kilo	k	One thousand times
$100 = 10^{2}$	hecto	h	One hundred times
$10 = 10^{1}$	deca	da	Ten times
$0.1 = 10^{-1}$	deci	d	One tenth of
$0.01 = 10^{-2}$	centi	c	One hundredth of
$0.000 = 10^{-3}$	milli	m	One thousandth of
$0.000\ 000 = 10^{-6}$	micro	μ(mu)	One millionth of
$0.000\ 000\ 000 = 10^{-9}$	nano	n	One billionth of
$0.000\ 000\ 000\ 000 = 10^{-12}$	pico	p	One trillionth of
$0.000\ 000\ 000\ 000\ 000 = 10^{-15}$	femto	f	One thousand trillionth of
$0.000\ 000\ 000\ 000\ 000\ 000 = 10^{-18}$	atto	a	One million trillionth of

Prefixes are to be used only with base units, except in the case of the SI mass unit, which contains the prefix symbol k. Multiples and submultiples of mass are formed by adding the prefixes to the word gram; for example, milligram (mg) instead of microkilogram (μkg).

The symbol of a prefix is considered to be combined with the unit symbol to which it is directly attached, forming with it a symbol for a new unit which can be provided with a positive or negative exponent and which can be combined with other unit symbols to form symbols for compound units. Compound prefixes should not be used; for example, write nm (namometre) instead of mμm.

EXAMPLES

$$1\ \text{cm}^3 = (10^{-2}\text{m})^3 = 10^{-6}\text{m}^3$$
$$1\ \mu\text{s}^{-1} = (10^{-6}\text{s})^{-1} = 10^{6}\text{s}^{-1}$$
$$1\ \text{mm}^2/\text{s} = (10^{-3}\text{m})^2/\text{s} = 10^{-6}\text{m}^2/\text{s}$$

Some units not in the SI system have such widespread use and play such an important role that they must be retained for general use. Those of interest to the field of fluid mechanics are:

Name	Symbol	Value in SI units
minute	min	1 min = 60 s
hour	h	1 h = 60 min = 3 600 s
day	d	1 d = 24 h = 86 400 s
degree	°	$1° = (\pi/180)$ rad
minute	′	$1' = (1/60)° = (\pi/10\,800)$ rad
second	″	$1'' = (1/60)' = (\pi/648\,000)$ rad
litre	l	$1\ \text{l} = 1\ \text{dm}^3 = 10^{-3}\ \text{m}^3$
metric ton	t	$1\ \text{t} = 10^3$ kg

Other units of interest that are to be temporarily accepted for international use are:

nautical mile		1 nautical mile = 1 852 m
knot		1 nautical mile per hour = (1 852/3 600) m/s
bar	bar	$1\ \text{bar} = 0.1\ \text{MPa} = 10^5$ Pa
standard atmosphere	atm	1 atm = 101 325 Pa

CGS units of interest with special names *that are not to be used internationally* are:

erg	erg	$1\ \text{erg} = 10^{-7}$ J
dyne	dyn	$1\ \text{dyn} = 10^{-5}$ N

poise	P	$1\ P = 1\ dyn \cdot s/cm^2 = 0.1\ Pa \cdot s$
stoke	St	$1\ St = 1\ cm^2/s = 10^{-4} m^2/s$

Other units of interest *that are not to be used internationally* are:

torr	torr	1 torr = (101 325)/(760) Pa
kilogram force	kgf	1 kgf = 9.806 65 N
calorie	cal	1 cal = 4.186 8 J

B-4 U.S. CUSTOMARY UNITS

In the United States the units of weights and lengths commonly employed are identical for practical purposes with the corresponding English units, but the capacity units differ from those now in use in the British Commonwealth, the U.S. gallon being defined as 231 cubic inches, and the bushel as 2,150.42* cubic inches, whereas the corresponding British Imperial units are, respectively, 277.42 cubic inches and 2,219.36 cubic inches.

RELATION TO SI UNITS

Length

By agreement in 1959 among the national standards laboratories of the English-speaking nations one yard was fixed as 0.9144 metres, whence one foot equals 0.9144/3, or 0.3048 metres, and one inch equals $0.3048 \times 100/12$, or 2.54 centimetres.

Time

The second used in the United States is identical to the SI second.

Mass

The same 1959 agreement that fixed the value of length of the metre also fixed the value of the pound mass (lbm) as 453.592 37 grams. This same value was also adopted by the Sixth International Steam Table Conference in 1967 and is the exact conversion for SI units.

---

* The standard U.S. bushel is the Winchester bushel, which is in cylindrical form, $18\frac{1}{2}$ in. in diameter and 8 in. deep. The exact capacity is $(\pi/4)(18.5)^2(8) = 684.5\ \pi$ cubic inches (2150.420 172 in^3).

Table B-1 *Conversion Factors*

To convert from		To	Multiply by		Equation no.
			For textbook calculations	*For precise conversion*	
acre		ft^2	*43,560		(B-1)
		m^2	4 047	$4.046\ 856\ 422 \times 10^3$	(B-2)
atmosphere (standard)		lbf/ft^2	2,116	2,116.216 624	(B-3)
		psia	14.70	14.695 948 78	(B-4)
		N/m^2	1.013×10^5	$*1.013\ 25 \times 10^5$	(B-5)
bar		lbf/ft^2	2,088	2,088.543 423	(B-6)
		psia	14.50	14.503 773 77	(B-7)
		N/m^2	$*1 \times 10^5$		(B-8)
barrel (petroleum, 42 gal)		ft^3	5.615	$*(231 \times 42/1728)$	(B-9)
		m^3	0.1590	$1.589\ 872\ 949 \times 10^{-1}$	(B-10)
British thermal unit (International Table)	Btu (energy)	ft-lbf	778.2	778.169 261 1	(B-11)
		J	1 055	$1.055\ 055\ 853 \times 10^3$	(B-12)
	Btu/lbm (specific energy)	ft-lbf/lbm	778.2	778.169 262 1	(B-13)
		J/kg		*2 326	(B-14)
	Btu/lbm-°R (specific entropy, specific heat and gas constant)	ft-lbf/lbm-°R	778.2	778.169 262 1	(B-15)
		$J/kg \cdot K$	4 187	$*4.186\ 8 \times 10^3$	(B-16)
calorie (international table)		ft-lbf	3.088	3.088 025 206	(B-17)
		J	4.187	*4.186 8	(B-18)
centipoise (dynamic viscosity)		$lbf\text{-}sec/ft^2$	2.089×10^{-5}	$2.088\ 543\ 423 \times 10^{-5}$	(B-19)
		$N \cdot s/m^2$	$*1 \times 10^{-3}$		(B-20)

cup	ft^3	8.355×10^{-3}	$*(231/1728 \times 16)$	(B-21)
	m^3	2.366×10^{-4}	$2.365\ 882\ 366 \times 10^{-4}$	(B-22)
day (mean solar)	s	*86,400		(B-23)
degree (angle)	rad	1.745×10^{-2}	$*(\pi/180)$	(B-24)
degree, Celsius	°F	$*t_F = 1.8t_C + 32$		(B-25)
	°R	$*T_R = 1.8t_C + 491.67$		(B-26)
	K	$*T_K = t_C + 273.15$		(B-27)
degree, Fahrenheit	°R	$*T_R = t_F + 459.67$		(B-28)
	°C	$*t_C = (t_F - 32)/1.8$		(B-29)
	K	$*T_K = (t_F + 459.67)/1.8$		(B-30)
degree, Rankine	°C	$*t_C = (T_R - 491.67)/1.8$		(B-31)
	K	$*T_K = T_R/1.8$		(B-32)
dyne	lbf	2.248×10^{-6}	$2.248\ 089\ 431 \times 10^{-6}$	(B-33)
	N	$*1 \times 10^{-5}$		(B-34)
erg	ft-lbf	7.376×10^{-8}	$7.375\ 621\ 492 \times 10^{-8}$	(B-35)
	J	$*1 \times 10^{-7}$		(B-36)
fathom	ft		*6	(B-37)
	m	1.829	*1.828 8	(B-38)
fluid ounce (U.S.)	ft^3	1.044×10^{-3}	$*(231/1728 \times 128)$	(B-39)
	m^3	2.957	$2.957\ 352\ 956 \times 10^{-6}$	(B-40)
foot	m	*0.3048		(B-41)
$\frac{\text{foot-pound force}}{\text{pound mass-degree Rankine}}$	J/kg·K	5.380	5.380 320 456	(B-42)
gallon (U.S. liquid)	ft^3	1.337×10^{-1}	$*(231/1728)$	(B-43)
	m^3	3.785×10^{-3}	$3.785\ 411\ 787 \times 10^{-3}$	(B-44)

*Exact conversion

Table B-1 *(continued)*

To convert from	To	Multiply by		Equation no.
		For textbook calculations	*For precise conversion*	
gill (U.S.)	ft^3	4.178×10^{-3}	$*(231/1728 \times 32)$	(B-45)
	m^3	1.183×10^{-4}	$1.182\ 941\ 183 \times 10^{-4}$	(B-46)
grain	lbm	*(1/7000)		(B-47)
	kg	6.480×10^{-5}	$*6.479\ 891 \times 10^{-5}$	(B-48)
gram	lbm	2.205×10^{-3}	$2.204\ 622\ 622 \times 10^{-3}$	(B-49)
gravity, standard acceleration	ft/sec^2	32.17	32.174 048 56	(B-50)
	m/s^2	9.807	*9.806 65	(B-51)
horsepower	ft-lbf/sec	*550		(B-52)
	W	745.7	745.699 871 7	(B-53)
hour (mean solar)	s	*3 600		(B-54)
inch	ft	*(1/12)		(B-55)
	m	*(2.54/100)		(B-56)
kilogram/cubic metre	$slugs/ft^3$	1.940×10^{-3}	$1.940\ 320\ 331 \times 10^{-3}$	(B-57)
	lbm/ft^3	6.243×10^{-2}	$6.242\ 796\ 057 \times 10^{-2}$	(B-58)
knot (international)	ft/sec	1.688	1.687 809 857	(B-59)
	m/s	*(1852/3600)		(B-60)
litre	ft^3	3.531×10^{-2}	$3.531\ 446\ 673 \times 10^{-2}$	(B-61)
	m^2	$*1 \times 10^{-3}$		(B-62)
mile (international, nautical)	ft	6,076	*(1852/0.3048)	(B-63)
	m	*1852		(B-64)
mile (U.S. statute)	ft	*5,280		(B-65)
	m	1609	*1609.344	(B-66)
minute (angle)	rad	2.909×10^{-4}	$*(\pi/180 \times 60)$	(B-67)

minute (mean solar)	s	*60		(B-68)
ounce-force (avoirdupois)	lbf	*(1/16)		(B-69)
	N	0.278 0	*$2.780\ 138\ 51 \times 10^{-1}$	(B-70)
ounce-mass (avoirdupois)	lbm	*(1/16)		(B-71)
	kg	2.834×10^{-2}	$2.834\ 952\ 313 \times 10^{-2}$	(B-72)
ounce (U.S. fluid)	ft^3	1.044×10^{-3}	*$(231/1728 \times 128)$	(B-73)
	m^3	2.957×10^{-5}	$2.957\ 352\ 958 \times 10^{-5}$	(B-74)
pascal	lbf/ft^2	2.089×10^{-2}	$2.088\ 543\ 423 \times 10^{-2}$	(B-75)
	N/m^2	*1		(B-76)
pascal-second (dynamic viscosity)	$lbf\text{-}sec/ft^2$	2.089×10^{-2}	$2.088\ 543\ 423 \times 10^{-2}$	(B-77)
	$N \cdot s/m^2$	*1		(B-78)
pint (U.S. liquid)	ft^3	1.671×10^{-2}	*$(231/1728 \times 8)$	(B-79)
	m^2	4.731×10^{-4}	$4.731\ 764\ 73 \times 10^{-4}$	(B-80)
poise (dynamic viscosity)	$lbf\text{-}sec/ft^2$	2.089×10^{-3}	$2.088\ 543\ 423 \times 10^{-3}$	(B-81)
	$N \cdot s/m^2$	*0.1		(B-82)
poundal	lbf	(1/32.17)	$3.108\ 095\ 017 \times 10^{-2}$	(B-83)
	N	0.138 3	$1.382\ 549\ 544 \times 10^{-1}$	(B-84)
pound-force (avoirdupois)	N	4.448	*4.448 221 615 260 5	(B-85)
pound-force/foot	N/m	14.59	$1.459\ 390\ 294 \times 10^{1}$	(B-86)
pound-force/square foot	psi	*(1/144)		(B-87)
	N/m^2	47.88	$4.788\ 025\ 897 \times 10^{1}$	(B-88)
pound-force/square inch	lbf/ft^2	*144		(B-89)
	N/m^2	6 895	$6.894\ 757\ 295 \times 10^{3}$	(B-90)
pound-force second/square foot (dynamic viscosity)	$N \cdot s/m^2$	47.88	$4.788\ 025\ 897 \times 10^{1}$	(B-91)

*Exact conversion

Table B-1 *(continued)*

To convert from	To	Multiply by		Equation no.
		For textbook calculations	*For precise conversion*	
$\frac{\text{pound-force second squared}}{\text{foot fourth}}$	kg/m^3	515.4	$5.153\ 788\ 180 \times 10^2$	(B-92)
pound-mass (avoirdupois)	slugs	(1/32.17)	$3.108\ 095\ 017 \times 10^{-2}$	(B-93)
	kg	0.453 6	$*4.535\ 923\ 7 \times 10^{-1}$	(B-94)
quart (U.S. liquid)	ft^3	3.342×10^{-2}	$*(231/1728 \times 4)$	(B-95)
	m^3	9.464×10^{-4}	$9.463\ 529\ 457 \times 10^{-4}$	(B-96)
second (angle)	rad	4.848×10^{-6}	$*(\pi/180 \times 3600)$	(B-97)
slug	lbm	32.17	32.174 048 56	(B-98)
	kg	14.59	*14.593 902 94	(B-99)
slug/cubic foot	lbm/ft^3	32.17	32.174 048 56	(B-100)
	kg/m^3	515.4	$5.153\ 788\ 186 \times 10^2$	(B-101)
stokes (kinematic viscosity)	ft^2/sec	1.076×10^{-3}	$1.076\ 391\ 042 \times 10^{-3}$	(B-102)
	m^2/s	$*1 \times 10^{-4}$		(B-103)
ton (long)	lbm	*2,240		(B-104)
	kg	1 016	$1.016\ 046\ 909 \times 10^3$	(B-105)
ton (short)	lbm	*2,000		(B-106)
	kg	907.2	$*9.071\ 847\ 4 \times 10^2$	(B-107)
torr (mm Hg at 0°C)	lbf/ft^2	2.784	2.784 495 557	(B-108)
	$lbf/in.^2$	1.934×10^{-2}	$1.933\ 677\ 470 \times 10^{-2}$	(B-109)
	N/m^2	133.3	*(101 325/760)	(B-110)
yard	ft	*3		(B-111)
	m	*0.914 4		(B-112)

*Exact conversion

REFERENCES

a. *Measuring Systems and Standards Organizations.* Edited by William K. Burton, American National Standards Institute, New York, 1969.

b. "A Metric America—A Decision Whose Time Has Come." Report to the Congress. National Bureau of Standards Special Publication 345. Washington, D.C., 1971.

c. "Definitions and Values," American Society of Mechanical Engineers Performance Test Code—PTC 2, New York, 1971.

d. "Rules for the Use of Units of the International System of Units." International Organization for Standardization Recommendation R-1000, ISO/R-1000-1969(E). Reprinted in New York; 1st ed. February 1969.

e. "The International System of Units (SI)," National Bureau of Standards Publication 330. Washington, D.C., 1972.

f. "Metric Practice Guide," American Society for Testing and Materials -E 380 -70. Philadelphia. Adopted by the American National Standards Institute as Z210.1—1971.

g. *ASME Orientation and Guide for Use of SI (Metric) Units,* 6th ed. American Society of Mechanical Engineers, New York, 1975.

Appendix C
Answers to Problems

Chapter 1

1 89.4 kPa
2 1 in. Hg abs
3 1 366°S
4 504.67°R, 7.22°C, 280.37 K
5 245.2 N
6 6.667 ft/sec^2
7 31.86 ft/sec^2
8 32.10 ft/sec^2
9 0.3814 lbf
10 5.860 slugs, 188.5 lbm, 85.50 kg
11 10^6 ft-lbf
12 558.0×10^6 ft-lbf
13 186.5 lbf
14 5.129 slugs/ft^3
15 Exact check
16 308.1 lbf/ft^3
17 8 822 N/m^3
18 6.200×10^{-4} m^3/kg
19 0.01700 ft^3/lbm
20 1.264, 1.265, 30.38°Be, −19.64°API (below lower limit of −1°)
21 0.002127 slugs/ft^3, 0.06843 lbf/ft^3
22 263.74 K, 7.025 kg/m^3
23 54,090 lbf/ft^2
24 1 063 kg/m^3
25 3,482 ft/sec, 4,742 ft/sec, 5,061 ft/sec
26 340.3 m/s
27 345.4×10^{-6} ft^2/sec, 627.2×10^{-6} lbf-sec/ft^2, 32.10×10^{-6} m^2/s, 0.030 03 N · s/m^2
28 0.151 5 N · m
29 1.097 mm (below surface)
30 Yes ($V_v = 29.70$ ft/sec)

Chapter 2

1 21.49 kN/m^3, 13.27 Mg/m^3
2 101.2 psia
3 99.96 kPa
4 108.0 psia
5 731.7 mm
6 21.13 ft
7 259.0 mm
8 616.2 mm
9 4°50′24″

10 500.3°F, 73.28 psia
11 −12.26°F, 6.76 psia
12 8.572×10^{-3} vs 8.567×10^{-3}
13 299.0 N
14 3,285 lbf, 1.8 in. below center line
15 11,700 lbf, 1.023 in. below center line
16 7 863 N
17 581,200 lbf, 31.11 ft below surface, and 9.335 ft to right of *ABC*.
18 69.58 in., 5.219 in.
19 1,850 psia
20 9.40 mm
21 Schedule 60
22 −27°16′48″ (reverse of slope in Fig. 2-34)
23 25 ft^3, 9,063 lbf, 252 lbf, 8,811 lbf
24 118.6 kPa
25 68.72 psia
26 148.7 rad/s
27 5.626 ft
28 2.750
29 53,880 tons
30 150.5 mm

Chapter 3

1 3.733 ft/sec, 36.00 gpm
2 80 m^3/s
3 94.25 m^3/s
4 94.25 m^3/s
5 1.259 ft/sec, 0.9178 ft/sec
6 1.25 ft/sec, 0.7120 ft^3/sec
7 $2/(a+1)(a+2)$, 0.8167
8 7.948 m/s, 0.404 6 m^3/s
9 10.09 ft/sec, 5.759 ft/sec
10 18.17 ft/sec, 12.19 ft/sec
11 18.17 ft/sec, 11.09 ft/sec
12 20.05 m/s
13 28.13 psia
14 2
15 1.045

Chapter 4

1 ¼ ft, 1 ft
2 2.828 m, 11.31 m

3 143.3 psia
4 91.10%, 228 120 kW
5 18,720 hp
6 3.385 N · m/kg
7 Proof
8 51.54 ft-lbf/lbm
9 −60 328 N · m/kg
10 −15,330 ft-lbf/lbm
11 −14 340 J/kg
12 604.16°F
13 14.10 Btu/lbm
14 814.7 kJ
15 307.9 Btu
16 2.840 MJ
17 30.15 knots
18 84.35 m/s
19 438.5 ft/sec
20 21.34 N · m/kg
21 50.17 Btu/lbm, 643 ft-lbf/lbm
22 4.823 kN
23 15,740 lbf
24 10,610 lbf, 5,958 hp, 23.59%
25 2 851 N, 955.2 kW
26 5.560 ft
27 0, 100.3 kW
28 23.39 psia
29 1 743 N

Chapter 5

1 (a) 260.96 K, 422,6 kPa, 323.9 m/s, 1, 0.896 8 kg/s
(b) 295.76 K, 655 kPa, 187.0 m/s, 0.5423, 0.708 3 kg/s
2 (a) 271.13 K, 435.1 kPa, 429.1 m/s, 1, 0.652 2 kg/s
(b) 298,68 K, 655 kPa, 251.8 m/s, 0.5592, 0.523 1 kg/s
3 1.011 in.
4 1.118 in.
5 61.66 psia, 2.056 lbm/sec, 1.615 lbm/sec
6 63.46 psia, 1.494 lbm/sec, 1.193 lbm/sec
7 54.37 m/s, 312.00 K, 372.9 kPa
8 298.1 m/s, 380.92 K, 600 kPa
9 (a) 1.156 × 1.471 in., 279.84°R, 2,050 ft/sec, 12.50 psia
(b) 1.156 × 1.334 in., 526.75°R, 703.2 ft/sec, 83.37 psia
10 (a) 1.695 × 2.098 in., 239.79°R, 4,456 ft/sec, 12.02 psia
(b) 1.695 × 1.948 in., 516.60°R, 1,635 ft/sec, 81.86 psia

11 32.56 mm, 173.51 K, 529.8 m/s, 2.006, 0.896 9 kg/s
12 32.52 mm, 172.78 K, 538.3 m/s, 2.008, 0.884 9 kg/s
13 0.5774, 460.94 K, 900.0 kPa, 248.5 m/s
14 0.5644, 421.68 K, 880.5 kPa, 302.0 m/s
15 2.646, 341.07°R, 16.88 psia, 2,396 ft/sec
16 3.607, 158.21°R, 8.43 psia, 6,527 ft/sec

Chapter 6

1 $\mu N/p$
2 E/p
3 7.501 ft, 1.817 ft/sec
4 0.483 6 kg/s
5 1/10
6 1/36.64
7 47.32 psia, 3.955 psi
8 1 kN
9 3.424 rpm, 102.6 rpm, 0.4906 psia
10 6.156 m
11 106.9 ft, 6.936 knots
12 54 m/s
13 $F = MLT^{-2}$, $M = PL^{-2}T^{3}$
14 ρV_o^2
15 $D^2(\Delta p/\rho)^{1/2}$
16 $\mathbf{C}_p$, $\mathbf{R}$, $(D\gamma/V^2\rho)$
17 Z/D, VT/D, $\mathbf{C}_p$, $\mathbf{M}$, $\mathbf{R}$
18 L/D, $\mathbf{R}$, $\mathbf{W}$
19 2.418×10^{-9}
20 0.003121
21 6 in., 33.75 ft-lbf/lbm, 3.56 bhp
22 32.880 MW, 106.9 rpm

Chapter 7

1 0.5033
2 0.3578
3 0.3110 ft, 0.01773 lbf
4 17.66 kN
5 13.82 kW
6 0.015, 0.4899
7 2.663 kW, 7.960 kW, 18.51 kW
8 5,248 lbf
9 0.070 90 m/s
10 211.8 ft/sec

11 210.8 m/s
12 0.3448
13 27.46 N
14 218.3 ft/sec
15 173.3 kN
16 59.57%
17 89.66 m/s
18 22.24 hp
19 13.64 rps
20 412.9 rpm, 132.4 hp, 37.80%

Chapter 8

1 20.48, 0.02070
2 1.165 lbf/ft^2
3 88.14 m
4 88.40×10^{-6} $lbf\text{-}sec/ft^2$
5 119.3 kN/m^2
6 6,341 lbm/hr
7 2.071×10^{-3} m^3/s
8 6-in. size
9 0.01894, 107×10^{-6} m
10 302.2 ft^3/hr
11 0.01017
12 2,468 lbm/hr
13 0.253 9 m^3/s
14 8.175 $lbf/in.^2$
15 147 kPa
16 60.47 psig
17 0.9907
18 205.5 gpm
19 2.091×10^{-3} m^3/s
20 14.47 lbm/sec
21 0.700 8 kg/s
22 41,288 lbm/hr
23 0.020 11 m^3/s
24 4.162 ft^3/sec, 0.05667
25 596.5 ft/sec
26 2.146 m

Chapter 9

1 5.787 ft/sec, 3.966 ft/sec
2 0.012 20 $m^{1/6}$

3 6.472 ft
4 1/23 538
5 h = 5.772 ft, b = 4.779 ft
6 h = 6.506 m, b = 3.487 m
7 1/3,909, 3.738 ft, subcritical
8 1/92.56
9 No
10 20.13%
11 4.212 ft
12 5.443 m
13 0.7573 ft^3/sec
14 0.302 7 m

Index

ABCDEFGHIJ–VB–79876